Cambridge, 25 April 84

Gustav Doetsch

Introduction to the Theory and Application of the Laplace Transformation

With 51 Figures
and a Table of Laplace Transforms

Translation by
Walter Nader

Springer-Verlag Berlin Heidelberg New York 1974

Gustav Doetsch
Professor Emeritus of Mathematics, University of Freiburg

Walter Nader
Assistant Professor, University of Alberta, Canada

Translation of the German Original Edition:
Einführung in Theorie und Anwendung der Laplace-Transformation
Zweite, neubearbeitete
und erweiterte Auflage (Mathematische Reihe, Band 24, Sammlung LMW)
Birkhäuser Verlag Basel, 1970

AMS Subject Classification (1970): 44A10, 42A68, 34A10, 34A25, 35A22, 30A84

ISBN 3-540-06407-9 Springer-Verlag Berlin Heidelberg New York
ISBN 0-387-06407-9 Springer-Verlag New York Heidelberg Berlin

ISBN 3-7643-0086-8 German edition Birkhäuser Verlag Basel

Library of Congress Catalog Card Number 73-10661. Printed in Germany.
Typesetting, printing and binding: Passavia Passau.

Preface

In anglo-american literature there exist numerous books, devoted to the application of the Laplace transformation in technical domains such as electrotechnics, mechanics etc. Chiefly, they treat problems which, in mathematical language, are governed by ordinary and partial differential equations, in various physically dressed forms. The theoretical foundations of the Laplace transformation are presented usually only in a simplified manner, presuming special properties with respect to the transformed functions, which allow easy proofs.

By contrast, the present book intends principally to develop those parts of the *theory* of the Laplace transformation, which are needed by mathematicians, physicists and engineers in their daily routine work, but in complete generality and with detailed, exact proofs. The *applications* to other mathematical domains and to technical problems are inserted, when the theory is adequately developed to present the tools necessary for their treatment.

Since the book proceeds, not in a rigorously systematic manner, but rather from easier to more difficult topics, it is suited to be read from the beginning as a textbook, when one wishes to familiarize oneself for the first time with the Laplace transformation.

For those who are interested only in particular details, all results are specified in "Theorems" with explicitly formulated assumptions and assertions.

Chapters 1–14 treat the question of convergence and the mapping properties of the Laplace transformation. The interpretation of the transformation as the mapping of one function space to another (original and image functions) constitutes the dominating idea of all subsequent considerations.

Chapters 14–22 immediately take advantage of the mapping properties for the solution of ordinary differential equations and of systems of such equations. In this part, especially important for practical applications, the concepts and the special cases occurring in technical literature are considered in detail. Up to this point no complex function theory is required.

Chapters 23–31 enter the more difficult parts of the theory. They are devoted to the complex inversion integral and its various evaluations (by deformation of the path of integration and by series developments) and to the Parseval equation. In these considerations the Fourier transformation is used as an auxiliary tool. Its principal properties are explained for this purpose. Also the question of the representability of a function as a Laplace transform is answered here.

Chapters 32–37 deal with a topic which is of special interest for both theory and application and which is commonly neglected in other books: that is the deduction of asymptotic expansions for the image function from properties of the original function, and conversely the passage from the image function to asymptotic expansions of the original function. In the latter case the inversion integral with angular path plays a decisive role; therefore its properties are developed, for the first time in the literature, in full detail. In technical problems this part provides the basis for the investigation of the behaviour of physical systems for large values of the time variable.

Chapter 38 presents the ordinary differential equation with polynomial coefficients. Here, again, the inversion integral with angular path is used for the contruction of the classical solution.

As examples of boundary value problems in partial differential equations, Chapter 39 treats the equation of heat conduction and the telegraph equation. The results of Chapters 35–37 are used to deduce the stationary state of the solutions, which is of special interest for engineering.

In Chapter 40 the linear integral equations of convolution type are solved. As an application the integral and the derivative of non-integral order in the intervall $(0, \infty)$ are defined.

Not only to procure a broader basis for the theory, but also to solve certain problems in practical engineering in a satisfactory manner, it is necessary to amplify the space of functions by the modern concept of distribution. The Laplace transformation can be defined for distributions in different ways. The usual method defines the Fourier transformation for distributions and on this basis the Laplace transformation. This method requires the limitation to "tempered" distributions and involves certain difficulties with regard to the definition of the convolution and the validity of the "convolution theorem". Here, however, a direct definition of the Laplace transformation is introduced, which is limited to distributions "of finite order". With this definition the mentioned difficulties do not appear; moreover it has the advantage that in this partial distribution space, a necessary und sufficient condition for the representability of an analytic function as a Laplace transform can be formulated, which is only sufficient but not necessary in the range of the previous mentioned definition.

The theory of distributions does not only make possible the legitimate treatment of such physical phenomena as the "impulse", but also the solution of a problem that has caused many discussions in the technical literature. When the initial value problem for a system of simultaneous differential equations is posed in the sense of classical mathematics, the initial values are understood as limits in the origin form the *right*. In general, this problem can be solved only if the initial values comply with certain "conditions of compatibility", a requirement which is fulfilled seldom in practice. Since, however, also in such a case the corresponding physical process ensues, someone mathematical description must exist. Such a description is possible, when 1. the functions are replaced by distributions, which are defined always over the whole axis and not only over the right half-axis, which is the domain of the initial value problem, and 2. the given initial values are understood as limits from the *left*; they originate, then, from the values of the unknowns in the left half-axis, i.e. from the *past* of the system, which agrees exactly with the physical intuition.

Instead of a general system, whose treatment would turn out very tedious, the system of first order for two unknown functions is solved completely with all details as a pattern, whereby already all essential steps are encountered.

This work represents the translation of a book which appeared in first edition in 1958 and in amplified second edition in 1970 in "Birkhäuser Verlag" (Basel und Stuttgart). The translation, which is based on the second edition, was prepared by Professor Dr. Walter Nader (University of Alberta, Edmonton, Canada) with extraordinary care. In innumerable epistolary discussions with the author, the translator has attempted to render the assertions of the German text in an adequate English structure. For his indefatigable endeavour I wish to express to Prof. Nader my warmest thanks.

Riedbergstrasse 8
D-7800 Freiburg i. Br.
Western Germany

GUSTAV DOETSCH

Table of Contents

1. Introduction of the Laplace Integral from Physical and Mathematical Points of View

The integral

$$\int_0^\infty e^{-st} f(t)\, dt$$

is known as the Laplace integral; t, the dummy variable of integration, scans the real numbers between 0 and ∞, and the parameter s may be real-valued or complex-valued. Should this integral converge for some values of s, then it defines a function $F(s)$:

$$\int_0^\infty e^{-st} f(t)\, dt = F(s)\,. \tag{1}$$

In Chapter 4 it will be shown how the correspondence between the functions $f(t)$ and $F(s)$ may be visualized as a "transformation", the Laplace transformation.

The Laplace integral is classified with mathematical objects like power series and Fourier series, which also describe functions by means of analytical expressions. Like these series, the Laplace integral was originally investigated in the pursuit of purely mathematical aims, and it was subsequently used in several branches of the sciences. Experience has demonstrated that with regard to possible applications, the Laplace integral excels these series. The Laplace integral serves as an effective tool, particularly in those branches that are of special interest not only to the mathematician but also to the physicist and to the engineer. This is, in part, due to the clear physical meaning of the Laplace integral, which will be explained in the sequel.

We begin with the well known representation of some function $f(x)$ in the finite interval $(-\pi, +\pi)$ by a *Fourier series*. Instead of the real representation

$$\frac{a_0}{2} + \sum_{n=1}^{\infty} (a_n \cos n\,x + b_n \sin n\,x) \tag{2}$$

employing the real oscillations $\cos n x$ and $\sin n x$, we prefer here, for practical reasons, the complex representation[1]

$$f(x) = \frac{1}{2\pi} \sum_{n=-\infty}^{+\infty} c_n\, e^{i n x}\,, \tag{3}$$

for which we combined the respective real oscillations, to compose the complex

[1] The factor $1/2\pi$ is included with (3) to establish complete analogy to formulae (5) and (12), which are usually written with this factor. The real oscillations $\cos n x$ and $\sin n x$ form a complete, orthogonal set for $n = 0, 1, 2, 3, \ldots$; for the complex oscillations $e^{i n x}$, we need $n = 0, \pm 1, \pm 2, \pm 3, \ldots$, to produce the complete, orthogonal set. It is for this reason that in (3) the summation of n extends between $-\infty$ and $+\infty$.

oscillations e^{inx}. The Fourier coefficients c_n are determined by means of the formula[2]

$$c_n = \int_{-\pi}^{+\pi} f(x)\, e^{-inx}\, dx\,. \tag{4}$$

The expansion (3) converges and represents $f(x)$[3] under quite general conditions, for instance, when $f(x)$ is composed of a finite number of monotonic pieces. From a physical point of view, Eq. (3) indicates that $f(x)$ may be constructed as a superposition of complex oscillations having frequencies $n = 0, \pm 1 \pm 2, \pm 3, \cdots$, the harmonic oscillations. The Fourier coefficients c_n are, in general, complex-valued, that is

$$c_n = r_n\, e^{i\varphi_n}.$$

With this, the nth term of the series (3) becomes

$$r_n\, e^{i(nx+\varphi_n)},$$

which shows that the oscillation of frequency n has *amplitude* r_n (when disregarding the factor $1/2\pi$), and the *initial phase angle* φ_n. In Physics, the totality of the amplitudes r_n together with the phase angles φ_n is called the spectrum of the physical phenomenon which is described by $f(x)$ in $(-\pi, +\pi)$. This spectrum is completely described by the sequence of the c_n, and we call the Fourier coefficients c_n the *spectral sequence* of $f(x)$.

Thus Eqs. (3) and (4) may be interpreted as follows.

By means of (4) *one obtains for the given function* $f(x)$ *the spectral sequence* c_n; *using these* c_n, *one can reconstruct* $f(x)$ *as a superposition of harmonic functions with frequencies* $n = 0, \pm 1, \pm 2, \pm 3, \cdots$, *as shown in* (3).

Nowadays, complex oscillations are used extensively in theoretical investigations, a development fostered by work in electrical engineering. Differentiation of real oscillations, $\sin nx$ and $\cos nx$, leads to an interchange of these; differentiation of a complex oscillation, e^{inx}, merely reproduces it. Hence it is more convenient to work with complex oscillations. Interpreting x as time, one envisages the complex oscillation $z = r_n e^{i(nx+\varphi_n)}$ as the motion of a point of the circle of radius r_n, centred at the origin of the complex z-plane; the point moves with constant angular velocity; that is the arc covered is proportional to x. For $n > 0$, this motion is in the mathematically positive sense about the origin; for $n < 0$, the motion is in the opposite, mathematically negative sense. That is, a complex oscillation with a *negative frequency* affords a meaningful physical interpretation, which is impossible with real oscillations. The orthogonal projections of the motion of the point into the real axis and into the imaginary axis respectively, produce the real component and the imaginary component of the complex oscillation. These two components correspond to the real oscillations $\cos nx$ and $\sin nx$, which are the ones actually observed in physical reality.

---

[2] For real-valued $f(x)$ we find $c_{-n} = \overline{c_n}$; hence c_0 is real-valued. Combining the conjugate terms for $-n$ and $+n$, and with $c_n = \pi(a_n - ib_n)$, we produce the conventionally employed form of the Fourier expansion: $$\frac{c_0}{2\pi} + \frac{1}{2\pi}\sum_{n=1}^{\infty} 2\,\Re\,(c_n\, e^{inx}) = \frac{a_0}{2} + \sum_{n=1}^{\infty}(a_n \cos nx + b_n \sin nx)\,.$$

[3] At discontinuity points x, where $f(x^-) \neq f(x^+)$, that is, where the function "jumps", the Fourier series converges to the mean of the limits: $[f(x^-) + f(x^+)]/2$.

If the independent variable represents time, then one is, in most cases, not interested in finite intervals, for time extends conceptually between $-\infty$ and $+\infty$. For the infinite interval $(-\infty, +\infty)$, the Fourier series (3) is to be replaced by the *Fourier integral*[4]

$$f(t) = \frac{1}{2\pi}\int_{-\infty}^{+\infty} F(y)\, e^{ity}\, dy, \tag{5}$$

employing the letter t instead of the letter x, to hint at the implied time. The function $F(y)$ in (5) is also determined by a Fourier integral

$$F(y) = \int_{-\infty}^{+\infty} f(t)\, e^{-iyt}\, dt. \tag{6}$$

Complex oscillations of all frequencies are involved in this case; one cannot construct $f(t)$ by merely superimposing a sequence of harmonic oscillations. Therefore, the sum in (3) had to be replaced by the integral in (5). The complex oscillation of frequency y is multiplied by the infinitesimal factor $F(y)dy$, which corresponds to the coefficient c_n of the Fourier series (3). Writing the generally complex $F(y)$ in the form

$$F(y) = r(y)\, e^{i\varphi(y)},$$

one finds:

$$f(t) = \frac{1}{2\pi}\int_{-\infty}^{+\infty} r(y)\, e^{i(yt+\varphi(y))}\, dy .$$

The complex oscillation of frequency y has the amplitude $r(y)dy = |F(y)|dy$ (again disregarding the factor $1/2\pi$), and the initial phase angle $\varphi(y) = \operatorname{arc} F(y)$.[5] For this reason, we call the function $F(y)$ the *spectral function* or the *spectral density* of $f(t)$.

For the finite interval $(-\pi, +\pi)$, we obtained the discrete spectrum; that is, the frequencies $n = 0, \pm 1, \pm 2, \pm 3, \cdots$. For the infinite interval $(-\infty, +\infty)$, we find a continuous spectrum for the frequencies y, with $-\infty < y < +\infty$.

We stated the formulae (5) and (6) formally, without regard to required conditions, the discussion of which we defer to Chapter 24. However, when comparing formulae (5) and (6) to formulae (3) and (4), we immediately detect a serious restriction: the spectral sequence c_n is meaningful for every integrable function $f(x)$; the spectral function $F(y)$ exists only if for t approaching $-\infty$, as well as $+\infty$, the function $f(t)$ behaves in a manner so that the integral (6) converges. This is not the case for some of the simplest and most commonly encountered functions. For instance, the integral (6) does not converge for $f(t) \equiv 1$, or $f(t) \equiv e^{i\omega t}$.

However, it is possible to overcome this difficulty. So far, we permitted t to vary between $-\infty$ and $+\infty$. Yet, for all cases of interest to the physicist, the

---

[4] As for the Fourier series, we obtain for the Fourier integral the mean of the limits, $[f(x^-) + f(x^+)]/2$, at those discontinuity points where $f(x^-) \neq f(x^+)$.

[5] We prefer the use of "arc z", arcus of z, for the angle of the vector representation of a complex number; this designation is more descriptive than the conventional "argument", which is also used to designate the argument of a function.

process under investigation begins at some specified instant, say $t = 0$, and accordingly t varies between 0 and $+\infty$, thus restricting the infinite interval to $(0, +\infty)$. This new situation may be considered as a special case simultaneously with the above unrestricted case, provided we define $f(t) = 0$ for $t < 0$. We find then for the spectral function:

$$F(y) = \int_0^\infty e^{-iyt} f(t)\, dt.$$

Nevertheless, the integral still does not converge for the above mentioned functions: 1, and $e^{i\omega t}$. However, we may now implement a fruitful modification: rather than studying some given function $f(t)$, we investigate the entire family of functions $e^{-xt}f(t)$, admitting all $x > X$, for some specified, fixed X. The spectral function of $e^{-xt}f(t)$ obviously depends upon both y and x, and we write, therefore,

$$F_x(y) = \int_0^\infty e^{-iyt}\left[e^{-xt} f(t)\right] dt. \tag{7}$$

For $x > 0$, the integral (7) converges for the above mentioned functions, 1 and $e^{i\omega t}$; indeed, it converges for all bounded functions. Moreover, the integral (7) converges for all functions that do not grow more strongly than e^{at} $(a > 0)$, provided we use $x > a$.[6]

Recalling the first modification, that is $f(t) \equiv 0$ for $t < 0$, we obtain, with the modified spectral function, and (5)[7]

$$\frac{1}{2\pi}\int_{-\infty}^{+\infty} e^{ity} F_x(y)\, dy = \begin{cases} 0 & \text{for } t < 0 \\ e^{-xt} f(t) & \text{for } t > 0. \end{cases} \tag{8}$$

Formulae (7) and (8) can be rewritten as follows:

(9) the spectral function of $e^{-xt}f(t)$: $F_x(y) = \int_0^\infty e^{-(x+iy)t} f(t)\, dt,$

(10) the time function $f(t)$: $\frac{1}{2\pi}\int_{-\infty}^{+\infty} e^{(x+iy)t} F_x(y)\, dy = \begin{cases} 0 & \text{for } t < 0 \\ f(t) & \text{for } t > 0. \end{cases}$

Quite naturally, a complex variable $(x + iy)$ appears in formulae (9) and (10); hence it becomes apparent that the function $F_x(y)$ does not depend separately upon x, and y, but simultaneously upon $x + iy$, and we may write $F(x + iy)$ instead of $F_x(y)$. The spectral functions corresponding to the range of values of the parameter x are associated in a manner that permits simultaneous representation by one function of a complex variable: $x + iy$. Usually, we represent the com-

---

[6] For the integral (6) with the lower limit $-\infty$, the factor e^{-xt} would aggravate the difficulties, since e^{-xt} grows as $t \to -\infty$.

[7] For $t = 0$, we obtain the average of $f(0^-) = 0$ and $f(0^+)$, that is, $f(0^+)/2$.

plex variable $(x + iy)$ by the letter z; here it is customary to use the letter s which fits well to the time variable t, t and s being neighbours in the alphabet:

$$s = x + i\,y\,.$$

In (10), where x actually designates a fixed value, we have $ds = dy/i$, and corresponding to the limits of integration $y = -\infty$, and $y = +\infty$, we obtain the new limits: $s = x - i\,\infty$, and $s = x + i\,\infty$. In the complex s-plane, the path of integration is the vertical line with constant abscissa x. Hence, we may rewrite formulae (9) and (10) as follows:

$$\int_0^\infty e^{-st} f(t)\,dt = F(s)\,, \tag{11}$$

$$\frac{1}{2\pi i}\int_{x-i\infty}^{x+i\infty} e^{ts} F(s)\,ds = \begin{cases} 0 & \text{for } t < 0 \\ f(t) & \text{for } t > 0\,. \end{cases} \tag{12}$$

Integral (11) is precisely the *Laplace integral* (1) which produces, for the given function $f(t)$, the function $F(s)$. Conversely, the integral (12) reproduces $f(t)$, using $F(s)$. For this reason, one could call (12) the *inverse* of (11).

We may now summarize the physical implications of formulae (11) and (12). *For the function $F(s)$, produced by means of the Laplace integral, consider the independent variable as complex valued*: $s = x + iy$. *With this*, $F(x + iy)$ *is the spectral function of the dampened function* $e^{-xt} f(t)$, *having* y *as frequency variable. The time function* $f(t)$ *can be reconstructed using formula* (12), *and* $F(x + iy)$.

Instead of the above physical considerations, one may follow strictly mathematical arguments. Attempting a generalization of the *power series*

$$\sum_{n=0}^{\infty} a_n z^n\,,$$

one can firstly replace the integer-valued sequence of exponents n by some arbitrary, increasing sequence of non-negative numbers λ_n. In general, the corresponding functions z^{λ_n} are not single-valued. The substitution $z = e^{-s}$ produces the terms $e^{-\lambda_n s}$, which are single-valued functions. In this manner, one generates the *Dirichlet series*:

$$\sum_{n=0}^{\infty} a_n e^{-\lambda_n s}\,.$$

It is common practice to designate the variable of the Dirichlet series by s. One further step is necessary for the intended generalization: we replace the discrete sequence λ_n by a continuous variable t, the sequence a_n by a function $f(t)$, and summation by integration. This leads to the Laplace integral

$$\int_0^\infty f(t)\,e^{-ts}\,dt\,.$$

Substituting at the onset of the above generalization for the power series a *Laurent series*

$$\sum_{n=-\infty}^{+\infty} a_n z^n\,,$$

one obtains through the same process of generalization the *two-sided Laplace integral,* which also serves as an important mathematical tool:

$$\int_{-\infty}^{+\infty} f(t)\, e^{-ts}\, dt\,.$$

The power series $\sum_{n=0}^{\infty} a_n z^n$ converges on a circular disc. Replacing the integers n by the generally real-valued λ_n, we have to consider the multi-valued behaviour of z^{λ_n}. Accordingly, the circular disc of convergence is to be envisaged as a portion of a multi-layered Riemann surface. The transformation $z = e^{-s}$ maps the infinitely many layers of this disc of convergence into a right half-plane of the s-plane. Indeed, a Dirichlet series converges in a *right half-plane*; the same property will be established for the Laplace integral.

On the circle $z = \varrho e^{i\vartheta}$ of fixed radius ϱ, we have:

$$\sum_{n=-\infty}^{+\infty} a_n z^n = \sum_{n=-\infty}^{+\infty} (a_n \varrho^n)\, e^{in\vartheta},$$

that is, on the circle, the Laurent series is a *Fourier series* of type (3). As an analogy, we find that on the vertical line $s = x + iy$, with constant abscissa x:

$$\int_{-\infty}^{+\infty} e^{-st} f(t)\, dt = \int_{-\infty}^{+\infty} e^{-iyt}\, [e^{-xt} f(t)]\, dt\,.$$

That is: on the vertical line, the two-sided Laplace integral is a *Fourier integral* (6). For power series, and the one-sided Laplace integral, the respective dummy variable of summation of the Fourier series, and of the Fourier integral, scans between 0 and $+\infty$. The study of Fourier series provides valuable information for the theory of power series, with particular regard to the behaviour of the power series on the boundary of the circular disc of convergence. Similarly, the Fourier integral is an important tool in the study of the Laplace integral.

Having thus introduced the Laplace integral as a natural generalization of the power series, we ought to suspect that known properties of the power series may pertain to the Laplace integral. That they do is, indeed, true to some extent. However, in some respects the Laplace integral behaves differently, and it consistently exhibits properties more complicated than those of the power series.

2. Examples of Laplace Integrals. Precise Definition of Integration

Laplace integrals of several, selected functions are evaluated in this Chapter, to develop a more intimate understanding of the Laplace integral.

When evaluating a Laplace integral of some function $f(t)$, we actually use $f(t)$ only for $0 \leqq t < +\infty$, hence it should be irrelevant, from the mathematical point of view, if and how $f(t)$ is defined for $t < 0$. However, some properties of the Laplace integral, particularly those which reflect a kinship to the Fourier integral, can be better understood if $f(t)$ is assigned the value zero for $-\infty < t < 0$.

1. $f(t) = u(t)$,

where $u(t)$ is defined as

$$u(t) = \begin{cases} 0 & \text{for } t \leqq 0 \\ 1 & \text{for } t > 0 . \end{cases}$$

The function $u(t)$ is called the *unit step function*, sometimes in electrical engineering the Heaviside unit function. We find, firstly:

$$\int_0^\omega e^{-st}\, dt = \frac{1}{s}\left(1 - e^{-s\omega}\right),$$

and with $\omega \to +\infty$, we obtain a limit if and only if $\Re s > 0$; it is

$$F(s) = \frac{1}{s} \quad \text{for } \Re s > 0 .$$

Using the explanation of Chapter 1, p. 4, we recognize $F(s) = F(x + iy) = (x + iy)^{-1}$ as the spectral function of $e^{-xt} u(t)$. Writing $s = r e^{i\varphi}$, we obtain

$$F(s) = r^{-1}\, e^{-i\varphi}.$$

The particular complex oscillation of $e^{-xt} u(t)$ of frequency y, has the amplitude r^{-1} and the initial phase angle $-\varphi$. That means that for growing $|y|$, the amplitude tends towards zero, and $|\varphi|$ tends towards $\pi/2$. The Laplace integral does not converge for $x = \Re s = 0$; that is, $u(t)$ does not have a spectral function. This also follows from the observation that with $x = 0$, for $y = 0$, that is for $s = 0$, $F(s)$ is meaningless.

$$2.\ f(t) = u(t - a) = \begin{cases} 0 & \text{for } t \leqq a \\ 1 & \text{for } t > a \end{cases} \qquad (a > 0),$$

the unit step function shifted to the right from $t = 0$ to $t = a$.

$$F(s) = \int_a^\infty e^{-st}\, dt = \frac{e^{-as}}{s} \quad \text{for } \Re s > 0 .$$

3. $f(t) = e^{\alpha t}$ (α arbitrary, complex).

$$F(s) = \int_0^\infty e^{-(s-a)t}\,dt = \frac{1}{s-a} \quad \text{for } \Re s > \Re a.$$

4. $\underline{f(t) \equiv \cosh kt} \equiv (1/2)[e^{kt} + e^{-kt}]$ (k arbitrary, complex).

Using

$$\int_0^\infty e^{-st}[f_1(t) + f_2(t)]\,dt = \int_0^\infty e^{-st} f_1(t)\,dt + \int_0^\infty e^{-st} f_2(t)\,dt$$

and the conclusions of example 3, one finds:

$$F(s) = \frac{1}{2}\left(\frac{1}{s-k} + \frac{1}{s+k}\right) = \frac{s}{s^2-k^2},$$

whereby we have to simultaneously restrict $\Re s > \Re k$, and $\Re s > -\Re k$; shortly $\Re s > |\Re k|$. For real k, this means $\Re s > |k|$.

5. $\underline{f(t) \equiv \sinh kt} \equiv (1/2)[e^{kt} - e^{-kt}]$ (k arbitrary, complex).

$$F(s) = \frac{1}{2}\left(\frac{1}{s-k} - \frac{1}{s+k}\right) = \frac{k}{s^2-k^2} \quad \text{for } \Re s > |\Re k|.$$

6. $\underline{f(t) \equiv \cos kt} \equiv (1/2)[e^{ikt} + e^{-ikt}]$ (k arbitrary, complex).

$$F(s) = \frac{1}{2}\left(\frac{1}{s-ik} + \frac{1}{s+ik}\right) = \frac{s}{s^2+k^2},$$

whereby we have to simultaneously restrict $\Re s > \Re(ik) = -\Im k$, and $\Re s > -\Re(ik) = \Im k$; shortly $\Re s > |\Im k|$. For real k, this means $\Re s > 0$.

7. $\underline{f(t) \equiv \sin kt} \equiv (1/2i)\,[e^{ikt} - e^{-ikt}]$ (k arbitrary, complex).

$$F(s) = \frac{1}{2}\left(\frac{1}{s-ik} - \frac{1}{s+ik}\right) = \frac{k}{s^2+k^2} \quad \text{for } \Re s > |\Im k|.$$

8. $\underline{f(t) \equiv t^a}$ (real $a > -1$).

The function t^a is multi-valued for non-integer a. Therefore, we specify for t^a the main branch; that is, for positive t, t^a is positive. Performing the integration in two sections, (0,1) and $(1, +\infty)$, we require for the existence of the first integral: *

$$\int_0^1 e^{-st}\, t^a\, dt$$

that $a > -1$. With this restriction, the first integral converges for all values of s. The second integral:

$$\int_1^\infty e^{-st}\, t^a\, dt$$

converges for $a \geqq 0$ if and only if $\Re s > 0$; if $-1 < a < 0$, then the second integral converges for $s = iy \neq 0$ also. For we have, with $s = iy$:

$$\int_1^\infty e^{-st}\, t^a\, dt = \int_1^\infty e^{-iyt}\, t^a\, dt = \int_1^\infty (\cos y\,t - i \sin y\,t)\, t^a\, dt\,.$$

Considering the integral

$$\int_1^\omega t^a \sin y\,t\, dt\,,$$

and letting the upper limit approach $+\infty$, not in a continuous manner, but in discrete steps through the zeroes $n\pi/y$ of $\sin yt$, we obtain the partial sums of an infinite series. The general term of this series:

$$\int_{n\pi/y}^{(n+1)\pi/y} t^a \sin y\,t\, dt$$

has the following properties: the terms have alternating sign, and tend monotonically, in absolute value, to zero (compare Fig. 1). Hence the convergence of the sequence of partial sums is guaranteed by Leibniz' criterion.

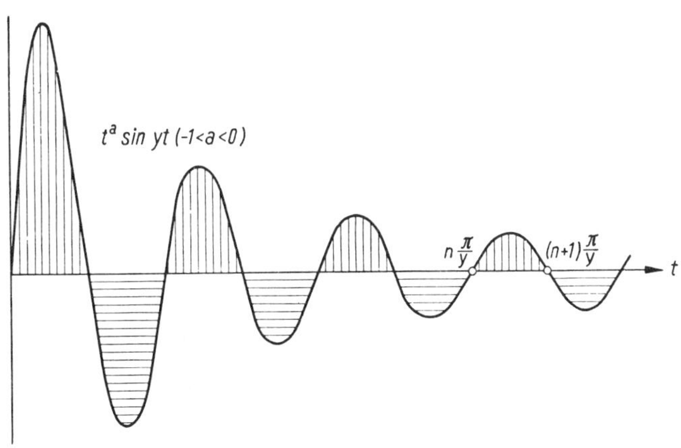

Figure 1

From this we can conclude that the integral with a continuously growing upper limit converges also. This last conclusion follows from the fact that for any ω between $n\pi/y$ and $(n + 1)\pi/y$:

$$\left| \int_1^{\omega} - \int_1^{n\pi/y} \right| \leqq \left| \int_{n\pi/y}^{(n+1)\pi/y} \right|,$$

and for $\omega \to \infty$, that is for $n \to \infty$, the right hand side tends towards zero. This last step in the above proof is by no means superfluous, for it may very well happen that some integral converges for a discretely growing upper limit, yet it diverges for the continuously growing upper limit. For instance, the limit

$$\lim_{n\to\infty} \int_0^{2n\pi} \sin t \, dt \quad (n = 1, 2, \cdots)$$

exists and is zero, since

$$\int_{2n\pi}^{2(n+1)\pi} \sin t \, dt = 0;$$

while the integral

$$\int_0^{\omega} \sin t \, dt$$

does not converge for continuously growing upper limit; instead, it oscillates between 0 and 2.

Extensive details have been supplied with the above proof, for in the sequel we shall, on several occasions, demonstrate convergence of some integral by comparison with a sequence; we can then refer to the details of the above proof.

The above conclusions regarding the sin-integral apply similarly to the cos-integral, and we conclude: The Laplace integral of t^a converges for $a \geqq 0$, provided $\Re s > 0$; it also converges for $-1 < a < 0$, provided $\Re s \geqq 0$, excepting $s = 0$. Next, we need to evaluate the integral. For positive, real s, we substitute $st = \tau$:

$$\int_0^{\infty} e^{-st} t^a \, dt = \frac{1}{s^{a+1}} \int_0^{\infty} e^{-\tau} \tau^a \, d\tau = \frac{\Gamma(a+1)}{s^{a+1}}.$$

This expression is inherently positive, hence we must use the main branch for s^{a+1}. For complex s and $\Re s > 0$, τ too is complex, and the path of integration, $0 \leqq t < +\infty$, is shifted into the ray from the origin through the point s towards ∞; that is, a ray in the right half-plane. This integral too is a well known representation of the Γ-function. For $-1 < a < 0$, the ray may coincide with the positive or the negative imaginary axis.

The experiences with these examples demonstrate the need for a well-defined concept of integration. This book is designed so that the reader who is familiar with Riemann integration only, can follow the development (except Chapter 30). Presuming a knowledge of Lebesgue integration, the statements remain essentially unchanged, although sometimes the expressions and the proofs could be simplified. Remarks regarding such modifications will be included for the benefit of those who are familiar with Lebesgue theory.

With regard to the upper limit, the Laplace integral is to be understood as an improper one; that is, we define[1]:

$$\int_0^\infty e^{-st} f(t)\, dt = \lim_{\omega \to \infty} \int_0^\omega e^{-st} f(t)\, dt. \tag{1}$$

We call a Laplace integral absolutely convergent, provided

$$\lim_{\omega \to \infty} \int_0^\omega |e^{-st} f(t)|\, dt \tag{2}$$

exists. Clearly, absolute convergence of a Laplace integral implies convergence in the sense of (1).

Obviously, we must require that the integral

$$\int_0^\omega e^{-st} f(t)\, dt$$

exists for all finite ω. However, the last example, $f(t) \equiv t^a$ for $-1 < a < 0$, shows that when using Riemann integration, it would be impractical to require $f(t)$ to be properly integrable in every finite interval and, consequently, bounded. In this example, $f(t)$ is, at $t = 0$, improperly integrable only; that is,

$$\lim_{\delta \to 0} \int_\delta^\omega f(t)\, dt$$

exists. Moreover, it is not sufficient to restrict our investigations to functions which require improper integration at $t = 0$ only, for we want to include in our studies functions which require *improper integration* at several points, such as

$$f(t) = \begin{cases} \dfrac{1}{\sqrt{t(1-t)}} & \text{for } 0 < t < 1 \\ 0 & \text{for } t \geqq 1, \end{cases}$$

---

[1] This is important in case the integral is interpreted as a Lebesgue integral. In the Lebesgue theory $\int_0^\infty$ can exist directly; that is, without resorting to the limiting process $\omega \to \infty$. In this case, the integral is eo ipso absolutely convergent. The interpretation of $\int_0^\infty$ as $\lim_{\omega \to \infty} \int_0^\omega$ is a generalization of the conventional Lebesgue integral. This generalization is explicitly admitted.

which necessitates improper integration at both $t = 0$ and $t = 1$. Functions of this type will be needed, for instance, in the theory of Bessel functions. We shall find that some theorems can be verified only under the stronger hypothesis that $f(t)$ is *absolutely* improperly integrable at those exceptional points. This means: suppose the interval (a, b) contains exactly one exceptional point t_0 with a neighbourhood in which $f(t)$ is unbounded; we require that $f(t)$ and, consequently, also $|f(t)|$ be properly integrable in every pair of subintervals $(a, t_0 - \varepsilon_1)$, $(t_0 + \varepsilon_2, b)$, with arbitrarily small $\varepsilon_1 > 0$, and $\varepsilon_2 > 0$. Then

$$\lim_{\varepsilon_1 \to 0} \int_a^{t_0 - \varepsilon_1} |f(t)| \, dt + \lim_{\varepsilon_2 \to 0} \int_{t_0 + \varepsilon_2}^{b} |f(t)| \, dt$$

should exist. For $t_0 = a$, we need consider only the second integral, and for $t_0 = b$, the first one only.

That is, in terms of Riemann integration, we require $f(t)$ to be properly integrable in every finite interval after, at most, a finite number of exceptional points with small neighbourhoods have been removed, and to be absolutely improperly integrable at those exceptional points. We state this briefly: *The function $f(t)$ be absolutely integrable in every finite interval* $0 \leqq t \leqq T$.[2] In the sequel, this prerequisite is tacitly assumed for every Laplace integral.

The factor e^{-st} is bounded in every finite interval; hence, $e^{-st} f(t)$ too is absolutely integrable in every finite interval.

Observe that a function which is improperly integrable at a finite point, need not be absolutely integrable, and, consequently, need not be Lebesgue integrable. We demonstrate this with the following example:

$$\int_0^1 \frac{\sin \frac{1}{t}}{t} \, dt .$$

The integral

$$\int_\varepsilon^1 \frac{\sin \frac{1}{t}}{t} \, dt$$

converges for $\varepsilon \to 0$. This is shown by means of the substitution $t = 1/u$, which transforms the integral into

$$\int_1^{1/\varepsilon} \frac{\sin u}{u} \, du ,$$

and for $1/\varepsilon$ growing discretely through the values $n\pi$ $(n = 1, 2, 3, \ldots)$, we obtain the partial sums of an alternating series, the terms of which tend monotonically towards zero (compare p. 10). However, the limit of

[2] One says, briefly: $f(t)$ is *locally* absolutely integrable. For the reader who is familiar with the Lebesgue theory, this means: $f(t)$ is Lebesgue integrable in every finite interval, and therefore eo ipso absolutely Lebesgue integrable.

$$\int_{\varepsilon}^{1}\left|\frac{\sin\frac{1}{t}}{t}\right|dt=\int_{1}^{1/\varepsilon}\left|\frac{\sin u}{u}\right|du$$

for $1/\varepsilon\to\infty$ does not exist, since

$$\int_{n\pi}^{(n+1)\pi}\left|\frac{\sin u}{u}\right|du$$

is of the order of $1/2n$; hence, for $n\to\infty$

$$\int_{0}^{n\pi}\left|\frac{\sin u}{u}\right|du$$

behaves like the partial sum of the diverging series

$$\frac{1}{2}\left(1+\frac{1}{2}+\frac{1}{3}+\cdots\right).$$

The theorems for Laplace integrals proved in this text, need not apply to functions like:

$$\frac{1}{t}\sin\frac{1}{t}\quad.$$

Remark: For all functions presented in this Chapter, the Laplace integral converges for some specified value of s. There are, of course, functions such that their Laplace integral would not converge for any value of s; for instance, $f(t)\equiv e^{t^2}$. In the above examples, the convergence of the Laplace integrals for some value of s follows from the fact that each of the example functions is dominated by some e^{pt} ($p>0$). *This is, by no means, a prerequisite for convergence*; a fact that is demonstrated by the example function on p. 18 which grows far more strongly than e^{pt}.

3. The Half-Plane of Convergence

Reviewing the examples of Chapter 2, we observe that for each of these functions the Laplace integral converges in a right half-plane. We shall show in this Chapter that this is generally true for Laplace integrals. Prior to that, we shall determine the domain of absolute convergence of a Laplace integral. For this, we shall need

Theorem 3.1. *A Laplace integral which converges absolutely at some point* s_0, *converges absolutely in the closed right half-plane*: $\Re s>\Re s_0$.

Proof: We utilize Cauchy's criterion of convergence: An integral

$$\int_0^\infty \varphi(t)\, dt$$

converges if and only if for every $\varepsilon > 0$, there is an ω, such that

$$\left| \int_{\omega_1}^{\omega_2} \varphi(t)\, dt \right| < \varepsilon \quad \text{for all} \quad \omega_2 > \omega_1 > \omega\,.$$

For $\Re s \geqq \Re s_0$, we find:

$$\int_{\omega_1}^{\omega_2} |\, e^{-st} f(t) \,|\, dt = \int_{\omega_1}^{\omega_2} |\, e^{-(s-s_0)t}\, e^{-s_0 t} f(t) \,|\, dt = \int_{\omega_1}^{\omega_2} e^{-\Re(s-s_0)t} \,|\, e^{-s_0 t} f(t) \,|\, dt$$

$$\leqq \int_{\omega_1}^{\omega_2} |\, e^{-s_0 t} f(t) \,|\, dt\,.$$

By hypothesis, the integral

$$\int_0^\infty |\, e^{-s_0 t} f(t) \,|\, dt$$

converges; hence, for every $\varepsilon > 0$, there is an ω, such that

$$\int_{\omega_1}^{\omega_2} |\, e^{-s_0 t} f(t) \,|\, dt < \varepsilon \quad \text{for all} \quad \omega_2 > \omega_1 > \omega.$$

It follows that

$$\int_{\omega_1}^{\omega_2} |\, e^{-st} f(t) \,|\, dt < \varepsilon \quad \text{for all} \quad \omega_2 > \omega_1 > \omega\,.$$

As an immediate consequence of Theorem 3.1, we derive

Theorem 3.2. *If the integral*

$$\int_0^\infty e^{-st} f(t)\, dt = F(s)$$

converges absolutely at s_0, *then* $F(s)$ *is bounded in the right half-plane:* $\Re s \geqq \Re s_0$.

Proof: For $\Re s \geqq \Re s_0$, one finds that

$$|F(s)| = \left| \int_0^\infty e^{-st} f(t)\, dt \right| \leqq \int_0^\infty e^{-\Re s\cdot t} \,|\, f(t) \,|\, dt \leqq \int_0^\infty e^{-\Re s_0 \cdot t} |\, f(t) \,|\, dt = \int_0^\infty |\, e^{-s_0 t} f(t) \,|\, dt.$$

By means of Theorem 3.1, we determine the domain of absolute convergence of a Laplace integral.

Theorem 3.3. *The exact domain of absolute convergence of a Laplace integral is either an open right half-plane:* $\Re s > \alpha$, *or else a closed right half-plane:* $\Re s \geqq \alpha$; *admitting the possibilities that* $\alpha = \pm\infty$.

Proof: For *real* s, three cases need be considered:

1. The integral converges absolutely for *every* real s; then, by Theorem 3.1, it converges for every complex s, and Theorem 3.3 is satisfied with $\alpha = -\infty$.

2. The integral converges absolutely for *no* real s; then, by Theorem 3.1, it converges absolutely for no complex s, and Theorem 3.3 is satisfied with $\alpha = +\infty$.

3. There is a real s where the integral converges absolutely, and a real s where the integral diverges absolutely. Let K_1 be the class of the real s_1 where the integral diverges absolutely, and let K_2 be the class of the real s_2 where the integral converges absolutely. This separation into classes implies a Dedekind cut: every real number belongs to precisely one of the two classes; the classes are non-empty; every real number s_1 of class K_1 is smaller than every real number s_2 of class K_2; for suppose there is an s_1 of K_1 which is larger than some s_2 of K_2 (equality is excluded by the definitions of K_1 and K_2), then, by Theorem 3.1, s_1 is a point of absolute convergence, contradicting the definition of K_1.

The Dedekind cut defines one finite, real number α. We now claim that for every *complex* s with $\Re s < \alpha$, the integral diverges absolutely. $\Re s < \alpha$ implies the existence of an s_1 in K_1, so that $\Re s < s_1 < \alpha$. Absolute convergence at s would imply absolute convergence at s_1, by Theorem 3.1. Also, for every s with $\Re s > \alpha$, the integral converges absolutely. Indeed, $\Re s > \alpha$ implies the existence of an s_2 in K_2, so that $\alpha < s_2 < \Re s$; absolute convergence at s_2, together with Theorem 3.1, guarantees absolute convergence at s (compare Fig. 2).

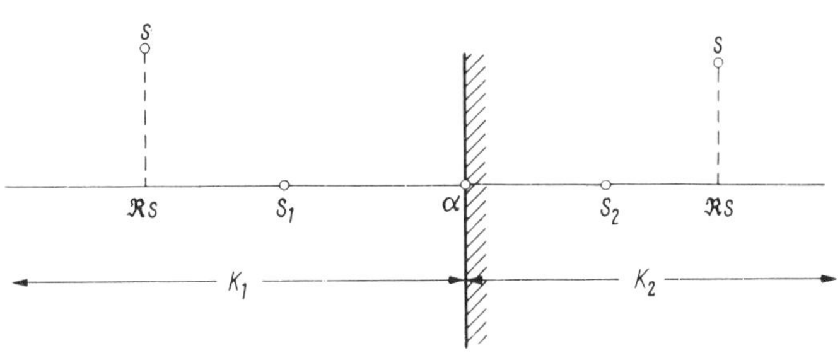

Figure 2

The straight line $\Re s = \alpha$ belongs either not at all, or else entirely, to the domain of absolute convergence, for Theorem 3.1 states that absolute convergence at one point of this line implies absolute convergence at every point of this line. Indeed, either possibility can be observed: for $f(t) \equiv 1/(1+t^2)$, we find $\alpha = 0$, and the Laplace integral converges absolutely for all s with $\Re s = 0$; for $f(t) \equiv 1$, we find $\alpha = 0$, and the integral diverges absolutely for all s with $\Re s = 0$.

The number α is called the *abscissa of absolute convergence* of the Laplace integral; the open half-plane $\Re s > \alpha$, or the closed half-plane $\Re s \geqq \alpha$, is referred to as the *half-plane of absolute convergence* of the Laplace integral.

The next Theorem is so important that it deserves the designation "Fundamental Theorem". It will be used to determine the domain of simple convergence[1] of a Laplace integral.

Theorem 3.4 (Fundamental Theorem). *If the Laplace integral*

$$\int_0^\infty e^{-st} f(t)\, dt$$

converges for $s = s_0$, *then it converges in the open half-plane* $\Re s > \Re s_0$, *where it can be expressed by the absolutely converging integral*

$$(s - s_0) \int_0^\infty e^{-(s-s_0)t}\, \varphi(t)\, dt$$

with

$$\varphi(t) = \int_0^t e^{-s_0\tau}\, f(\tau)\, d\tau.$$

Supplementary Remark: The same conclusion is valid for a Laplace integral which does not converge at s_0 *(that is, the limit of* $\varphi(t)$ *as* $t \to \infty$ *does not exist), provided* $\varphi(t)$ *is bounded:* $|\varphi(t)| \leqq M$, *for* $t \geqq 0$.

Proof: Using integration by parts,[2] together with $\varphi(0) = 0$, one finds:

$$(1) \qquad \int_0^\omega e^{-st} f(t)\, dt = \int_0^\omega e^{-(s-s_0)t}\, e^{-s_0 t} f(t)\, dt =$$

$$= e^{-(s-s_0)\omega}\, \varphi(\omega) + (s - s_0) \int_0^\omega e^{-(s-s_0)t}\, \varphi(t)\, dt.$$

If the Laplace integral converges at s_0, then $\varphi(t)$ has a limit F_0 for $t \to \infty$. Moreover, the integral $\varphi(t)$ is continuous for $t \geqq 0$. Hence $\varphi(t)$ is bounded: $|\varphi(t)| \leqq M$ for $t \geqq 0$.[3] Consequently, for $\Re s > \Re s_0$ the limits

$$\lim_{\omega \to \infty} e^{-(s-s_0)\omega}\, \varphi(\omega) = 0$$

---

[1] An integral that converges, but does not converge absolutely, is called *conditionally* converging. We call a converging integral *simply* converging if the question regarding absolute or conditional convergence is avoided.

[2] Here, and often in the sequel, we use the "generalized" integration by parts: If

$$U(t) = A + \int_a^t u(\tau)\, d\tau, \quad V(t) = B + \int_a^t v(\tau)\, d\tau,$$

then

$$\int_a^b U(t)\, v(t) dt = U(t)\, V(t) \Big|_a^b - \int_a^b u(t)\, V(t)\, dt.$$

We require neither $u = U'$ nor $v = V'$.

[3] For sufficiently large $t > T$, $\varphi(t)$ differs but little from the limit F_0; hence, $\varphi(t)$ is bounded. In the finite interval $0 \leqq t \leqq T$, $\varphi(t)$ is continuous and, consequently, bounded. Thus, $\varphi(t)$ is bounded for $t \geqq 0$.

and
$$\lim_{\omega\to\infty}\int_0^{\omega} e^{-(s-s_0)t}\,\varphi(t)\,dt=\int_0^{\infty} e^{-(s-s_0)t}\,\varphi(t)\,dt$$
exist. The integral converges absolutely, since
$$\int_0^{\infty} |\,e^{-(s-s_0)t}\,\varphi(t)\,|\,dt \leq M\int_0^{\infty} e^{-\Re(s-s_0)t}\,dt.$$
From (1), we find for $\omega\to\infty$:
$$\int_0^{\infty} e^{-st} f(t)\,dt=(s-s_0)\int_0^{\infty} e^{-(s-s_0)t}\,\varphi(t)\,dt \quad \text{for } \Re s>\Re s_0.$$

Observe that in the proof we actually used no more than $|\varphi(t)|\leq M$.

For many proofs, valuable aid is derived from the possibility that any (possibly only conditionally converging) Laplace integral *can be expressed by an absolutely converging integral.*

Using Theorem 3.4, we derive, in a manner similar to the one employed in the verification of Theorem 3.3:

Theorem 3.5. *The exact domain of simple convergence of a Laplace integral is a right half-plane: $\Re s>\beta$, possibly including none of, or part of, or all of the line $\Re s=\beta$; admitting the possibilities that $\beta=\pm\infty$.*

For the function $f(t)\equiv 1/(1+t^2)$, we find $\beta=0$, and the entire line $\Re s=\beta$ belongs to the domain of simple convergence. For the function $f(t)\equiv 1/(1+t)$, we find $\beta=0$, and the integral diverges for $s=0$. However, it converges for $s=iy$ $(y\neq 0)$, since
$$\int_0^{\infty} e^{-iyt}\,\frac{1}{1+t}\,dt=\int_0^{\infty}\frac{\cos yt}{1+t}\,dt-i\int_0^{\infty}\frac{\sin yt}{1+t}\,dt,$$

and either integral converges for $y\neq 0$ (compare p. 9). For the function $f(t)\equiv 1$, we find $\beta=0$, and no point of the line $\Re s=0$ belongs to the domain of simple convergence.

We call the number β the *abscissa of convergence* of the Laplace integral; the open half-plane $\Re s>\beta$ is referred to as *half-plane of convergence* of the Laplace integral; the line $\Re s=\beta$ is called the *line of convergence.*

Obviously, it suffices to *investigate real numbers* in the search for the numbers α and β.

In Chapter 1, we developed the Laplace integral as a continuous analogue of the power series. The domain of convergence of a power series is a circular disc, possibly including points of the boundary. In the interior of the disc, the power series converges absolutely. When written in the form $\sum\limits_{n=0}^{\infty} a_n e^{-sn}$, the power series converges in a right half-plane; in the interior of this half-plane it converges absolutely. That is, for a power series, the domain of simple convergence and the domain of absolute convergence coincide, with the possible exception of points on the boundary. This is not generally true for the Laplace integral. In support of this statement, we present an example of a Laplace integral which *converges everywhere, yet nowhere absolutely*. Define the function $f(t)$[4] as follows:[5]

$$f(t) = \begin{cases} 0 & \text{for } 0 \leqq t < \log\log 3 = a \\ (-1)^n \exp\left(\frac{1}{2} e^t\right) & \text{for } \log\log n \leqq t < \log\log(n+1) \quad (n = 3, 4, \cdots). \end{cases}$$

This integral diverges absolutely for all s, since

$$\int_a^\infty | e^{-st} f(t) | \, dt = \int_a^\infty \exp\left(-\Re s \cdot t + \frac{1}{2} e^t\right) dt$$

and, however large $\Re s$ may be selected, $e^t/2$ ultimately grows more strongly than $\Re s \cdot t$.

To establish simple convergence, it is sufficient to consider real s. We investigate $f(t)$ in an interval of constant sign, and form the integral:

$$I_n = \int_{\log\log n}^{\log\log(n+1)} \exp\left(-s\,t + \frac{1}{2} e^t\right) dt = \int_n^{n+1} \frac{(\log x)^{-s-1}}{x^{1/2}} \, dx,$$

using the substitution $e^t = \log x$ to produce the right integral. For each positive or negative s, beyond a certain point, the integrand decreases monotonically to 0. Hence, from a certain n onwards, we have: $I_{n+1} < I_n$, and $I_n \to 0$ as $n \to \infty$. Letting, for the integral

$$\int_0^\omega e^{-st} f(t) \, dt, \tag{2}$$

the upper limit of integration approach $+\infty$, not in a continuous manner, but in discrete steps, through the values of $\log\log n$, for $n = 3, 4, 5, \ldots$, we generate the series:

$$-I_3 + I_4 - I_5 + - \cdots,$$

whose convergence is guaranteed by Leibniz' criterion for alternating series. Using arguments similar to those on p. 10, we conclude that (2) also converges for the continuously increasing upper limit of integration.

Hence, we actually encounter $\alpha \neq \beta$, in which case, necessarily, $\beta < \alpha$, since absolute convergence implies convergence. For this case we have a *band of conditional convergence* $\beta < \Re s < \alpha$.

---

[4] In the second line, we need $\log\log n \geqq 0$, which necessitates $n \geqq e = 2.71 \cdots$. It is for this reason that $f(t)$ is defined separately between 0 and $\log\log 3$.

[5] The form $\exp(x)$ is used instead of e^x, whenever the latter representation creates typographical difficulties.

4. The Laplace Integral as a Transformation

The Laplace integral of some function converges in a right half-plane, provided that it converges at some point. In this case, a function $F(s)$ is defined by the Laplace integral:

$$\int_0^\infty e^{-st} f(t)\,dt = F(s). \tag{1}$$

One may say that a correspondence is established between $f(t)$ and $F(s)$ by means of the Laplace integral. This correspondence may be interpreted as a transformation which transforms the function $f(t)$ into the function $F(s)$. In this sense, we call the correspondence the **Laplace transformation,** which is expressed by the symbol $\mathfrak{L}$:

$$\mathfrak{L}\{f(t)\} = F(s). \tag{2}$$

That is, the transformation $\mathfrak{L}$, when acting upon the function $f(t)$, produces the function $F(s)$, the *Laplace transform of* $f(t)$. For brevity, we shall write $\mathfrak{L}$-transformation instead of Laplace transformation.

The representation (2) should be interpreted like the notation of a function, $\varphi(x) = y$, which indicates: with the argument x we relate the value y by means of the function φ.

Using modern terminology, $\mathfrak{L}$ may be called an *operator* that produces the function $F(s)$ when acting upon the function $f(t)$. This operator $\mathfrak{L}$ has the following properties:
it is "additive" (or "distributive"):

$$\mathfrak{L}\{f_1 + f_2\} \doteq \mathfrak{L}\{f_1\} + \mathfrak{L}\{f_2\}$$

and "homogeneous":

$$\mathfrak{L}\{a f\} = a\,\mathfrak{L}\{f\} \quad (\alpha \text{ an arbitrary constant});$$

hence, it is "linear":

$$\mathfrak{L}\{a_1 f_1 + a_2 f_2\} = a_1\,\mathfrak{L}\{f_1\} + a_2\,\mathfrak{L}\{f_2\} \quad (\alpha_1, \alpha_2 \text{ arbitrary constants}).$$

The operation indicated by $\mathfrak{L}$ is integration, hence $\mathfrak{L}$ is classified as an integral operator, and we call the $\mathfrak{L}$-transformation an *integral transformation.*

A correspondence may be interpreted as a *mapping*. Envisage the correspondence, or transformation, performed by some apparatus like a photographic camera which produces an image of the original. From this interpretation originates the terminology: **original function** for $f(t)$, and **image function** for $F(s)$.

In modern mathematics, a suggestive approach results from considering the totality of specific objects as a set of points in some abstract space. In an abstract space, one may conveniently visualize logical relations by means of concepts borrowed from geometry. In this sense, one defines as the **original (function) space** the totality of all functions $f(t)$ for which the Laplace integral converges at some point, and which, therefore, may be encountered as original functions. Similarly, one defines as the **image (function) space,** the totality of all functions that may occur as image functions of the $\mathfrak{L}$-transformation.

In the sequel, lower case letters will be reserved to denote original functions, and the corresponding upper case letters will be used to represent the respective image functions; for instance, f and F, φ and Φ. Exceptions to this practical convention have to be made for those symbols which are already specified by prior usage, like $\Gamma(t)$. The chosen notation conveniently simplifies the presentation of the theory. In applications, one must occasionally yield to established practices.

Sometimes, one wants to indicate the point s, at which the $\mathfrak{L}$-transform is to be evaluated. This can be accomplished by either of the following notations:

$$\mathfrak{L}\{f; s\} \quad \text{or} \quad \mathfrak{L}\{f\}_s.$$

The relation (2) may be expressed more compactly by the **symbol of correspondence** ○—●, which is defined by

$$f(t) \circ\!\!-\!\!\bullet F(s) \text{ or equivalently } F(s) \bullet\!\!-\!\!\circ f(t),$$

read: "$f(t)$ is correlated with $F(s)$", or "$f(t)$ corresponds to $F(s)$".

In all those problems where actual evaluation of the integral (1) is not attempted, the reader is advised to de-emphasize the explicit definition of the $\mathfrak{L}$-transformation by an integral, and to use the operator $\mathfrak{L}$ or the symbol of correspondence: ○—●. In applications, we will primarily be concerned with the *properties* of the $\mathfrak{L}$-transformation represented by these symbols, disregarding the integral by which this transformation is defined. For, in a similar manner, when working with integrals, one usually thinks of rules and properties rather than of the definition of the integral as the limit of a sum.

5. The Unique Inverse of the Laplace Transformation

In the previous Chapters, we have consistently considered the $\mathfrak{L}$-transformation as the correspondence which relates with each original function its corresponding image function, obviously in a unique manner. This relation may be looked at in the inverse orientation; that is, one may begin with some specific image function and seek the corresponding original functions. This inverse transformation will be designated as $\mathfrak{L}^{-1}$-transformation. Clearly, this inverse transformation cannot be unique, for two original functions that differ at a finite number of points, nevertheless have the same image function. Indeed, this conclusion may be carried even further. For this purpose, we introduce the nullfunction $n(t)$, which is characterized by the property that its definite integral vanishes identically for all upper limits:[1]

(1) $$\int_0^t n(\tau)\, d\tau = 0 \quad \text{for all } t \geqq 0,$$

[1] When using Lebesgue integration, a function satisfies the condition (1) of the nullfunction if and only

For such a nullfunction one concludes, using integration by parts:

$$\int_0^\omega e^{-st}\, n(t)\, dt = e^{-st}\int_0^t n(\tau)\, d\tau \Big|_0^\omega + s\int_0^\omega e^{-st}\, dt \int_0^t n(\tau)\, d\tau = 0,$$

hence

$$\lim_{\omega\to\infty}\int_0^\omega e^{-st}\, n(t)\, dt = \mathfrak{L}(n) = 0.$$

With this, we have demonstrated that any nullfunction may be added to an original function without affecting the corresponding image function; hence, the inverse transformation cannot be single-valued. Fortunately, with the above observation we have accounted for all possibilities, for we can state the

Theorem 5.1 (Uniqueness Theorem). *Two original functions, whose image functions are identical (in a right half-plane), differ at most by a nullfunction.*

In the Lebesgue theory it is customary to consider two functions as being equivalent, provided they differ only by a nullfunction. In this sense, the $\mathfrak{L}^{-1}$-transformation is unique.

Clearly, Theorem 5.1 is equivalent to the statement: If $\mathfrak{L}\{f\} = F(s) \equiv 0$; then $f(t)$ is a nullfunction. It is interesting that the same conclusion may be reached from the weaker hypothesis: $\mathfrak{L}\{f\} = F(s)$ assumes the value zero on a sequence of points that are located at equal intervals along a line parallel to the real axis. To establish this more powerful statement, we shall need

Theorem 5.2. *Let $\psi(x)$ be a continuous function, and suppose that the moments of every order of $\psi(x)$ on the finite interval (a, b) vanish, that is:*

$$\int_a^b x^\mu\, \psi(x)\, dx = 0 \quad \textit{für } \mu = 0, 1, \cdots;$$

then: $\psi(x) \equiv 0$ in (a, b).

Proof: Without loss of generality, we may restrict $\psi(x)$ to be real-valued. For a complex-valued $\psi(x)$, the conclusion may then be applied separately to show the simultaneous vanishing of the real part and the imaginary part. The Weierstrass approximation theorem guarantees, for every $\delta > 0$, the existence of a polynomial $p_\delta(x)$ such that in the finite interval (a, b) the continuous function $\psi(x)$ differs from $p_\delta(x)$ by at most δ, hence

$$\psi(x) = p_\delta(x) + \delta\, \vartheta(x) \quad \text{with} \quad |\vartheta(x)| \leqq 1 \quad \text{for } a \leqq x \leqq b\,.$$

---

if the function assumes the value zero almost everywhere, that is everywhere except on a set of measure zero. Clearly, for any nullfunction $n(t)$, $e^{-st}n(t)$ is also a nullfunction, and $\mathfrak{L}\{n\} = 0$.
The criterion (1) for a nullfunction was chosen since it is also applicable to Riemann integration, while a function which is zero almost everywhere need not be Riemann integrable; a fact which is demonstrated by the example: $n(t) \equiv 1$ for rational t, and $n(t) \equiv 0$ for irrational t.

Multiplying this equation by $\psi(x)$, and then integrating between a and b, yields

$$\int_c^b \psi^2(x)\,dx = \int_a^b p_\delta(x)\,\psi(x)\,dx + \delta\int_a^b \vartheta(x)\,\psi(x)\,dx\,.$$

The first integral of the right hand side is a linear combination of moments of $\psi(x)$ on (a, b); it is zero by hypothesis. The remainder of the equation indicates:

$$\int_a^b \psi^2(x)\,dx \leqq \delta\int_a^b |\,\psi(x)\,|\,dx\,.$$

Let us suppose that $\psi(x)$ is not identically zero in (a, b). Then there exists a point, and by continuity of $\psi(x)$ an entire neighbourhood of this point, where $|\psi(x)| > 0$. Whence

$$\int_a^b \psi^2(x)\,dx > 0\,, \text{ and } \int_a^b |\,\psi(x)\,|\,dx > 0\,,$$

and we may divide by the latter non-zero number, to find

$$\delta \geqq \int_a^b \psi^2(x)\,dx : \int_a^b |\,\psi(x)\,|\,dx > 0\,,$$

thus producing a contradiction, since δ may be selected to be arbitrarily small. Hence $\psi(x) \equiv 0$ in (a, b).

This Theorem 5.2 is used in the verification of

Theorem 5.3. *If $\mathfrak{L}\{f\} = F(s)$ vanishes on an infinite sequence of points that are located at equal intervals along a line parallel to the real axis:*

$$F(s_0 + n\,\sigma) = 0 \quad (\sigma > 0\,, n = 1, 2, \cdots) \tag{2}$$

s_0 *being a point of convergence of $\mathfrak{L}\{f\}$; then it follows that $f(t)$ is a nullfunction.*

Remark: Observe the interesting implication of Theorem 5.3: An image function which vanishes on a sequence of equidistant points along a line parallel to the real axis, vanishes identically.

Proof of Theorem 5.3: Invoking the Fundamental Theorem 3.4, for $\Re s > \Re s_0$, we find

$$F(s) = (s - s_0)\int_0^\infty e^{-(s-s_0)t}\,\varphi(t)\,dt$$

with

$$\varphi(t) = \int_0^t e^{-s_0\tau}\,f(\tau)\,d\tau\,;$$

hence,

$$F(s_0 + n\,\sigma) = n\,\sigma\int_0^\infty e^{-n\sigma t}\,\varphi(t)\,dt\,.$$

By hypothesis (2):

$$\int_0^\infty e^{-n\sigma t}\,\varphi(t)\,dt = 0 \quad \text{for } n = 1, 2, \cdots$$

Employing the substitution

$$e^{-\sigma t} = x\,,\; t = -\frac{\log x}{\sigma}\,,\; \varphi\left(-\frac{\log x}{\sigma}\right) = \psi(x)\,,$$

we rewrite the last equation:

$$\frac{1}{\sigma}\int_0^1 x^{n-1}\,\psi(x)\,dx = 0 \quad \text{for } n = 1, 2, \cdots$$

or

$$\int_0^1 x^{\mu}\,\psi(x)\,dx = 0 \qquad \text{for } \mu = 0, 1, \cdots$$

To make $\psi(x)$ continuous in the interval $0 \leqq x \leqq 1$,[2] define:

$$\psi(0) = \lim_{t\to\infty} \varphi(t) = F(s_0),\; \psi(1) = \varphi(0) = 0\,.$$

Thus, we may apply Theorem 5.2, and we observe that

$$\text{(3)} \qquad \psi(x) \equiv 0\,,\ \text{that is } \varphi(t) = \int_0^t e^{-s_0\tau} f(\tau)\,d\tau \equiv 0\,.$$

Integration by parts produces:

$$e^{-s_0 t}\int_0^t f(\tau)\,d\tau + s_0\int_0^t e^{-s_0\tau}\,d\tau\int_0^\tau f(u)\,du \equiv 0\,.$$

The integrand of the second integral is continuous, hence we can differentiate this integral, and consequently also the first integral. Differentiation yields:

$$-s_0\,e^{-s_0 t}\int_0^t f(\tau)\,d\tau + e^{-s_0 t}\frac{d}{dt}\int_0^t f(\tau)\,d\tau + s_0\,e^{-s_0 t}\int_0^t f(u)\,du \equiv 0$$

or

$$\frac{d}{dt}\int_0^t f(\tau)\,d\tau \equiv 0\,.$$

---

[2] The function $f(t)$ need not be continuous; however, we do require a continuous function $\psi(x)$ to satisfy the hypotheses of Theorem 5.2. It is for this reason that we did not start directly with

$$F(s_0 + n\,\sigma) = \int_0^\infty e^{-n\sigma t}\,[e^{-s_0 t}\,f(t)]\,dt = 0$$

in the above proof, but instead, followed the detour involving $\varphi(t)$.

With $\int_0^t f(\tau)\,d\tau = 0$ for $t = 0$, it follows that

$$\int_0^t f(\tau)\,d\tau \equiv 0\,,$$

that is, $f(t)$ is a nullfunction.[3]

A non-trivial image function $F(s) = \mathfrak{L}\{f\}$ may nevertheless have infinitely many zeros along a line parallel to the real axis, provided these are not spaced at equal intervals along this line. This is demonstrated by two examples:

(4) $$\mathfrak{L}\left\{\frac{1}{\sqrt{\pi t}}\cos\frac{1}{t}\right\} = \frac{1}{\sqrt{s}}\,e^{-\sqrt{2s}}\cos\sqrt{2s}\,,$$

(5) $$\mathfrak{L}\left\{\frac{1}{\sqrt{\pi t}}\sin\frac{1}{t}\right\} = \frac{1}{\sqrt{s}}\,e^{-\sqrt{2s}}\sin\sqrt{2s}\,.$$

With the aid of Theorem 5.3, we strengthen Theorem 5.1, and obtain

Theorem 5.4 (Strengthened Uniqueness Theorem). *Two original functions, whose image functions assume equal values on an infinite sequence of points that are located at equal intervals along a line parallel to the real axis, differ at most by a nullfunction.*

For applications, one often needs to establish exact equality of original functions. This may be accomplished with the aid of

Theorem 5.5. *Two original functions which differ by a nullfunction, are exactly equal at those points where both functions are either continuous from the left, or continuous from the right.*

Suppose f_1 and f_2 are continuous from the left at the point t, then $n = f_1 - f_2$ is also continuous from the left at this point t. Hence, $n(t)$ may be obtained by differentation of $\int_0^t n(\tau)\,d\tau \equiv 0$ from the left; this yields zero.

As a consequence of Theorem 5.3, we establish

Theorem 5.6. *A Laplace transform $F(s) \not\equiv 0$ cannot be periodic.*

Proof: Suppose $F(s)$ is periodic with complex period σ; that is, in the right half-plane of convergence:

$$F(s) = F(s+\sigma) \qquad (\sigma \text{ a complex constant}),$$

then

$$\int_0^\infty e^{-st} f(t)\,dt - \int_0^\infty e^{-(s+\sigma)t} f(t)\,dt = \int_0^\infty e^{-st}\left(1 - e^{-\sigma t}\right) f(t)\,dt = 0\,.$$

[3] In the Lebesgue theory, (3) immediately implies that $e^{-st} f(t)$ is a nullfunction and, furthermore, that $f(t)$ is a nullfunction, since $e^{-s_0 t} \neq 0$.

Hence $(1 - e^{-\sigma t}) f(t)$ is a nullfunction. The factor $(1 - e^{-\sigma t})$ has the zeros: for non-imaginary σ, there is exactly one zero at $t = 0$; for purely imaginary σ, there are zeros given by the real-valued sequence $t = n 2\pi i/\sigma$ with $n = 0, 1, 2, 3, \cdots$. In the Lebesgue theory one can immediately conclude that $f(t)$ is a nullfunction. When using Riemann integration, one can establish the same conclusion, for we have

$$\int_0^t (1 - e^{-\sigma t}) f(t)\, dt \equiv 0,$$

or

$$\int_0^t f(\tau)\, d\tau = \int_0^t e^{-\sigma\tau} f(\tau)\, d\tau \quad \text{for } t \geqq 0, \tag{6}$$

hence, with

$$\int_0^t f(\tau)\, d\tau = \varphi(t),$$

and using integration by parts:

$$\varphi(t) = \int_0^t e^{-\sigma\tau} f(\tau)\, d\tau = e^{-\sigma t}\, \varphi(t) + \sigma \int_0^t e^{-\sigma\tau}\, \varphi(\tau)\, d\tau,$$

that is:

$$(1 - e^{-\sigma t})\, \varphi(t) = \sigma \int_0^t e^{-\sigma\tau}\, \varphi(\tau)\, d\tau. \tag{7}$$

The function $\varphi(t)$ is continuous, hence the right hand side can be differentiated, consequently also the left hand side, and $\varphi(t)$ also, with the possible exception of the zeros of $(1 - e^{-\sigma t})$, which we enumerated above. Differentiation of (7) produces:

$$\sigma\, e^{-\sigma t}\, \varphi(t) + (1 - e^{-\sigma t})\, \varphi'(t) = \sigma\, e^{-\sigma t}\, \varphi(t);$$

hence,

$$\varphi'(t) = 0$$

with the possible exception of the zeroes of $(1 - e^{-\sigma t})$. For the continuous function $\varphi(t)$ with $\varphi(0) = 0$, it follows that $\varphi(t) \equiv 0$. Thus $f(t)$ is a nullfunction, and $F(s) \equiv 0$.

Theorem 5.6, when applied to the function $e^{-\alpha s}$ (α arbitrary, complex) with period $2\pi i/\alpha$, shows that $e^{-\alpha s}$ is not an image function.

6. The Laplace Transform as an Analytic Function

On p. 5 we developed the Laplace integral as a continuous analogue of the power series. In this Chapter, we shall demonstrate that a Laplace integral, like a power series, always represents an analytic function.

Theorem 6.1. *A $\mathfrak{L}$-transform is an analytic function in the interior of its half-plane of convergence, $\Re s > \beta$, that is, it has derivatives of all orders. The derivatives are obtained by differentiation under the integral symbol:*

$$F^{(n)}(s) = (-1)^n \int_0^\infty e^{-st}\, t^n f(t)\, dt = (-1)^n \mathfrak{L}\{t^n f(t)\}.$$

The derivatives too are $\mathfrak{L}$-transforms.

Proof: It suffices to verify the Theorem for $n = 1$; the general conclusion follows by iteration. For an interior point of the half-plane of convergence s we must show that

(1) $$\lim_{h\to 0} \frac{F(s+h) - F(s)}{h} = -\int_0^\infty e^{-st}\, t f(t)\, dt.$$

The analogous steps in the corresponding proof for the power series are easy, for the power series converges absolutely in the interior of the domain of convergence. However, this cannot be generally assumed for the $\mathfrak{L}$-integral. Thus, we need resort to the Fundamental Theorem 3.4, and represent $F(s)$ by the absolutely converging integral

(2) $$F(s) = (s - s_0) \int_0^\infty e^{-(s-s_0)t}\, \varphi(t)\, dt \quad (\Re s > \Re s_0)$$

with

(3) $$\varphi(t) = \int_0^t e^{-s_0\tau} f(\tau)\, d\tau,$$

where s_0 is a point of convergence of $\mathfrak{L}\{f\}$. We select this point s_0 in the following manner: Since $\Re s > \beta$, the abscissa of convergence, we can set $\Re s - \beta = 3\xi > 0$, and[1] we specify:

$$s_0 = \beta + \xi,$$

hence:

(4) $$\Re(s - s_0) = 2\xi > 0.$$

Formally differentiating (2) under the integral symbol, we find

(5) $$\Psi(s) = \int_0^\infty e^{-(s-s_0)t}\, \varphi(t)\, dt - (s - s_0) \int_0^\infty e^{-(s-s_0)t}\, t\, \varphi(t)\, dt.$$

[1] In case $\beta = -\infty$, replace β in the proof by any real number to the left of s.

Now, we claim that

$$\lim_{h\to 0} \frac{F(s+h)-F(s)}{h} = \Psi(s).$$

For the verification, we a priori restrict

$$|h| < \xi \tag{6}$$

and form, using the expression (2) for $F(s)$,

$$D(h) = \frac{F(s+h)-F(s)}{h} - \Psi(s) =$$

$$= \frac{1}{h}\left\{(s+h-s_0)\int_0^\infty e^{-(s+h-s_0)t}\,\varphi(t)\,dt - (s-s_0)\int_0^\infty e^{-(s-s_0)t}\,\varphi(t)\,dt\right\}$$

$$-\int_0^\infty e^{-(s-s_0)t}\,\varphi(t)\,dt + (s-s_0)\int_0^\infty e^{-(s-s_0)t}\,t\,\varphi(t)\,dt$$

$$= \int_0^\infty e^{-(s-s_0)t}\,(e^{-ht}-1)\,\varphi(t)\,dt + (s-s_0)\int_0^\infty e^{-(s-s_0)t}\left(\frac{e^{-ht}-1}{h}+t\right)\varphi(t)\,dt.$$

Exploiting the restriction (6), we can establish the following bound

$$|e^{-ht}-1| = \left|-\frac{ht}{1!}+\frac{(ht)^2}{2!}-\frac{(ht)^3}{3!}+\cdots\right| \leqq |h|\,t\left(1+\frac{|h|t}{1!}+\frac{|h|^2t^2}{2!}+\cdots\right)$$

$$= |h|\,t\,e^{|h|t} \leqq |h|\,t\,e^{\xi t},$$

$$\left|\frac{e^{-ht}-1}{h}+t\right| = \left|\frac{ht^2}{2!}-\frac{h^2t^3}{3!}+\frac{h^3t^4}{4!}-+\cdots\right|$$

$$\leqq |h|\,t^2\left(1+\frac{|h|t}{1!}+\frac{|h|^2t^2}{2!}+\cdots\right) = |h|\,t^2 e^{|h|t} \leqq |h|\,t^2 e^{\xi t}.$$

The function $\varphi(t)$ is continuous for $t \geqq 0$, and $\lim_{t\to\infty}\varphi(t) = F(s_0)$, hence

$$|\varphi(t)| \leqq M \quad \text{for } t \geqq 0. \tag{7}$$

Because of (4), we obtain

$$|D(h)| \leqq |h|\,M\int_0^\infty e^{-2\xi t}\,t\,e^{\xi t}\,dt + |s-s_0|\,|h|\,M\int_0^\infty e^{-2\xi t}\,t^2 e^{\xi t}\,dt$$

$$= |h|\,M\left\{\int_0^\infty e^{-\xi t}\,t\,dt + |s-s_0|\int_0^\infty e^{-\xi t}\,t^2\,dt\right\} = |h|\,\text{const.}$$

Consequently, $D(h) \to 0$, for $h \to 0$, which implies that $F'(s) = \Psi(s)$.

Now, we seek a simpler representation for $\Psi(s)$. Using integration by parts, and (3), we find:

$$\int_0^t \tau[e^{-s_0\tau} f(\tau)]\,d\tau = \tau\,\varphi(\tau)\Big|_0^t - \int_0^t \varphi(\tau)\,d\tau = t\,\varphi(t) - \int_0^t \varphi(\tau)\,d\tau\,,$$

that is:

$$t\,\varphi(t) = \int_0^t [\varphi(\tau) + \tau\, e^{-s_0\tau} f(\tau)]\,d\tau\,, \tag{8}$$

and, because of (8),

$$-(s - s_0)\int_0^\infty e^{-(s-s_0)t}\,[t\,\varphi(t)]\,dt$$

$$= e^{-(s-s_0)t}\,t\,\varphi(t)\Big|_0^\infty - \int_0^\infty e^{-(s-s_0)t}\,[\varphi(t) + t\,e^{-s_0 t} f(t)]\,dt$$

$$= -\int_0^\infty e^{-(s-s_0)t}\,\varphi(t)\,dt - \int_0^\infty e^{-st}\,t\,f(t)\,dt\,, \tag{9}$$

since for $t \to \infty$, $e^{-(s-s_0)t}t\varphi(t) \to 0$, because of (4) and (7). Substitution of (9) into (5) yields for $\Psi(s)$ the expression (1).

Remark: Theorem 6.1 could also be established by firstly verifying it for $\mathfrak{L}$-integrals with finite upper limits T, and then invoking the limiting process: $T \to \infty$. For this purpose, one requires deeper theorems of both Lebesgue theory and theory of functions. In contrast, the above verification employs elementary means only.

The finite $\mathfrak{L}$-integral exists for every s. Thus, one may formulate the more specialized

Theorem 6.2. *A "finite" $\mathfrak{L}$-transform*

$$F(s) = \int_0^T e^{-st} f(t)\,dt$$

is an entire function.

Although the $\mathfrak{L}$-transform does behave like a power series insofar as it represents an analytic function, it does, nevertheless, differ essentially from a power series in several respects:

1. Every function which is analytic on a circular disc can be expressed by a power series. However, not every function which is analytic in a right half-plane

can be represented by a $\mathfrak{L}$-transform. This is demonstrated by the example e^{-as} (see p. 25).

2. A power series converges exactly on the largest circular disc about the centre of the power series expansion, on which the represented function is analytic; that is, there is at least one singular point on the boundary of the circular disc of convergence. By contrast, *on the line of convergence of a $\mathfrak{L}$-transform, there need not be a singular point of the function.* Indeed, the function may be analytic in a half-plane which extends beyond the half-plane of convergence, possibly in the entire plane. This is demonstrated by the following example. Let $f(t)$ be defined by

$$f(t) = -\pi e^t \sin \pi e^t = \frac{a}{dt} (\cos \pi e^t).$$

Using integration by parts twice, one obtains:

$$\int_0^{\omega} e^{-st} (-\pi e^t \sin \pi e^t)\, dt$$

$$= 1 + e^{-s\omega} \cos \pi e^{\omega} + \frac{s}{\pi} e^{-(s+1)\omega} \sin \pi e^{\omega} + \frac{s(s+1)}{\pi} \int_0^{\omega} e^{-(s+1)t} \sin \pi e^t \, dt .$$

When considering the limit for the right hand side as $\omega \to \infty$, we observe that the second term converges exactly for $\Re s > 0$, and that all other terms certainly have limits for $\Re (s+1) > 0$; hence $\mathfrak{L}\{f\}$ converges exactly for $\Re s > 0$, and

$$\int_0^{\infty} e^{-st} (-\pi e^t \sin \pi e^t)\, dt = 1 - \frac{s(s+1)}{\pi^2} \int_0^{\infty} e^{-(s+2)t} (-\pi e^t \sin \pi e^t)\, dt,$$

which may be written as a functional equation:

$$F(s) = 1 - \frac{s(s+1)}{\pi^2} F(s+2).$$

With the aid of the latter, one can continue $F(s)$, analytically and free of singularities from the half-plane $\Re s > 0$ into the half-plane $\Re s > -2$, from this half-plane into the half-plane $\Re s > -4$, and so forth; thus, $F(s)$ is analytic in the entire plane.

This example demonstrates that the half-plane of convergence of the $\mathfrak{L}$-integral need not be the largest plane in which the $\mathfrak{L}$-transform is an analytic function. For this reason, we define:

Let χ be the lower limit of the real x so that $F(s)$ is analytic for $\Re s > x$. Then we call χ the *holomorphy abscissa*,[2] and the half-plane $\Re s > \chi$ the *holomorphy half-plane* for $F(s) = \mathfrak{L}\{f\}$. We have $\chi \leqq \beta$; cases with $\chi < \beta$ are actually encountered.

[2] The term "holomorphic" is used occasionally instead of "analytic", particularly if the function is not considered as a whole, but locally. For instance, "$F(s)$ is holomorphic at s_0" means "$F(s)$ is analytic in some neighbourhood of s_0". Thus, we can replace the rather ugly term analyticity by the term holomorphy.

Theorem 6.1 is of extreme importance for the theory of the $\mathfrak{L}$-transformation. The original function need be defined for real t only; it is largely unrestricted. The image function, however, belongs to the distinguished class of analytic functions, so that one can employ the powerful theory of complex functions in the study of image functions.

7. The Mapping of a Linear Substitution of the Variable

To explain the intentions which govern several of the subsequent Chapters let us employ an analogy. In Chapter 4 we compared the $\mathfrak{L}$-transformation with some apparatus which *maps* two function spaces into each other. Alternatively, one could compare the process of this transformation with the translation between two languages. The latter establishes a correspondence between the words of two languages, while the $\mathfrak{L}$-transformation relates the functions of two function spaces. The first requisite for a translation between languages is a collection of correspondences between individual words, that is, a *dictionary*. Similarly, for the $\mathfrak{L}$-transformation we need a collection of correspondences between individual functions, a table of transforms. However, a dictionary is insufficient when attempting a translation between two languages. For instance, verbs are conjugated in the first language, and one must know how to perform the corresponding operation in the second language. Also, words are joined in the first language to form sentences, and the translator must know how to construct the appropriate form of the sentence in the second language. In short, for a translation between two languages, one requires not only a dictionary but also a *grammar*, the rules of which tell us how some operations performed on the words of the first language are reflected in the second language. The same principle applies to the $\mathfrak{L}$-transformation. When subjecting some original function to a specified operation, like differentiation, then there is a corresponding operation that acts upon the image function. Similarly, when one forms a sum, or a product, or some other association of several original functions, then the corresponding image functions are to be combined in a certain manner. That is, here we are interested in correspondences of operations rather than those of functions.

In the sequel, we shall derive these "*grammatical rules*" for the $\mathfrak{L}$-transformation. We begin with the simplest modification, replacing the independent variable of the original function by a linear substitution, and we seek the corresponding operation on the image function. We start with two special cases.

Given some original function $f(t)$, we form a new original function $f_1(t)$ by replacing the independent variable t by at:

$$f_1(t) = f(a\,t)\,.$$

In general, this substitution has meaning only for $a > 0$, since $f(t)$ need be defined for $t \geqq 0$ only. One finds:

$$\mathfrak{L}\{f_1(t)\} = \int_0^\infty e^{-st} f(at)\,dt = \frac{1}{a}\int_0^\infty e^{-(s/a)u} f(u)\,du = \frac{1}{a}F\left(\frac{s}{a}\right).$$

The graph of f_1 is obtained from the graph of f by a similarity deformation (shrinking of the t extension) with ratio $1:a$; for instance for $a = 2$, the abscissa of the graph is shrunk to one half. Therefore the above conclusion is called "Similarity Theorem":

Theorem 7.1 (Similarity Theorem). *We have:*

$$f(at) \circ\!\!-\!\!\bullet \frac{1}{a} F\left(\frac{s}{a}\right) \quad \textit{for } a > 0.$$

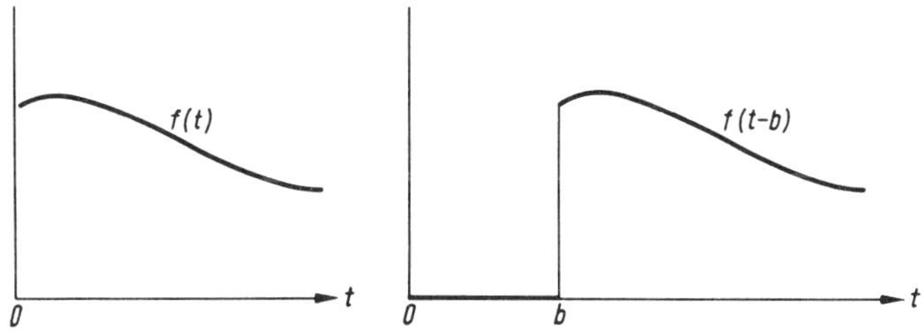

Figure 3

As a second case we consider the change of $f(t)$ to

$$f_2(t) = f(t - b) \text{ where } b > 0.$$

The original function $f(t)$ being defined for $t \geqq 0$ only, $f_2(t)$ has meaning only for $t > b$. For the evaluation of the $\mathfrak{L}$-transform of $f_2(t)$, we must define $f_2(t)$ in the interval $0 \leqq t < b$. This is accomplished by assigning the value zero to the function $f_2(t)$ in this interval $0 \leqq t < b$.
Hence:

$$f_2(t) = \begin{cases} 0 & \text{for } 0 \leqq t < b \\ f(t - b) & \text{for } t \geqq b. \end{cases}$$

In geometric terms, we have translated or shifted the graph of f to the right through the distance b, and we used the section of the t-axis between 0 and b to complete the graph of f_2 (compare Fig. 3).

The above presented explanation of f_2 follows trivially, provided we agree to generally assign to any original function $f(t)$ the value zero for $t < 0$. The function f_2 can be expressed in terms of function f more conveniently and compactly by the use of the shifted unit step function introduced as example 2 in Chapter 2:

$$f_2(t) = f(t - b)\,u(t - b).$$

The $\mathfrak{L}$-transform of $f_2(t)$ is found by:

$$\mathfrak{L}\{f_2\} = \int_b^\infty e^{-st} f(t-b)\, dt = e^{-bs} \int_0^\infty e^{-su} f(u)\, du = e^{-bs} F(s)\,.$$

Theorem 7.2 (First Shifting Theorem). *We have:*

$$f(t-b)\, u(t-b) \circ\!\!-\!\!\bullet\, e^{-bs}\, F(s) \quad \textit{for } b > 0\,.$$

With the Theorems 7.1, and 7.2, we established the grammatical rules that tell us what operation on $F(s)$ results from the simplest operation on $f(t)$: the similarity transformation, and the translation or shifting. We shall apply these in several practical applications:

Had we considered in example 6 of Chapter 2 merely the special case

$$\cos t \circ\!\!-\!\!\bullet \frac{s}{s^2+1}\,,$$

we could now use the Similarity Theorem 7.1, to find the more general result with $a > 0$:

$$\cos a\,t \circ\!\!-\!\!\bullet \frac{1}{a}\, \frac{s/a}{(s/a)^2+1} = \frac{s}{s^2+a^2}\,.$$

Indeed, the derived result is correct for arbitrary complex a, as shown in Chapter 2. In general, one should be cautious when using the Similarity Theorem 7.2 for negative a, and more so for complex a. Erroneous conclusions may result, although the $f(at)$ may have meaning. For instance, the Bessel function

$$J_0(t) = \sum_{n=0}^{\infty} \frac{(-1)^n}{(n!)^2} \left(\frac{t}{2}\right)^{2n}$$

has the $\mathfrak{L}$-transform (compare p. 266):

$$\mathfrak{L}\{J_0(t)\} = \frac{1}{\sqrt{1+s^2}} \quad \text{for } \Re s > 0\,.$$

When using $a = -1$ with the Similarity Theorem 7.1, one would produce:

$$J_0(-t) \circ\!\!-\!\!\bullet -\frac{1}{\sqrt{1+s^2}}\,.$$

We find, in fact, when using $J_0(-t) = J_0(t)$:

$$J_0(-t) \circ\!\!-\!\!\bullet \frac{1}{\sqrt{1+s^2}}\,.$$

We utilize the First Shifting Theorem 7.2 to find the $\mathfrak{L}$-transform of the function:

$$f(t) = \begin{cases} \sin t & \text{for } 0 \leqq t \leqq 2\pi \\ 0 & \text{for } t > 2\pi\,. \end{cases}$$

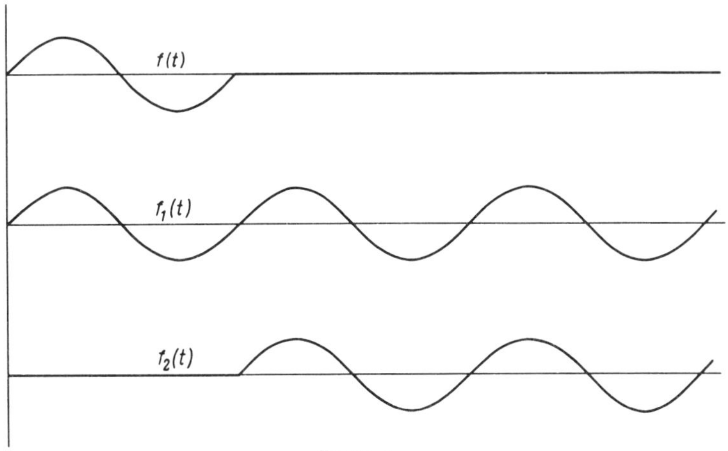

Figure 4

As shown in Fig. 4, $f(t)$ may be represented as the difference of the function

$$f_1(t) = \sin t \ ,$$

defined for $t \geqq 0$, and the function

$$f_2(t) = \begin{cases} 0 & \text{for } 0 \leqq t \leqq 2\pi \\ \sin(t - 2\pi) = \sin t & \text{for } t > 2\pi \ , \end{cases}$$

which is generated by translating the function $f_1(t)$ to the right through the distance 2π. The $\mathfrak{L}$-transform of the function $f(t)$ is therefore, using Theorem 7.2:

$$F(s) = \frac{1}{s^2+1} - \frac{1}{s^2+1} e^{-2\pi s} = \frac{1-e^{-2\pi s}}{s^2+1} \ .$$

Consider some arbitrary *periodic function* $p(t)$, having period ω, which has the $\mathfrak{L}$-transform $P(s)$. Define the function $p_\omega(t)$ to be equal to $p(t)$ within its first period, $0 \leqq t < \omega$, and to assume the value zero outside this interval. Employing the method of the example above, we can find the $\mathfrak{L}$-transform of $p_\omega(t)$:

(1) $$P_\omega(s) = P(s)\left(1 - e^{-\omega s}\right) .$$

Naturally, one may invert formula (1) to produce formula (2), by means of which one can determine the $\mathfrak{L}$-transform $P(s)$ of the periodic function $p(t)$, using the "finite" $\mathfrak{L}$-transform $P_\omega(s)$ of the first period:

(2) $$P(s) = \frac{1}{1-e^{-\omega s}} \int_0^\omega e^{-st} p(t)\, dt \, .$$

So far, we have started with some original function and found the corresponding image function. This association is unique. In Chapter 5, we have shown by Theorem 5.1 that every image function has an essentially unique original function. Thus, we could employ the First Shifting Theorem 7.2 inversely; for instance, whenever we encounter an image function of the type $F_2(s) = e^{-bs}F(s)$, with $F(s) \bullet\!\!-\!\!\circ f(t)$, then the original function of $F_2(s)$ is the function $f(t)$ shifted through the distance b to the right, with $f(t) = 0$ for $t < 0$. In this manner we can immediately find the original function of

$$\frac{1}{s(s+1)}\, e^{-s};$$

for we have:

$$\frac{1}{s(s+1)} = \frac{1}{s} - \frac{1}{s+1} \circ\!\!-\!\!\bullet\ 1 - e^{-t},$$

hence

$$\frac{1}{s(s+1)}\, e^{-s} \circ\!\!-\!\!\bullet \begin{cases} 0 & \text{for } 0 \leqq t < 1 \\ 1 - e^{-(t-1)} & \text{for } t \geqq 1. \end{cases}$$

Combining the conclusions of the special cases expressed by Theorems 7.1, and 7.2, we produce a general linear substitution. For this purpose consider:

$$f_0(t) = f(at - b) \quad \text{with } a > 0,\, b > 0,$$

setting now:

$$f_0(t) = 0 \quad \text{for } at - b < 0\text{, that is, for } \quad t < \frac{b}{a}.$$

Instead of successively applying Theorem 7.1 and Theorem 7.2, we evaluate the $\mathfrak{L}$-transform of $f_0(t)$ directly:

$$\mathfrak{L}\{f_0\} = \int_{b/a}^{\infty} e^{-st} \; f(at - b)\; dt = \frac{1}{a}\, e^{-(b/a)s} \int_0^{\infty} e^{-(s/a)u} \; f(u)\, du = \frac{1}{a}\, e^{-(b/a)s} F\left(\frac{s}{a}\right).$$

Theorem 7.3. *We have:*

$$f(at - b) \circ\!\!-\!\!\bullet \frac{1}{a}\, e^{-(b/a)s} F\left(\frac{s}{a}\right) \quad \textit{for } a > 0,\ b > 0,$$

provided $f(t) = 0$ *for* $t < 0$.

Replacing, in Theorem 7.3, $-b$ by the positive number $+b$, we obtain the function $f(at + b)$ which is defined for all $t \geqq 0$. However, this new function does not encompass all values of $f(t)$, only those with $t \geqq b$. A corresponding rule can be derived, which is not used as often as Theorem 7.3, although it does find applications, for instance with difference equations. We derive:

$$\mathfrak{L}\{f(at+b)\} = \int_0^\infty e^{-st} f(at+b)\,dt = \frac{1}{a} e^{(b/a)s} \int_b^\infty e^{-(s/a)u} f(u)\,du$$

$$= \frac{1}{a} e^{(b/a)s} \left\{ \int_0^\infty e^{-(s/a)t} f(t)\,dt - \int_0^b e^{-(s/a)t} f(t)\,dt \right\},$$

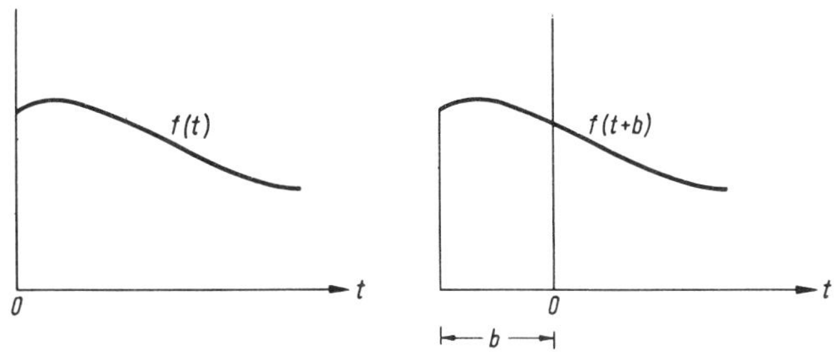

Figure 5

involving the $\mathfrak{L}$-transform of $f(t)$, and also the "finite" $\mathfrak{L}$-transform

$$F_b(s) = \int_0^b e^{-st} f(t)\,dt.$$

Thus, we have established

Theorem 7.4. *We have:*

$$f(at+b) \circ\!\!-\!\!\bullet \frac{1}{a} e^{(b/a)s} \left\{ F\left(\frac{s}{a}\right) - F_b\left(\frac{s}{a}\right) \right\} \quad \textit{for}\ a > 0,\ b > 0.$$

Selecting $a = 1$, as shown in Fig. 5, we specialize Theorem 7.4 to obtain

Theorem 7.5 (Second Shifting Theorem). *We have:*

$$f(t+b) \circ\!\!-\!\!\bullet e^{bs} \{F(s) - F_b(s)\} \quad \textit{for}\ b > 0.$$

Linear substitutions are not restricted to the independent variable of the original function, they may also be used with the independent variable of the image function. That is, we are interested in $F_1(s) = F(cs + d)$. The constant c must be positive so that $(cs + d)$, like s, varies in a right half-plane; d may assume any complex value. Suppose $F(s)$ has meaning for $\Re s > \beta$, then $F_1(s)$ has meaning for

$$\Re(cs + d) > \beta, \text{ that is, for } \Re s > \frac{\beta - \Re d}{c}.$$

We find:

$$F_1(s) = \int_0^\infty e^{-(cs+d)t} f(t)\,dt = \frac{1}{c} \int_0^\infty e^{-(cs+d)u/c} f\left(\frac{u}{c}\right) du = \frac{1}{c} \int_0^\infty e^{-st} \left[e^{-(d/c)t} f\left(\frac{t}{c}\right) \right] dt.$$

Hence, with the aid of the Uniqueness Theorem 5.1:

Theorem 7.6. *We have:*

$$F(cs+d) \bullet\!\!-\!\!\circ \frac{1}{c}\, e^{-(d/c)t}\, f\left(\frac{t}{c}\right) \quad \textit{for } c>0, \quad \textit{and complex } d.$$

Most frequently, we use the special case with $c = 1$.

Theorem 7.7 (Damping Theorem). *We have:*

$$F(s+d) \bullet\!\!-\!\!\circ e^{-dt}\, f(t) \quad \textit{for complex } d.$$

The designation Damping Theorem is actually justifiable only for $d > 0$: A shift of the image function through the positive distance d indicates that the corresponding original function is to be multiplied by the damping coefficient e^{-dt}, which converges to zero as $t \to \infty$.

Theorem 7.7, when applied to $t^a \circ\!\!-\!\!\bullet \Gamma(a+1)/s^{a+1}$, generates the new correspondence:

$$e^{-dt}\, t^a \circ\!\!-\!\!\bullet \frac{\Gamma(a+1)}{(s+d)^{a+1}} \quad \text{for} \quad a > -1, \ \text{complex } d.$$

Here, for a few simple problems, we developed the technique of reading the modifications of an original function, like translation or damping, *through inspection of the modifications of the corresponding image function.* It is our aim to develop and enrich this technique in the sequel. Recalling the comparison of the original space and the image space to two languages, this faculty corresponds to the facility of a bilingual person who can grasp the meaning of a sentence in one language without prior explicit and tedious translation into the other tongue. This technique will prove to be particularly important in all such cases where the original function is not explicitly known, when, nevertheless, we can deduce information from the image function.

8. The Mapping of Integration

In Chapter 7, we investigated how a trivial and elementary operation, like linear transformation of the independent variable of the original function, affects the corresponding image function. Here, for the first time, we shall investigate the effect of a transcendental operation, that is, integration of the original function, upon the corresponding image function.

Theorem 8.1 (Integration Theorem). *Define*

$$\varphi(t) = \int_0^t f(\tau)\, d\tau .$$

If $\mathfrak{L}\{f\}$ converges for some real $s = x_0 > 0$, then $\mathfrak{L}\{\varphi\}$ converges for $s = x_0$, and we have:

$$\mathfrak{L}\{\varphi\} = \frac{1}{s}\,\mathfrak{L}\{f\}\text{, that is } \Phi(s) = \frac{1}{s}\,F(s) \text{ for } s = x_0 \text{ and } \Re s > x_0.$$

Moreover,[1]

$$\varphi(t) = o\left(e^{x_0 t}\right) \quad \text{as} \quad t \to \infty,$$

hence $\mathfrak{L}\{\varphi\}$ converges absolutely for $\Re s > x_0$.

Remark: Observe that x_0 is restricted to $x_0 > 0$. Theorem 8.1 may fail for $x_0 \leqq 0$.

Proof of Theorem 8.1: We use de L'Hospital's rule:

Let $g(z)$ and $h(z)$ be differentiable functions for $z > Z$; suppose that $h(z)$ is real valued, and that $h(z) \to +\infty$ as $z \to +\infty$,[2] and that $h'(z) \neq 0$. If

$$\lim_{z\to\infty} \frac{g'(z)}{h'(z)} = A$$

holds, then

$$\lim_{z\to\infty} \frac{g(z)}{h(z)} = A$$

(for real $g(z)$, $A = \pm\infty$ are admissible).

We define:

$$\psi(z) = \int_0^z e^{-x_0 t}\,\varphi(t)\,dt,$$

$$g(z) = e^{x_0 z}\,\psi(z), \quad h(z) = e^{x_0 z}.$$

The function $\varphi(t)$ is continuous, hence $g(z)$ is differentiable for $z > 0$; $h(z)$ is also differentiable; the function $h(z)$ is real-valued for real x_0; $h(z) \to +\infty$ as $z \to +\infty$,

---

[1] $\varphi(t) = o(g(t))$ as $t \to \infty$, means

$$\frac{\varphi(t)}{g(t)} \to 0 \quad \text{as} \quad t \to \infty,$$

presuming $g(t) \neq 0$ as $t > T$.

[2] The requirement: $h(t) \to \infty$ as $z \to \infty$ is essential. De L'Hospital's rule is incorrect without this hypothesis, as shown by the counter example:

$$g(z) = 1 - e^{-2z}, \qquad g'(z) = 2e^{-2z};$$

$$h(z) = 1 - e^{-z}, \qquad h'(z) = e^{-z};$$

here:

$$\frac{g(z)}{h(z)} \to 1, \qquad \frac{g'(z)}{h'(z)} \to 0 \quad \text{as } z \to \infty.$$

since $x_0 > 0$; and $h'(z) = x_0 e^{x_0 t} \neq 0$. Using the generalized integration by parts, we form:

$$\frac{g'(z)}{h'(z)} = \frac{e^{x_0 z}(x_0 \psi + \psi')}{x_0 e^{x_0 z}} = \frac{1}{x_0}(x_0 \psi + \psi')$$

$$= \frac{1}{x_0}\left[x_0 \int_0^z e^{-x_0 t} \varphi(t)\, dt + e^{-x_0 z} \varphi(z)\right]$$

$$= \frac{1}{x_0}\left[- e^{-x_0 t} \varphi(t)\Big|_0^z + \int_0^z e^{-x_0 t} f(t)\, dt + e^{-x_0 z} \varphi(z)\right]$$

$$= \frac{1}{x_0} \int_0^z e^{-x_0 t} f(t)\, dt.$$

$\mathfrak{L}\{f\}$ converges for $s = x_0$, hence $g'(z)/h'(z)$ tends towards $F(x_0)/x_0$ as $z \to +\infty$. By de L'Hospital's rule, $g(z)/h(z)$ has the same limit:

$$\frac{g(z)}{h(z)} = \psi(z) = \int_0^z e^{-x_0 t} \varphi(t)\, dt \to \frac{1}{x_0} F(x_0) \quad \text{as } z \to \infty.$$

This implies:

$$\Phi(x_0) = \frac{1}{x_0} F(x_0),$$

and we have demonstrated the conclusion of Theorem 8.1, concerning the existence and expression of $\mathfrak{L}\{\varphi\}$ for $s = x_0$. The hypotheses are certainly satisfied for all real $x > x_0$, hence:

$$\Phi(x) = \frac{1}{x} F(x) \quad \text{for } x > x_0.$$

The functions $\Phi(s)$ and $F(s)$ are both analytic for $\Re s > x_0$. Consequently one may continue this functional equation into the half-plane $\Re s > x_0$.

Rewriting the equation

$$\lim_{z \to \infty} \frac{g(z)}{h(z)} = \lim_{z \to \infty} \frac{g'(z)}{h'(z)}$$

explicitly, we obtain:

$$\lim_{z \to \infty} \psi(z) = \lim_{z \to \infty} \frac{1}{x_0} [x_0 \psi(z) + \psi'(z)];$$

hence,

$$\lim_{z\to\infty} \psi'(z) = 0,$$

or

$$\lim_{z\to\infty} e^{-x_0 z}\, \varphi(z) = 0,$$

which verifies the last of the conclusions of Theorem 8.1.

The order property $\varphi(t) = o(e^{x_0 t})$, which occurs here incidentally, is quite revealing. From the convergence of $\mathfrak{L}\{f\}$ at $s = x_0$, one can draw no conclusion regarding the intensity of growth of $f(t)$ as $t \to \infty$. Remember the example on p. 29 with which we demonstrated that $e^{-x_0 t} f(t)$ need not tend towards zero for $t \to \infty$. One can, however, predict that the integral $\varphi(t)$ of $f(t)$ grows less strongly than $e^{x_0 t}$. This latter property can be used profitably on occasion.

Supplementary Remark: Suppose $\mathfrak{L}\{f\}$ converges at a *complex* $s = s_0$, with $\Re s_0 \geqq 0$, then $\mathfrak{L}\{\varphi\}$ converges and is equal to $(1/s)\, \mathfrak{L}\{f\}$ for $\Re s > \Re s_0$; for there exists a real $x_0 > 0$, with $\Re s_0 < x_0 < \Re s$, such that $\mathfrak{L}\{f\}$ converges for $s = x_0 > 0$. Theorem 8.1 guarantees the conclusion.

The Extended Convolution Theorem 10.4 provides another proof of Theorem 8.1 for complex s_0. However, when using Riemann integration, this verification is restricted to functions $f(t)$ of the class $\mathfrak{J}_0$, which will be introduced and explained on p. 45 (compare also p. 52).

9. The Mapping of Differentiation

Using Theorem 8.1, we shall derive, in this Chapter, Theorem 9.1, which provides the image of differentiation. The latter will prove extremely useful in practical applications of the $\mathfrak{L}$-transformation. A few introductory remarks will aid the subsequent development.

The functions $f(t)$ that are to be investigated are defined and differentiable for $t > 0$; the derivative need not exist for $t = 0$: possibly, $f(t)$ is not defined for $t = 0$, or $f(t)$ is not differentiable at $t = 0$, although $f(t)$ is defined at $t = 0$, as, for instance, shown by the two functions: $f(t) = 1$ for $t > 0$, with $f(0) = 0$; and $f(t) = 2t^{1/2}$ for $t \geqq 0$, so that $f'(t) = t^{-1/2}$. To guarantee the existence of $\mathfrak{L}\{f'\}$, we require at least the existence of $\lim_{t\to+0} f(t) = f(0^+)$, since $\mathfrak{L}\{f'\}$ has meaning only if f' is integrable in every finite interval; in particular,

$$\int_0^1 f'(\tau)\, d\tau$$

must exist. It is given by

$$\lim_{t\to+0} \int_t^1 f'(\tau) d\tau = \lim_{t\to+0} [f(1) - f(t)] = f(1) - \lim_{t\to+0} f(t).$$

Hence,

$$\lim_{t\to+0} f(t) = f(0^+)$$

exists. Replacing, in Theorem 8.1,

$$f(t) \quad \text{by} \quad f'(t) \quad \text{and} \quad \varphi(t) \quad \text{by} \quad \int_0^t f'(\tau)\,d\tau = f(t) - f(0^+),$$

and with

$$\mathfrak{L}\{f(0^+)\} = \frac{f(0^+)}{s} \quad \text{for } \Re s > 0$$

one obtains

Theorem 9.1 (Differentiation Theorem). *If $f(t)$ is differentiable for $t > 0$, and $\mathfrak{L}\{f'\}$ converges for some real $x_0 > 0$; then the limit $f(0^+)$ exists, and $\mathfrak{L}\{f\}$ too converges for $s = x_0$. We have the relation*

$$\mathfrak{L}\{f'\} = s\,\mathfrak{L}\{f\} - f(0^+) \quad \textit{for} \quad s = x_0, \quad \textit{and for } \Re s > x_0.$$

Moreover:

$$f(t) = o\left(e^{x_0 t}\right) \quad \textit{as} \quad t \to \infty;$$

hence $\mathfrak{L}\{f\}$ converges absolutely for $\Re s > x_0$.

Remark: Recalling the remark pertaining to Theorem 8.1, we re-emphasize the importance of the hypothesis $x_0 > 0$.

In applications involving differential equations, a generalization of Theorem 9.1 is of importance. In the proof of Theorem 9.1, we merely used: If $f'(t)$ exists for $t > 0$, and it is integrable in every finite interval $0 \leq t \leq T$, it follows that

$$f(t) = f(0^+) + \int_0^t f'(\tau)\,d\tau.$$

One encounters problems involving some function $f(t)$ which is not differentiable for $t > 0$; nevertheless, a function $f^{(1)}(t)$ exists, so that:

$$f(t) = f(0^+) + \int_0^t f^{(1)}(\tau)\,d\tau.$$

For instance, the function

$$f(t) = \begin{cases} 0 & \text{for } 0 \leqq t < 1 \\ t-1 & \text{for } t \geqq 1 \end{cases}$$

is not differentiable at $t = 1$. However, with

$$f^{(1)}(t) = \begin{cases} 0 & \text{for } 0 \leqq t < 1 \\ 1 & \text{for } t \geqq 1 \end{cases}$$

the required relation can be satisfied.

The function $f^{(1)}(t)$ may be called the *generalized derivative*[1] of $f(t)$. Clearly, Theorem 9.1 is valid if $f'(t)$ is replaced by $f^{(1)}(t)$.

Theorem 9.2. *If $f(t)$ has the generalized derivative $f^{(1)}(t)$, and $\mathfrak{L}\{f^{(1)}\}$ converges for some real $x_0 > 0$; then $\mathfrak{L}\{f\}$ too converges for $s = x_0$. We have*

$$\mathfrak{L}\{f^{(1)}\} = s\,\mathfrak{L}\{f\} - f(0+) \quad \textit{for } s = x_0 \textit{ and } \Re s > x_0.$$

Moreover,

$$f(t) = o\left(e^{x_0 t}\right) \quad \textit{as } t \to \infty;$$

hence $\mathfrak{L}\{f\}$ converges absolutely for $\Re s > x_0$.

Now, if $f(t)$ is differentiable twice for $t > 0$, and $\mathfrak{L}\{f''\}$ converges for some $x_0 > 0$, then we may apply Theorem 9.1 to f' instead of f, and we find that $f'(0+)$ exists, that $\mathfrak{L}\{f'\}$ exists for $s = x_0$, and that

$$\mathfrak{L}\{f''\} = s\,\mathfrak{L}\{f'\} - f'(0+) \quad \text{for } s = x_0 \text{ and } \Re s > x_0. \tag{1}$$

Applying Theorem 9.1 once more, we conclude that $f(0+)$ exists, that $\mathfrak{L}\{f\}$ converges for $s = x_0$, and that

$$\mathfrak{L}\{f'\} = s\,\mathfrak{L}\{f\} - f(0+) \quad \text{for } s = x_0 \text{ and } \Re s > x_0. \tag{2}$$

Combining (1) with (2), we produce:

$$\mathfrak{L}\{f''\} = s^2\,\mathfrak{L}\{f\} - f(0+)\,s - f'(0+) \quad \text{for } s = x_0 \text{ and } \Re s > x_0.$$

Moreover, we have the order relations

$$f'(t) = o\left(e^{x_0 t}\right) \quad \text{and} \quad f(t) = o\left(e^{x_0 t}\right) \quad \text{as } t \to \infty.$$

By repeated applications of the above process, one establishes

---

[1] In the Lebesgue theory, the fact that $f(t)$ is the integral of a function $f^{(1)}(t)$ is equivalent to the property: $f(t)$ is totally continuous. Hence, $f(t)$ is almost everywhere differentiable in the conventional sense, and almost everywhere $f'(t)$ equals $f^{(1)}(t)$.

Theorem 9.3. *If $f(t)$ is differentiable n times for $t > 0$,*[2] *and $\mathfrak{L}\{f^{(n)}\}$ converges for some real $x_0 > 0$; then the limits*

$$\lim_{t\to+0} f(t) = f(0^+), \quad \lim_{t\to+0} f'(t) = f'(0^+), \cdots, \quad \lim_{t\to+0} f^{(n-1)}(t) = f^{(n-1)}(0^+)$$

exist, and $\mathfrak{L}\{f\}$ too converges for $s = x_0$. We have

$$\mathfrak{L}\{f^{(n)}\} = s^n \mathfrak{L}\{f\} - f(0^+)\, s^{n-1} - f'(0^+)\, s^{n-2} - \cdots - f^{(n-1)}(0^+) \tag{3}$$

for $s = x_0$ and for $\Re s > x_0$. Moreover,

$$f(t) = o\left(e^{x_0 t}\right), \quad f'(t) = o\left(e^{x_0 t}\right), \cdots, \quad f^{(n-1)}(t) = o\left(e^{x_0 t}\right) \quad \text{as } t \to \infty;$$

hence $\mathfrak{L}\{f\}, \mathfrak{L}\{f'\}, \cdots, \mathfrak{L}\{f^{(n-1)}\}$ converge absolutely for $\Re s > x_0$.

For the subsequent verification of the Addendum to Theorem 9.3 we require the generally useful

Lemma. *If $f(x)$ is differentiable in $a < x \leqq b$, and if*[3]

$$\lim_{x\to a+0} f(x) = l \quad \text{and} \quad \lim_{x\to a+0} f'(x) = l'$$

exist, then upon assigning to $f(a)$ the value l, $f'(a)$ exists and equals l'.

Proof: Define $f(a) = l$; then $f(x)$ is differentiable in the interior of the interval $a \leqq x \leqq b$, and $f(x)$ is continuous at the end points of the interval. The mean value theorem of differentiation guarantees for every x, with $a < x \leqq b$, the existence of an intermediate abscissa ξ, $a < \xi < x$, so that

$$\frac{f(x) - f(a)}{x - a} = f'(\xi).$$

For $x \to a + 0$, ξ also tends towards a; by hypothesis the right hand side tends to l', and, consequently, also the left hand side, that is $f'(a) = l'$.

---

[2] The function $f(t)$ may have a generalized n^{th} derivative; that is, a function $f^{(n)}(t)$ exists, so that

$$f^{(n-1)}(t) = f^{(n-1)}(0^+) + \int_0^t f^{(n)}(\tau)\, d\tau.$$

[3] The hypothesis $f(x) \to l$ is redundant; it follows from $f'(x) \to l'$. The function $f'(x)$ has a limit for $x \to a + 0$, hence $f'(x)$ is bounded in a neighbourhood of a, $a < x \leqq c$; consequently $f'(x)$ is Lebesgue integrable, and $\int_a^c f'(x)\,dx$ exists. We have

$$\int_a^c f'(u)\, du = \lim_{x\to a+0} \int_x^c f'(u)\, du = \lim_{x\to a+0} [f(c) - f(x)],$$

which confirms the existence of $\lim_{x\to a+0} f(x)$.

Now suppose, $f(t)$ is n times differentiable not only for $t > 0$, but also at $t = 0$, then $f(t), f'(t), \cdots, f^{(n-1)}(t)$ are continuous at $t = 0$; hence the limits $f(0+), f'(0+), \cdots, f^{(n-1)}(0+)$ may be replaced by the values $f(0), f'(0), \cdots, f^{(n-1)}(0)$. For the more general case, that is when $f^{(n)}(t)$ exists only for $t > 0$, the above substitution is also permissible, provided we assign *to* $f(t)$ at $t = 0$ the value $f(0+)$, thereby possibly altering $f(t)$ at $t = 0$.

For $n = 1$, the conclusion is trivial. If in Theorem 9.3 we have $n \geqq 2$, then $f(0+)$ and $f'(0+)$ exist. Assigning for $f(0)$ the value $f(0+)$, by the above Lemma, $f'(0)$ exists and equals $f'(0+)$. For $n \geqq 3$, one concludes in an analogous manner that $f''(0)$ exists, and equals $f''(0+)$, and so forth. Thus, we verify the

Addendum to Theorem 9.3. *If we assign for* $f(0)$ *the value* $f(0+)$, *then the derivatives* $f'(0), \cdots, f^{(n-1)}(0)$ *exist. It follows that* $f(t), f'(t), \cdots, f^{(n-1)}(t)$ *are continuous for* $t \geqq 0$, *whence formula* (3) *may be re-written thus:*

$$\mathfrak{L}\{f^{(n)}\} = s^n \mathfrak{L}\{f\} - f(0)\, s^{n-1} - f'(0)\, s^{n-2} - \cdots - f^{(n-1)}(0). \tag{4}$$

An alternate presentation of formula (4) *is:*

$$\mathfrak{L}\{f^{(n)}\} = s^n \mathfrak{L}\left\{f(t) - \left[f(0) + \frac{f'(0)}{1!}\, t + \cdots + \frac{f^{(n-1)}(0)}{(n-1)!}\, t^{n-1}\right]\right\}. \tag{5}$$

The expression in the square brackets consists of the first n terms of the Taylor expansion of $f(t)$ about the centre $t = 0$. The function in the swung brackets, and its first $(n - 1)$ derivatives vanish at $t = 0$, the nth derivative agrees with that of $f(t)$.

Theorem 9.3 establishes a remarkable fact; it implies that n-fold differentiation, a *transcendental* operation, in the original space corresponds to an *algebraic* operation in the image space: the multiplication by s^n, and the subtraction of a polynomial in s, the coefficients of which are given by the so-called initial values of the original function.

In this context, we recall Theorem 6.1, which may be interpreted as a grammatical rule for the interpretation of the nth derivative of some image function by an algebraic operation which acts upon the corresponding original function.

Theorem 9.4. *The n-fold differentiation of the image function corresponds to the multiplication of the original function by* $(-t)^n$:

$$F^{(n)}(s) \bullet\!\!-\!\!\circ\, (-t)^n f(t).$$

10. The Mapping of the Convolution

In Chapters 7, 8, and 9, we studied the effect of operations involving *one* function such as, for instance, differentiation. In this Chapter, we shall investigate operations which involve more than one function, like sum or product of functions. The formula for the sum of functions

$$\mathfrak{L}\{f_1+f_2\}=\mathfrak{L}\{f_1\}+\mathfrak{L}\{f_2\}$$

is immediately obvious. However, the image of the product of two original functions, $f_1 \cdot f_2$ say, is quite complicated, thus necessitating that its study be deferred to Chapter 31.

By contrast, the operation of the multiplication of two image functions, $F_1 \cdot F_2$, corresponds to an operation on the respective original functions, f_1 and f_2, which is relatively simple. Recalling the interpretation of the $\mathfrak{L}$-integral as a continuous generalization of the power series, one could conjecture the sought correspondence. Forming the product of two convergent power series,

$$\varphi_1(z) = \sum_{n=0}^{\infty} a_n z^n \quad \text{and} \quad \varphi_2(z) = \sum_{n=0}^{\infty} b_n z^n ,$$

one finds the new power series

$$\varphi_1(z)\, \varphi_2(z) = \varphi(z) = \sum_{n=0}^{\infty} c_n z^n ,$$

whose coefficients c_n are found, using the a_n and the b_n, by means of the formula:

$$c_n = \sum_{\nu=0}^{n} a_\nu b_{n-\nu} .$$

Thus, one should expect that the result of the multiplication of two $\mathfrak{L}$-integrals,

$$F_1(s) = \int_0^\infty e^{-st} f_1(t)\, dt \quad \text{and} \quad F_2(s) = \int_0^\infty e^{-st} f_2(t)\, dt ,$$

is another $\mathfrak{L}$-integral

$$F_1(s)\, F_2(s) = F(s) = \int_0^\infty e^{-st} f(t)\, dt ,$$

the original function of which is similarly composed, using the respective original functions, f_1 and f_2, by means of the formula:

$$f(t) = \int_0^t f_1(\tau)\, f_2(t-\tau)\, d\tau . \tag{1}$$

Accordingly, the product of two image functions, $F_1 \cdot F_2$, would correspond to the combination of the respective original functions, f_1 and f_2, as shown by formula (1).

The combination (1) is called the *convolution (Faltung)* of the functions f_1 and f_2; it occurs in many branches of mathematics and theoretical physics, especially in the theory of differential equations. In several aspects, the convolution behaves like a product, whence it is symbolically written as a product, replacing the dot of the product by the asterisk of the convolution, and we call $f_1(t)$ and $f_2(t)$ the factors of the convolution:

$$(2) \qquad \int_0^t f_1(\tau) f_2(t-\tau)\, d\tau = f_1 * f_2.$$

We shall substantiate the above conjecture, although under hypotheses more stringent than those required for the power series. This is due to two difficulties:

1. Power series converge absolutely in the interior of the domain of convergence; thus one may multiply power series term by term, and then rearrange the terms to collect terms with like powers of z. By contrast, the absolute convergence of $\mathfrak{L}$-integrals in the interior of the domain of convergence is not guaranteed.

2. The expression for the c_n is a finite sum; it always exists. However, the integral (1) for $f(t)$ need not exist. For instance, consider the functions

$$f_1(t) = t^{-1/2}, \quad f_2(t) = |1-t|^{-1/2},$$

for which the convolution $f(t)$ by (2) does not exist for $t = 1$, since

$$f(1) = \int_0^1 \tau^{-1/2} |1-(1-\tau)|^{-1/2}\, d\tau = \int_0^1 \tau^{-1}\, d\tau.$$

Consequently, we shall have to request properties concerning the absolute convergence of the $\mathfrak{L}$-integrals; moreover, we shall have to restrict the admissible original functions to a specific class of functions so that the existence of the integral (2) is assured.[1] We form the

Class $\mathfrak{J}_0$ *of those absolutely integrable functions* $f(t)$, *which are bounded in every finite interval that does not include the origin:* $0 < T_1 \leqq t \leqq T_2$.

Thus, when using Riemann integration, zero is the only point where a $\mathfrak{J}_0$-function may be merely improperly integrable.

Let f_1 and f_2 designate $\mathfrak{J}_0$-functions, and let t be some positive, fixed abscissa; then

$$|f_1(\tau)| \leqq M_1, \quad |f_2(\tau)| \leqq M_2 \text{ for } \frac{t}{2} \leqq \tau \leqq t,$$

and, therefore:

$$\left| \int_0^{t/2} f_1(\tau) f_2(t-\tau)\, d\tau \right| \leqq M_2 \int_0^{t/2} |f_1(\tau)|\, d\tau,$$

$$\left| \int_{t/2}^{t} f_1(\tau) f_2(t-\tau)\, d\tau \right| \leqq M_1 \int_0^{t/2} |f_2(u)|\, du,$$

that is, both parts of the integral (2) exist; consequently $f_1 * f_2$ does exist.

[1] When using Lebesgue integration, the last restriction may be omitted, since $f_1 * f_2$ exists almost everywhere, and it too is Lebesgue integrable.

If we request, moreover, absolute convergence of the respective $\mathfrak{L}$-integrals, then we can verify the conjectured property.

Theorem 10.1 (Convolution Theorem). *If $\mathfrak{L}\{f_1\}$ and $\mathfrak{L}\{f_2\}$ converge absolutely for $s = s_0$, and if f_1 and f_2 belong to the class $\mathfrak{J}_0$; then $\mathfrak{L}\{f_1 * f_2\}$ converges absolutely for $s = s_0$, and we have*

$$\mathfrak{L}\{f_1 * f_2\} = \mathfrak{L}\{f_1\} \cdot \mathfrak{L}\{f_2\} \quad \text{for } \Re s \geqq \Re s_0 .$$

Proof: We define

$$f_1(t) = f_2(t) = 0 \quad \text{for } t < 0 ; \tag{3}$$

then, for $s = s_0$:

$$\mathfrak{L}\{f_1\} \cdot \mathfrak{L}\{f_2\} = \int_{-\infty}^{+\infty} e^{-s_0\tau} f_1(\tau)\, d\tau \cdot \int_{-\infty}^{+\infty} e^{-s_0 u} f_2(u)\, du . \tag{4}$$

The second integral is a constant, hence it may be taken under the first integration symbol. Also, introducing the new variable t by the substitution: $u = t - \tau$, we may re-write (4) as a repeated integral:

$$\int_{-\infty}^{+\infty} e^{-s_0\tau} f_1(\tau) \left[\int_{-\infty}^{+\infty} e^{-s_0(t-\tau)} f_2(t-\tau)\, dt \right] d\tau .$$

The integrals converge absolutely; hence we may commute the order of integration, leading to other repeated integrals which converge absolutely:

$$\int_{-\infty}^{+\infty} e^{-s_0 t} \left[\int_{-\infty}^{+\infty} f_1(\tau) f_2(t-\tau)\, d\tau \right] dt .$$

As a consequence of (3), we have:

$$\int_{-\infty}^{+\infty} f_1(\tau) f_2(t-\tau)\, d\tau = \begin{cases} \int_0^t f_1(\tau) f_2(t-\tau)\, d\tau & \text{for } t \geqq 0 \\ 0 & \text{for } t < 0 . \end{cases}$$

Thus, the Theorem 10.1 is verified for $s = s_0$. For $\Re s \geqq \Re s_0$ all hypotheses are necessarily satisfied, and the Theorem holds for the specified range of values.

From this Theorem we may immediately deduce some important consequences. Earlier, it was hinted that the convolution shares several properties with the product. Indeed, the convolution satisfies the *commutative* law, for the substitution $t = u - \tau$ yields

$$f_1 * f_2 = \int_0^t f_1(\tau) f_2(t-\tau)\, d\tau = \int_0^t f_1(t-u) f_2(u)\, du = f_2 * f_1 .$$

Moreover, the convolution satisfies the *associative* law:

$$(f_1 * f_2) * f_3 = f_1 * (f_2 * f_3) \ .$$

That is, for more than two factors, it is immaterial in which succession the convolutions are performed. Hence it suffices to write $f_1 * f_2 * f_3$.

Proof: Upon modifying the original functions, so that from a certain, positive abscissa T onwards these functions are assigned the value zero, while, for brevity, retaining the functional notation, we obtain absolute convergence of the $\mathfrak{L}$-integrals $\mathfrak{L}\{f_i\}$, $i = 1$, 2, and 3, for every s. The same holds true for $\mathfrak{L}\{f_1 * f_2\}$, by Theorem 10.1. Repeated application of Theorem 10.1 yields:

$$\mathfrak{L}\{(f_1 * f_2) * f_3\} = \mathfrak{L}\{f_1 * f_2\} \cdot \mathfrak{L}\{f_3\} = \mathfrak{L}\{f_1\} \cdot \mathfrak{L}\{f_2\} \cdot \mathfrak{L}\{f_3\}.$$

The identical expression is obtained for $\mathfrak{L}\{f_1 * (f_2 * f_3)\}$, hence

$$\mathfrak{L}\{(f_1 * f_2) * f_3\} = \mathfrak{L}\{f_1 * (f_2 * f_3)\}.$$

The Uniqueness Theorem 5.1 shows that

$$(f_1 * f_2) * f_3 = f_1 * (f_2 * f_3) + \text{nullfunction.}$$

In Theorem 10.2 we shall demonstrate that a convolution is continuous for $t > 0$; therefore, by Theorem 5.5, the nullfunction is identically zero. The convolutions of the modified functions coincide with those of the given functions for $0 < t \leqq T$, and the conclusion is verified for $0 < t \leqq T$; the conclusion is generally true for all $t > 0$, since T was arbitrarily chosen, and for $t = 0$ all convolutions are zero.

Now, we supply a theorem which is important for many applications.

Theorem 10.2. *Let f_1 and f_2 be $\mathfrak{J}_0$-functions; then the convolution $f_1 * f_2$ is continuous for $t > 0$.*

Proof: Let t be a fixed, positive number. We have to demonstrate that, for $\delta \to 0$,

$$D(t, \delta) = \int_0^{t+\delta} f_1(\tau) f_2(t + \delta - \tau) \, d\tau - \int_0^t f_1(\tau) f_2(t - \tau) \, d\tau$$

converges towards zero. Without loss of generality, we may assume δ to be positive, since the verification for negative δ is analogous to the one presented here. Also, we restrict, a priori,

$$0 < \delta \leqq 1 \, .$$

Let t_0 designate some fixed, positive number, the precise choice of which we defer; restricting, however,

$$0 < t_0 \leqq \frac{t}{2} \, .$$

Next, we decompose $D(t, \delta)$ in the following manner:

$$D(t, \delta) = \int_0^{t_0} f_1(\tau)\,[f_2(t+\delta-\tau) - f_2(t-\tau)]\,d\tau$$

$$+ \int_{t_0}^{t} f_1(\tau)\,[f_2(t+\delta-\tau) - f_2(t-\tau)]\,d\tau$$

$$+ \int_t^{t+\delta} f_1(\tau)\,f_2(t+\delta-\tau)\,d\tau = I_1 + I_2 + I_3.$$

For the integral I_1, the smallest argument of f_2 is $(t - t_0)$, and the largest is $(t + \delta)$; that is, the arguments vary, at most, in the interval $(t/2, t + 1)$. In this interval, the $\mathfrak{J}_0$-function f_2 is bounded: $|f_2| \leqq M_2$.

For the integral I_2, the argument of f_1 is in the interval (t_0, t). The $\mathfrak{J}_0$-function f_1 is bounded in this interval by a bound that depends upon the choice of t_0: $|f_1| \leqq m_1(t_0)$.

For the integral I_3, the argument of f_1 varies in the interval $(t, t + \delta)$, that is, at most, in the interval $(t, t + 1)$. In this interval we have $|f_1| \leqq M_1$.

With these individual bounds, we can produce the bound:

$$|D(t, \delta)| \leqq 2\,M_2 \int_0^{t_0} |f_1(\tau)|\,d\tau + m_1(t_0) \int_0^{t-t_0} |f_2(u+\delta) - f_2(u)|\,du$$

$$+ M_1 \int_0^{\delta} |f_2(u)|\,du.$$

For any given $\varepsilon > 0$, we first select t_0 sufficiently small, to make

$$2\,M_2 \int_0^{t_0} |f_1(\tau)|\,d\tau < \frac{\varepsilon}{3}.$$

With this t_0, $m_1(t_0)$ is also specified. Next, we choose $\delta_0 > 0$, so that simultaneously, for $0 < \delta < \delta_0$,[2]

$$m_1(t_0) \int_0^{t-t_0} |f_2(u+\delta) - f_2(u)|\,du < \frac{\varepsilon}{3},$$

---

[2] We use the theorem: If $f(t)$ is absolutely integrable, then for $\delta \to 0$,

$$\int_0^T |f(t+\delta) - f(t)|\,dt \to 0.$$

This important theorem will be needed on p. 144 also. It is well known in the Lebesgue theory. Surprisingly, it cannot be found in textbooks on Riemann integration. For a proof of the theorem for Riemann integrals see the author's book "Theorie und Anwendung der Laplace Transformation", Berlin 1937, pp. 399, 400 (Dover Publications 1942).

and

$$M_1 \int_0^\delta |f_2(u)|\, du < \frac{\varepsilon}{3}.$$

Hence, after combining the last three inequalities, we find that

$$|D(t, \delta)| < \varepsilon \quad \text{for } 0 < \delta < \delta_0.$$

Special attention is called to the fact that continuity of the convolution has been established merely for $t > 0$. Indeed, the convolution need not be continuous at $t = 0$: $f_1 * f_2$ is zero at $t = 0$, although $f_1 * f_2$ need not tend towards zero, as $t \to 0$. This is demonstrated by the example:

$$t^{-1/2} * t^{-1/2} = \int_0^t \tau^{-1/2}(t-\tau)^{-1/2}\, d\tau = \int_0^1 u^{-1/2}(1-u)^{-1/2}\, du$$

$$(5) \qquad = B\left(\frac{1}{2}, \frac{1}{2}\right) = \frac{\Gamma\left(\frac{1}{2}\right)\Gamma\left(\frac{1}{2}\right)}{\Gamma(1)} = \pi \quad \text{for } t > 0.$$

This convolution has the constant value π for $t > 0$. Hence, as $t \to 0$, the limit of the convolution is π.

However, one can verify the following Theorem:

Theorem 10.3. *If one of the two $\mathfrak{J}_0$-functions, f_1 or f_2, is bounded in a neighbourhood of zero, say $|f_1(t)| \leqq M_1$, for $0 \leqq t \leqq T$; then the convolution $f_1 * f_2$ is continuous for $t \geqq 0$, that is, including $t = 0$.*

For we have, with $0 < t \leqq T$:

$$\left| \int_0^t f_1(\tau) f_2(t-\tau)\, d\tau \right| \leqq M_1 \int_0^t |f_2(u)|\, du \to 0 \quad \text{as } t \to 0.$$

In the proof of Theorem 10.1, we used the hypothesis that *both $\mathfrak{L}\{f_1\}$ and $\mathfrak{L}\{f_2\}$ converge absolutely*. A counterexample (compare p. 56) demonstrates that this hypothesis cannot be disregarded entirely. We shall show, however, that the absolute convergence of one of the two $\mathfrak{L}$-integrals is sufficient; a relaxation of the hypothesis that is of great importance for practical applications. This is actually an analogy to the theorem by Mertens in the theory of series: Multiplying an absolutely converging series $\sum_0^\infty a_n$ term by term with a simply convergent series $\sum_{n=0} b_n$, and collecting the terms of the product along the Cauchy diagonal, that is,

$$c_n = \sum_{\nu=0}^{n} a_\nu b_{n-\nu},$$

we obtain a converging series $\sum_{n=0}^{\infty} c_n$ which equals $\sum_{n=0}^{\infty} a_n \cdot \sum_{n=0}^{\infty} b_n$.

Both Theorem 10.1 and the associative property of the convolution, the proof of which required Theorem 10.1, are needed in the proof of the Extended Convolution Theorem 10.4. Thus, the proof of the more restricted Theorem 10.1 was not superfluous.

Theorem 10.4 (Extended Convolution Theorem). *If $\mathfrak{L}\{f_1\}$ converges absolutely for $s = s_0$, and $\mathfrak{L}\{f_2\}$ converges simply for $s = s_0$, f_1 and f_2 being $\mathfrak{J}_0$-functions, then $\mathfrak{L}\{f_1 * f_2\}$ converges simply for $s = s_0$, and we have*

$$\mathfrak{L}\{f_1 * f_2\} = \mathfrak{L}\{f_1\} \cdot \mathfrak{L}\{f_2\} \quad \textit{for } s = s_0, \textit{ and for } \Re s > \Re s_0.$$

Proof: It suffices to establish Theorem 10.4 for $s_0 = 0$, since the application of the conclusion so derived, to $e^{-s_0 t} f_1(t)$ and $e^{-s_0 t} f_2(t)$, together with

$$(e^{-s_0 t} f_1) * (e^{-s_0 t} f_2) = \int_0^t e^{-s_0 \tau} f_1(\tau)\, e^{-s_0(t-\tau)} f_2(t-\tau)\, d\tau$$

(6) $$= e^{-s_0 t} (f_1 * f_2)$$

establishes the unrestricted case.

For $s_0 = 0$, the hypotheses become:

$$\int_0^\infty |f_1(\tau)|\, d\tau = \lim_{t\to\infty} \int_0^t |f_1(\tau)|\, d\tau \quad \text{and} \quad \int_0^\infty f_2(\tau)\, d\tau = \lim_{t\to\infty} \int_0^t f_2(\tau)\, d\tau$$

exist. The conclusion to be verified is:

$$\lim_{t\to\infty} \int_0^t f_1 * f_2(\tau)\, d\tau = \lim_{t\to\infty} \int_0^t f_1(\tau)\, d\tau \cdot \lim_{t\to\infty} \int_0^t f_2(\tau)\, d\tau.$$

Using convolution notation, the integral $\int_0^t f(\tau)\, d\tau$ may be written compactly as $f * 1$. We re-write the conclusion, using this compact notation:

(7) $$\lim_{t\to\infty} f_1 * f_2 * 1 = \lim_{t\to\infty} f_1 * 1 \cdot \lim_{t\to\infty} f_2 * 1.$$

First, we treat the special case:

$$F_2(0) = \lim_{t\to\infty} f_2 * 1 = 0,$$

for which we must show:

$$\lim_{t\to\infty} f_1 * f_2 * 1 = 0.$$

For this special case, with $\varepsilon > 0$, we can select a T, so that

$$|f_2 * 1| < \varepsilon \quad \text{for } t \geqq T.$$

Moreover, $f_2 * 1$ is a continuous function with a limit for $t \to \infty$; hence $f_2 * 1$ is bounded for all t:

$$|f_2 * 1| < M \quad \text{for } t \geqq 0.$$

For $t \geqq T$, we obtain the estimate:

$$|f_1 * f_2 * 1| \leqq \left| \int_0^T f_1(t-\tau)\,[f_2 * 1](\tau)\,d\tau \right| + \left| \int_T^t f_1(t-\tau)\,[f_2 * 1](\tau)\,d\tau \right|$$

(8)
$$\leqq M \int_{t-T}^{t} |f_1(u)|\,du + \varepsilon \int_0^\infty |f_1(u)|\,du.$$

The integral $\int_0^\infty |f_1(u)|\,du$ exists; hence, by the Cauchy criterion of convergence,

(9)
$$\int_{t_1}^{t_2} |f_1(u)|\,du < \varepsilon \quad \text{for all pairs of values } t_2 > t_1 \geqq T,$$

possibly after increasing T. For $t \geqq 2T$, we find $t - T \geqq T$, and we may use the estimate (9) with the inequality (8), to find that

$$|f_1 * f_2 * 1| \leqq \varepsilon \left(M + \int_0^\infty |f_1(u)|\,du \right) \quad \text{for } t \geqq 2T.$$

The expression in the round brackets is a constant, hence

$$\lim_{t \to \infty} f_1 * f_2 * 1 = 0,$$

and the conclusion (7) is true for this special case with $F_2(0) = 0$.

Now, if $F_2(0) \neq 0$, then

$$\mathfrak{L}\{f_2(t) - F_2(0)\,e^{-t}\}$$

converges for $s = 0$, and $\Re s > 0$; it is equal to

$$\mathfrak{L}\{f_2\} - \frac{F_2(0)}{s+1}.$$

Both $\mathfrak{L}\{f_1\}$ and $\mathfrak{L}\{e^{-t}\}$ converge absolutely at $s = 0$, whence by Theorem 10.1

$$\mathfrak{L}\{f_1 * (F_2(0)\, e^{-t})\}_{s=0} = \mathfrak{L}\{f_1\}_{s=0} \cdot \mathfrak{L}\{F_2(0)\, e^{-t}\}_{s=0} = \mathfrak{L}\{f_1\}_{s=0} \cdot \left(\frac{F_2(0)}{s+1}\right)_{s=0}$$

$$= \mathfrak{L}\{f_1\}_{s=0} \cdot F_2(0) = \mathfrak{L}\{f_1\}_{s=0} \cdot \mathfrak{L}\{f_2\}_{s=0}.$$

Addition of this last equation to the prior one yields:

$$\mathfrak{L}\{f_1 * f_2\} = \mathfrak{L}\{f_1\} \cdot \mathfrak{L}\{f_2\} \quad \text{for } s = 0.$$

As demonstrated at the onset of this proof, the conclusion for $s = 0$ may now be generalized for arbitrary s_0. For $\Re s > \Re s_0$, all hypotheses of the Theorem are necessarily satisfied, and Theorem 10.4 holds for $\Re s > \Re s_0$.

Notice that Theorem 10.4, unlike Theorem 10.1, does not claim absolute convergence of $\mathfrak{L}\{f_1 * f_2\}$; also, the conclusions are not verified for $\Re s \geqq \Re s_0$, but merely for $s = s_0$, and for $\Re s > \Re s_0$.

When setting $\int_0^t f(\tau)\, d\tau = f * 1$, Theorem 8.1, for the special case of $\mathfrak{J}_0$-functions, is contained in Theorem 10.4. We now understand the reason for the condition requested with Theorem 10.1, $x_0 > 0$, since $\mathfrak{L}\{1\} = 1/s$ converges (absolutely) for $\Re s > 0$ only.

In applications, it is often necessary to differentiate convolution integrals. First of all, observe that $f_1 * f_2$ need not be differentiable for all values of t, a fact that is easily demonstrated by the example $f * 1 = \int_0^t f(\tau)\, d\tau$. This may fail to have a derivative, or even a one-sided derivative, at some points. There is a well-known rule for the differentiation of an integral with respect to a parameter, here t, which appears in the limits of integration and in the integrand. However, the hypotheses that are usually employed in the derivation of this rule[3] are quite restricting; hence, this rule is insufficient for our purpose. It is for this reason that we derive a special Theorem for the differentiation of the convolution $f_1 * f_2$. In practical applications, one often encounters functions that exhibit kinks. That is, there are points where both one-sided derivatives, from the left and from the right respectively, do exist, but differ numerically. Thus, we approach the problem from a more general point of view, including the concepts of left derivative and right derivative.

Theorem 10.5. *Let $f_1(t)$ be differentiable for $t > 0$, and let $f'_1(t)$ and $f_2(t)$ belong to the class of $\mathfrak{J}_0$-functions.*

Then, at those points $t > 0$, where f_2 is continuous from the right (or from the left), $f(t) = f_1 * f_2$ is differentiable from the right (or from the left). We have

$$f'(t) = f_1' * f_2 + f_1(0+)\, f_2(t). \tag{10}$$

---

[3]
$$\frac{d}{d\alpha} \int_{h_1(\alpha)}^{h_2(\alpha)} f(x, \alpha)\, dx = \int_{h_1(\alpha)}^{h_2(\alpha)} \frac{\partial f(x,\alpha)}{\partial \alpha}\, dx + h_2'(\alpha)\, f(h_2(\alpha), \alpha) - h_1'(\alpha)\, f(h_1(\alpha), \alpha),$$

provided both, $h'_1(\alpha)$ and $h'_2(\alpha)$, are continuous in an interval $\alpha_1 \leqq \alpha \leqq \alpha_2$, and $\partial f/\partial \alpha$ is continuous in the area of the α, x-plane which is bounded by the straight lines $\alpha = \alpha_1$, $\alpha = \alpha_2$, and the curves $x = h_1(\alpha)$, $x = h_2(\alpha)$.

For the special case with $f_1(0^+) = 0$, the hypothesis concerning the continuity of $f_2(t)$ is superfluous, and the (conventional) derivative $f'(t)$ exists for all $t > 0$.[4]

Remark: Observe that f_1 need not be differentiable at $t = 0$, as in the example: $f_1(t) = t^{1/2}$. – The function f'_1 is integrable, hence $\lim_{t \to +0} f_1(t) = f_1(0^+)$ exists (compare p. 40).

Proof of Theorem 10.5: We have[5]

$$f_1(\tau) = \int_0^\tau f'_1(u)\, du + f_1(0^+),$$

hence

$$f(t) = \int_0^t f_2(t-\tau) \left[\int_0^\tau f'_1(u)\, du + f_1(0^+) \right] d\tau$$

$$= \int_0^t f_2(t-\tau)\, d\tau \int_0^\tau f'_1(u)\, du + f_1(0^+) \int_0^t f_2(u)\, du.$$

For bounded functions f'_1 and f_2, one may, without difficulty, convert the iterated integral into a double integral over the triangle $0 \leqq u \leqq \tau \leqq t$ in the τu-plane (see Fig. 6), for this double integral exists, since the integrand is the product of two factors, each depending upon one of the two variables of integration only. The functions f'_1 and f_2, being $\mathfrak{J}_0$-functions, are absolutely integrable. Thus the above conversion is also legitimate in case one or both of these $\mathfrak{J}_0$-functions are unbounded in some neighbourhood of the origin. It follows that:

$$\int_0^t f_2(t-\tau)\, d\tau \int_0^\tau f'_1(u)\, du = \iint_{0 \leqq u \leqq \tau \leqq t} f_2(t-\tau) f'_1(u)\, d\tau\, du.$$

Using the transformation

$$\begin{cases} \tau = \quad -y + t \\ u = x - y \end{cases} \qquad \text{or} \qquad \begin{cases} x = -\tau + u + t \\ y = -\tau \qquad + t \end{cases}$$

with the Jacobian $J = 1$, we obtain the new double integral

$$\iint f_2(y) f'_1(x-y)\, dx\, dy,$$

---

[4] Formula (10) may be re-written in a more detailed manner: designating the derivative from the left by f'_-, and the derivative from the right by f'_+, we obtain:

$$f'_-(t) = f'_1 * f_2 + f_1(0^+) f_2(t^-), \qquad f'_+(t) = f'_1 * f_2 + f_1(0^+) f_2(t^+).$$

[5] For the specified hypotheses, and with $\delta > 0$,

$$\int_\delta^\tau f'_1(u)\, du = f_1(\tau) - f_1(\delta),$$

hence, for $\delta \to 0$,

$$\int_0^\tau f'_1(u)\, du = f_1(\tau) - f_1(0^+).$$

which is to be evaluated over the triangle $0 \leqq y \leqq x \leqq t$ in the xy-plane (see Fig. 6). This integral may be written as an iterated integral:

$$\int_0^t dx \int_0^x f_1'(x-y)\, f_2(y)\, dy,$$

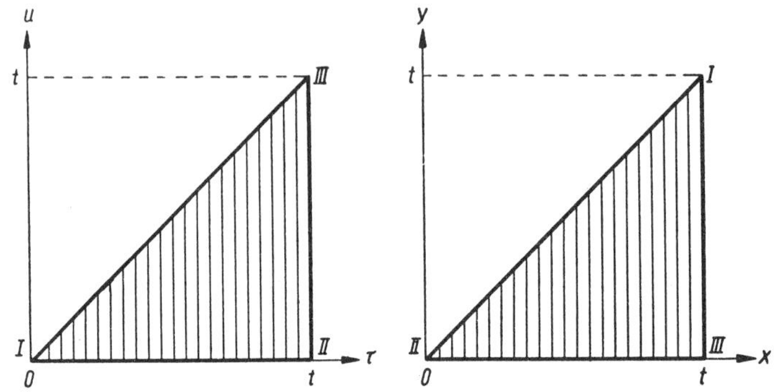

Figure 6

for the inner integral is a convolution of $\mathfrak{J}_0$-functions, the existence of which is assured.[6] Thus, we find for $f(t)$:

$$f(t) = \int_0^t dx \int_0^x f_1'(x-y)\, f_2(y)\, dy + f_1(0^+) \int_0^t f_2(u)\, du.$$

In this expression, we have t only in the upper limit of integration. For the integral with the dummy variable x, the integrand is a convolution of $\mathfrak{J}_0$-functions, which is continuous for $x > 0$ (not necessarily at $x = 0$, though); hence, this integral is differentiable at every point $t > 0$; the derivative is $f'_1 * f_2$. The integral with the dummy variable u, which actually contributes only if $f_1(0^+) \neq 0$, has, if f_2 is continuous at $t > 0$ from the right (or left), the right (or left) derivative $f_2(t)$.

Formula (10) may fail for $t = 0$. This is demonstrated by the following example: $f_1(t) = t^{1/2}$, $f_2(t) = t^{-1/2}$. The convolution of these functions is:

$$f(t) = f_1 * f_2 = \int_0^t \tau^{1/2}(t-\tau)^{-1/2}\, d\tau = t \int_0^1 u^{1/2}(1-u)^{-1/2}\, du$$

$$= t\, B\left(\frac{3}{2}, \frac{1}{2}\right) = t\, \frac{\Gamma\left(\frac{3}{2}\right) \Gamma\left(\frac{1}{2}\right)}{\Gamma(2)} = \frac{1}{2}\, \pi t,$$

which holds also for $t = 0$. For $t \geqq 0$, the derivative of this convolution is $\pi/2$. When using formula (10), one finds (compare (5)):

$$f'(t) = \frac{1}{2}\, t^{-1/2} * t^{-1/2} = \begin{cases} 0 & \text{for } t = 0 \\ \pi/2 & \text{for } t > 0, \end{cases}$$

[6] When using Lebesgue integration, we defend these conversions more easily by means of Fubini's theorem.

the correct answer for $t > 0$, an incorrect answer for $t = 0$.

Notice that although the convolution $f_1 * f_2$ is symmetric in f_1 and f_2, the derived formula for the derivative of the convolution is not symmetric. This reflects the asymmetric hypotheses of Theorem 10.5.

11. Applications of the Convolution Theorem: Integral Relations

The $\mathfrak{L}$-transformation permits the transformation of the convolution, a complicated integral representation, into a simple algebraic product. This facility can be utilized to produce simple proofs of integral relations which are otherwise difficult to verify.

1. *The n-fold iterated integral* (n a natural number)

$$\varphi_n(t) = \int_0^t d\tau_n \int_0^{\tau_n} d\tau_{n-1} \cdots d\tau_2 \int_0^{\tau_2} f(\tau_1)\, d\tau_1$$

can be written, using convolution notation,

$$\varphi_n(t) = f(t) * \underbrace{1}_{1} * \underbrace{1}_{2} * \cdots * \underbrace{1}_{n} = f * 1^{*n}.$$

Applying the process that was used in the proof of the associative property of the convolution, we alter $f(t)$ temporarily by assigning to it the value zero for all t beyond some arbitrary, fixed point, to guarantee the convergence of $\mathfrak{L}\{f\}$, and we find by the Convolution Theorem 10.1:

$$\mathfrak{L}\{\varphi_n\} = \mathfrak{L}\{f\} \cdot \mathfrak{L}\{1\}^n = \mathfrak{L}\{f\} \cdot \frac{1}{s^n}$$

$$= \mathfrak{L}\{f\} \cdot \mathfrak{L}\left\{\frac{t^{n-1}}{(n-1)!}\right\} = \mathfrak{L}\left\{f * \frac{t^{n-1}}{(n-1)!}\right\}.$$

We retained, for brevity, the symbols f and φ_n for the altered functions. Utilizing the continuity of the original function φ_n, and upon invoking the Uniqueness Theorem 5.5, one finds for the iterated integral the representation by the simple integral

$$\varphi_n(t) = f * \frac{t^{n-1}}{(n-1)!} = \frac{1}{(n-1)!} \int_0^t f(\tau)\,(t-\tau)^{n-1}\, d\tau. \tag{1}$$

2. The above presented derivation serves as an example of a technique that can frequently be employed; to determine the convolution of two original functions, one forms the product of the respective image functions, and seeks the original

function of the product. As another fine example of this method consider the *Bessel function*

$$J_0(t) = \sum_{n=0}^{\infty} \frac{(-1)^n}{(n!)^2} \left(\frac{t}{2}\right)^{2n}$$

which has, for $\Re s > 0$, the absolutely converging $\mathfrak{L}$-transform

$$\mathfrak{L}\{J_0\} = \frac{1}{\sqrt{s^2+1}}$$

(compare p. 266). By Theorem 10.1, we find for $\Re s > 0$:

$$J_0 * J_0 \circ\!\!-\!\!\bullet \frac{1}{s^2+1} \bullet\!\!-\!\!\circ \sin t,$$

hence

(2) $$J_0(t) * J_0(t) = \sin t.$$

The remarkable formula (2) also serves as the announced example by means of which we can demonstrate that the Convolution Theorem 10.1 may fail if we admit two original functions which have merely conditionally converging $\mathfrak{L}$-transforms. We shall show (see p. 269) that[1]

$$J_0(t) = \sqrt{\frac{2}{\pi t}} \cos\left(t - \frac{\pi}{4}\right) + O\left(t^{-3/2}\right) \quad \text{as} \quad t \to \infty,$$

and $\mathfrak{L}\{J_0\}$ converges absolutely for $\Re s > 0$; for $\Re s = 0$ it converges conditionally (compare proof on p. 9). However, $\mathfrak{L}\{J_0 * J_0\} = \mathfrak{L}\{\sin t\}$ does not converge for $s = 0$.

3. A large number of similar examples could easily be produced, The next example once more demonstrates the almost trivial connection in the image space that corresponds to a rather complicated one in the original space.

The function

$$\psi(x, t) = \frac{x}{2\sqrt{\pi}\, t^{3/2}} e^{-x^2/4t}$$

holds a fundamental position in the theory of *heat conduction* (compare p. 283). We have the correspondence

$$\psi(x, t) \circ\!\!-\!\!\bullet e^{-x\sqrt{s}} \quad (x > 0).$$

---

[1] If $g(t) \geqq 0$, then $f(t) = O(g(t))$ as $t \to \infty$, is equivalent to

$$|f(t)| \leqq k g(t) \text{ for } t \geqq T,$$

where k and T represent constants.

For the image function, we use the algebraic additivity theorem:

$$e^{-x_1\sqrt{s}} \cdot e^{-x_2\sqrt{s}} = e^{-(x_1+x_2)\sqrt{s}}.$$

Hence, for the respective original functions, we find the transcendental additivity theorem:

(3) $$\psi(x_1, t) * \psi(x_2, t) = \psi(x_1 + x_2, t) \quad (x_1 > 0, x_2 > 0).$$

This is, indeed, a rather complicated expression when written explicitly. In the footnote on p. 283 we explain the importance of the relation (3) to the theory of heat conduction. A direct verification of the relation (3) would occupy several pages. By comparison, the above verification of the relation (3) by means of the $\mathfrak{L}$-transformation seems like magic.

4. For the above example 2, we multiplied two $\mathfrak{L}$-transforms of the *Bessel function* J_0. Now, we go the other way and split the $\mathfrak{L}$-transform of the Bessel function J_0 into two factors:

$$J_0(t) \circ\!\!-\!\!\bullet \frac{1}{(s^2+1)^{1/2}} = \frac{1}{(s+i)^{1/2}} \frac{1}{(s-i)^{1/2}}.$$

With

$$\frac{1}{(s+a)^{1/2}} \bullet\!\!-\!\!\circ \frac{1}{\Gamma(1/2)} t^{-1/2} e^{-at} \quad \left(\Gamma\left(\frac{1}{2}\right) = \sqrt{\pi}\right)$$

one finds, by the Convolution Theorem 10.1,

$$J_0(t) = \frac{1}{\pi} \left(t^{-1/2} e^{-it}\right) * \left(t^{-1/2} e^{it}\right)$$

$$= \frac{1}{\pi} \int_0^t \tau^{-1/2} e^{-i\tau} (t-\tau)^{-1/2} e^{i(t-\tau)} d\tau = \frac{1}{\pi} e^{it} \int_0^t \tau^{-1/2} (t-\tau)^{-1/2} e^{-2i\tau} d\tau$$

$$= \frac{1}{\pi} e^{it} \int_0^1 u^{-1/2} (1-u)^{-1/2} e^{-2itu} du = \frac{1}{\pi} \int_0^1 e^{it(1-2u)} \left[u(1-u)\right]^{-1/2} du.$$

Using the first substitution

$$1 - 2u = v, \text{that is, } u = \frac{1-v}{2}, \quad 1 - u = \frac{1+v}{2}$$

one obtains a representation of J_0 by the *finite Fourier integral*

(4) $$J_0(t) = \frac{1}{\pi} \int_{-1}^{+1} e^{itv} (1-v^2)^{-1/2} dv.$$

By means of the second substitution

$$v = \cos\varphi, \quad 1 - v^2 = \sin^2\varphi$$

we produce the so-called *Poisson integral*

(5) $$J_0(t) = \frac{1}{\pi} \int_0^{\pi} e^{it\cos\varphi} d\varphi = \frac{2}{\pi} \int_0^{\pi/2} \cos(t\cos\varphi)\, d\varphi.$$

12. The Laplace Transformation of Distributions

The physicist needs to introduce certain concepts which enable him to mathematically describe physical phenomena. One of these concepts is the "impulse" δ which is supposed to mathematically represent a shock like, for instance, the impact of a hammer in mechanics, or a large voltage increase of exceedingly short duration in electrical engineering. Concepts of this nature cannot be comprehended within the frame of the classical theory of functions. However the modern theory of distributions embraces these concepts in a consistent manner. Moreover, this new theory avoids many of the difficulties of classical analysis. Thus, we must extend the theory of the Laplace-transformation, which in the previous Chapters has been developed for functions only, to distributions. This necessitates a knowledge of the foundations of the theory of distributions which is understood here in the sense of L. Schwartz. The termini and theorems of the theory of distributions employed here are compiled and presented in an Appendix, organized in 22 statements; in the text we shall refer to this Appendix whenever desirable, citing App. and the No. of the statement of specific interest.

The $\mathfrak{L}$-transformation, as defined in Chapter 4, refers to "right-sided" functions which need be defined for $t \geqq 0$ only. When considering the entire real line $-\infty < t < +\infty$, one must define these functions to be zero in $t < 0$. Correspondingly, only those distributions will be considered which, in the open half line $t < 0$, are equal to the trivial function distribution 0 (compare App. No. 10). Equivalently, one can characterize these distributions as those which have the support in the right half-line $t \geqq 0$ (see App. No. 11). We shall designate with $\mathscr{D}'_+$ the totality of these "right-sided" distributions. All locally integrable functions defined on $t \geqq 0$, which are zero for $t < 0$, may be interpreted as function-distributions (compare App. No. 9) and are as such included in $\mathscr{D}'_+$.

The classical $\mathfrak{L}$-transformation cannot be defined for all right-sided functions; similarly, within the frame of Schwartz' theory, one does not define the $\mathfrak{L}$-transformation for all distributions in $\mathscr{D}'_+$, but only for a subspace of $\mathscr{D}'_+$. This subspace may be chosen in various ways.[1] The method of selection used here is based upon the following concept:

A distribution T is said to be of finite order if T is a distribution-derivative of a function $h(t)$ which is continuous on R^1: $-\infty < t < \infty$, that is: $T = D^k h(t)$. If k designates the smallest non-negative integer, allowing this equality, then T is said to be of order k.

A Remark: It is a fact that *every* distribution is *locally* a distribution-derivative of finite order of some continuous function.[2] Explicitly: Consider some fixed, bounded and closed interval $I: a \leqq t \leqq b$; let $\mathscr{D}_I$ designate the subspace of $\mathscr{D}$ (see App. No. 5) which is composed of all those functions $\varphi(t)$ which have their support (see App. No. 3) entirely in I. Then for

---

[1] A definition based upon the Fourier transformation of distributions can be found in the book by L. Schwartz [1], Chapter VIII, cited in the Appendix on p. 313 An alternative definition based upon a subspace of "tempered distributions" is introduced by L. Schwartz [2], Chapter VI, and by Zemanian, Chapter 8.

[2] Compare the book by Zemanian, p. 86.

every distribution T there exists a function, which is continuous on R^1, and a smallest integer number $k \geqq 0$, so that for every $\varphi(t) \in \mathscr{D}_I$ (see App. No. 13):

$$\langle T, \varphi\rangle = (-1)^k \int_{-\infty}^{+\infty} h(t)\, \varphi^{(k)}(t)\, dt = (-1)^k \langle h(t), \varphi^{(k)}(t)\rangle = \langle D^k h(t), \varphi(t)\rangle .$$

For distributions of finite order, this conclusion holds for every $\varphi(t)$ in $\mathscr{D}$.

An Example: The distribution δ, the impulse, is defined as the distribution-derivative of the unit step function $u(t)$: $\delta = Du(t)$ (compare App. No. 15). However, this in itself is insufficient to establish δ as a distribution of finite order, since $u(t)$ is not continuous on R^1. Upon integrating $u(t)$, we obtain the unit ramp function:

(1) $$h(t) = 0 \quad \text{for} \quad t < 0, \quad h(t) = t \quad \text{for} \quad t \geqq 0,$$

which is continuous on R^1. At $t = 0$, this ramp function fails to have a derivative; however, the distribution-derivative $Dh(t)$ does exist, since

$$\langle Dh, \varphi\rangle = -\langle h, \varphi'\rangle = -\int_0^\infty t\, \varphi'(t)\, dt = -t\, \varphi(t)\Big|_0^\infty + \int_0^\infty \varphi(t)\, dt$$

$$= 0 + \int_0^\infty \varphi(t)\, dt = \int_{-\infty}^{+\infty} u(t)\, \varphi(t)\, dt = \langle u, \varphi\rangle,$$

hence $Dh(t) = u(t)$. It follows that

(2) $$D^2 h(t) = Du(t) = \delta,$$

and we have shown that δ is of finite order; it is of order 2. Consequently, all $D^n \delta = \delta^{(n)}(t)$ are also of finite order:

(3) $$\delta^{(n)}(t) = D^{n+2} h(t).$$

Theorem 12.1. *Suppose the function $f(t)$ is defined and locally integrable in $-\infty < t < +\infty$; and set, for some arbitrary fixed value a, $h(t) = \int_a^t f(\tau)\, d\tau$. Then $f(t) = Dh(t)$. The function $h(t)$ is continuous, hence $f(t)$ is of finite order; it is of order 0 if $f(t)$ is continuous, otherwise of order* 1.

Proof:[3] We have

$$\langle Dh, \varphi\rangle = -\langle h, \varphi'\rangle = -\int_{-\infty}^{+\infty} h(\tau)\, \varphi'(\tau)\, d\tau .$$

The limits of integration are, in fact, finite. We can apply the generalized rule of

[3] $h(t)$ is an integral, hence it is totally continuous. According to the Lebesgue theory, $h(t)$ is differentiable almost everywhere and, almost everywhere, $h'(t) = f(t)$.

integration by parts (compare the footnote[2] on p. 16); the contributions from the two ends of the interval vanish, since φ vanishes at these ends. Thus we obtain:

$$\langle D h, \varphi \rangle = \int_{-\infty}^{+\infty} f(\tau)\, \varphi(\tau)\, d\tau = \langle f, \varphi \rangle ,$$

this implies $Dh = f$.

From all distributions of finite order we select those which are derived from continuous functions which satisfy the two additional requirements:

(4) $$h(t) = 0 \quad \text{for } t < 0,$$

(5) $\mathfrak{L}\{h(t)\}$ converges absolutely for $\Re s > \sigma$, the value σ depending upon h.

The totality of the distributions selected according to the stated criteria form a space $\mathscr{D}'_0$ which, obviously, is a subspace of $\mathscr{D}'_+$.

For the distributions of $\mathscr{D}'_0$ we define the $\mathfrak{L}$-transform in the following manner: *If $T \in \mathscr{D}'_0$ and $T = D^k h(t)$; then we define the $\mathfrak{L}$-transform of T by*

(6) $$\mathfrak{L}\{T\} = s^k\, \mathfrak{L}\{h(t)\} = s^k H(s) = F(s).$$

$\mathfrak{L}\{T\}$ exists for $\Re s > \sigma$ where it represents an analytic function $F(s)$.

Theorem 12.2. *If some function $f(t)$ has a $\mathfrak{L}$-transform $F(s)$ in the classical sense, then $\mathfrak{L}\{f\}$ also exists in the sense of the theory of distributions and is equal to $F(s)$.*

Proof: The function $f(t)$ is defined for $t \geqq 0$; for $t < 0$ we define it to be 0. The function $h(t) = \int_0^t f(\tau)\, d\tau$ has two properties. It is zero for $t < 0$, and it is continuous for all t. $\mathfrak{L}\{h(t)\}$ exists by Theorem 8.1; it is given by $F(s)/s$. By Theorem 12.1 we have $f = Dh$; hence when considering f as a distribution we find, by (6),

$$\mathfrak{L}\{f\} = s\, \mathfrak{L}\{h\} = s \frac{1}{s} F(s) = F(s).$$

Definition (6) implies that the *inverse $\mathfrak{L}$-transformation for distributions* is *unique* in the same sense at that for functions. From $\mathfrak{L}\{T\} = 0$ we conclude, by (6), that $\mathfrak{L}\{h(t)\} = 0$. Hence, by the argumentation of Chapter 5, $h(t)$ is a null function. We conclude that two distributions with identical $\mathfrak{L}$-transforms differ by the null distribution.

13. The Laplace Transforms of Several Special Distributions

1. $\underline{T = \delta}$

The distribution δ is of finite order for it is equal to $D^2 h(t)$ where $h(t)$ designates the continuous unit ramp function defined by (12.1), which satisfies both conditions (12.4,5), the second condition with $\sigma = 0$. Hence δ has a $\mathfrak{L}$-transform, which is given by

(1) $$\mathfrak{L}\{\delta\} = s^2 \mathfrak{L}\{h(t)\} = s^2 \mathfrak{L}\{t\} = s^2 \frac{1}{s^2} = 1 \quad \text{for } \Re s > 0.$$

The constant function does not occur in the image space of functions.

2. $\underline{T = \delta^{(n)}}$ $(n \geq 1)$

We have (compare (12.3)):

$$\delta^{(n)} = D^n \delta = D^{n+2} h(t),$$

hence

(2) $$\mathfrak{L}\{\delta^{(n)}\} = s^{n+2} \mathfrak{L}\{h(t)\} = s^n \quad \text{for } \Re s > 0.$$

The occurrence of powers of s with positive integer-valued exponents is remarkable, since the image space of functions contains merely powers of s with negative exponents.

3. $\underline{T = \delta(t-a)}$ $(a > 0)$

We have (compare App. No. 17)

$$\delta(t-a) = Du(t-a) = D^2 h(t-a).$$

The function $h(t-a)$ satisfies the conditions (12.4,5), the second condition with $\sigma = 0$, for $a > 0$ (these conditions cannot be satisfied for $a < 0$). Thus, by Theorem 7.2:

(3) $$\mathfrak{L}\{\delta(t-a)\} = s^2 \mathfrak{L}\{h(t-a)\} = s^2 e^{-as} \mathfrak{L}\{h(t)\} = e^{-as} \quad \text{for } \Re s > 0.$$

Thus the exponential function, which cannot be the $\mathfrak{L}$-transform of a classical function (compare the remarks towards the end of Chapter 5), is entered into the image space, at least for positive a.

Pseudofunctions

The pseudofunctions of App. No. 22 have the form $D^n f(t)$, where $f(t)$ is continuous at all points t except $t = 0$. The function

$$\int_0^t f(\tau)\, d\tau \cdot u(t)$$

is continuous everywhere, including at the point $t = 0$. We use (compare App. No. 14):

$$f(t) = \frac{d}{dt}\int_0^t f(\tau)\,d\tau = D\int_0^t f(\tau)\,d\tau\,;$$

hence the pseudofunction is given by

$$D^{n+1}\int_0^t f(\tau)\,d\tau\,;$$

it is of finite order.

The $\mathfrak{L}$-transform of the latter is, by definition (12.6), using Theorem 8.1,

$$\mathfrak{L}\left\{D^{n+1}\int_0^t f(\tau)\,d\tau\right\} = s^{n+1}\mathfrak{L}\left\{\int_0^t f(\tau)\,d\tau\right\} = s^{n+1}\frac{1}{s}\mathfrak{L}\left\{f\right\} = s^n\,\mathfrak{L}\left\{f\right\}.$$

This demonstrates that rule (12.6) may immediately be formally applied to the original definition of the pseudofunction.

In the sequel we shall use the correspondence

$$\mathfrak{L}\left\{\log t\right\} = -\frac{\log s + C}{s} \qquad (C = \text{Eulers constant}).$$

$$4.\ \underline{\text{Pf}\,[t^{-1}u(t)]} = D\,[\log t \cdot u(t)]$$

(4) $$\mathfrak{L}\{\text{Pf}\,[t^{-1}u(t)]\} = s\,\mathfrak{L}\{\log t\} = -\log s - C$$

$$5.\ \underline{\text{Pf}\,[t^{-2}u(t)]} = -D^2\,[\log t \cdot u(t)] - \delta'(t)$$

(5) $$\begin{aligned}\mathfrak{L}\{\text{Pf}\,[t^{-2}u(t)]\} &= -s^2\,\mathfrak{L}\{\log t\} - s = s(\log s + C) - s\\ &= s(\log s + C - 1)\end{aligned}$$

$$6.\ \underline{\text{Pf}\,[t^{-n}u(t)]} = \frac{(-1)^{n-1}}{(n-1)!}\{D^n\,[\log t \cdot u(t)] + (\psi(n) + C)\,\delta^{(n-1)}(t) \qquad (n \geqq 1)$$

(6) $$\begin{aligned}\mathfrak{L}\{\text{Pf}\,[t^{-n}u(t)]\} &= \frac{(-1)^{n-1}}{(n-1)!}\left\{-s^n\,\frac{\log s + C}{s} + (\psi(n) + C)\,s^{n-1}\right\}\\ &= \frac{(-1)^n}{(n-1)!}\,s^{n-1}(\log s - \psi(n))\end{aligned}$$

$$7.\ \underline{\text{Pf}\,[t^{-\lambda}u(t)]} = D^n\left[\frac{(-1)^n}{(\lambda-1)\ldots(\lambda-n)}\,t^{-\lambda+n}\,u(t)\right]$$

($\lambda > 1$, not an integer, $-\lambda + n > -1$, n an integer)

$$\mathfrak{L}\{\mathrm{Pf}\,[t^{-\lambda}\,u(t)]\} = s^n \frac{(-1)^n}{(\lambda-1)\dots(\lambda-n)} \frac{\Gamma(-\lambda+n+1)}{s^{-\lambda+n+1}}$$

$$= \frac{\Gamma(-\lambda+n+1)}{(-\lambda+1)\dots(-\lambda+n)\,s^{-\lambda+1}}.$$

Because of:

$$\Gamma(-\lambda+n+1) = (-\lambda+n)\dots(-\lambda+1)\,\Gamma(-\lambda+1)$$

one finds:

(7) $$\mathfrak{L}\{\mathrm{Pf}\,[t^{-\lambda}\,u(t)]\} = \frac{\Gamma(-\lambda+1)}{s^{-\lambda+1}}.$$

The last expression formally agrees with the correspondence (Chapter 2, example 8).

(8) $$\mathfrak{L}\{t^{\lambda}\} = \frac{\Gamma(\lambda+1)}{s^{\lambda+1}} \qquad (\lambda > -1).$$

For all λ which are not integer-valued, one may use the same formula (8), except that for $\lambda < -1$ the function t^{λ} must be replaced by the pseudofunction $\mathrm{Pf}\,[t^{\lambda}u(t)]$.

By contrast, the formula (6) for negative integer-valued exponents differs considerably from the corresponding formula (9) for positive integer-valued exponents:

(9) $$\mathfrak{L}\left\{t^n\right\} = \frac{\Gamma(n+1)}{s^{n+1}} = \frac{n\,!}{s^{n+1}} \qquad (n = 0,1,\dots).$$

This is not surprising since (9) cannot have an analogous expression for $n = -1$, $-2, \dots$, because $\Gamma(n+1)$ becomes ∞ for such n.

When studying the right hand sides of the formulae (7), (8), (9), one observes that only the powers $1, s, s^2, \dots$ are missing. This reflects the fact that, by (1) and (2), these powers do not correspond to powers of t; instead they are the transforms of $\delta, \delta', \delta'', \dots$. When starting with the powers $s^{-\alpha}$ and searching for the corresponding originals, f_α, one finds, for the family of the f_α, the definition:

(10) $$f_\alpha = \begin{cases} \dfrac{t^{\alpha-1}}{\Gamma(\alpha)}\,u(t) & \text{for } \alpha > 0 \\ \mathrm{Pf}\,\dfrac{t^{\alpha-1}}{\Gamma(\alpha)}\,u(t) & \text{for } \alpha < 0,\ \alpha \neq -1, -2, \dots \\ \delta^{(n)} & \text{for } \alpha = -n = 0, -1, -2, \dots \end{cases}$$

Hence, we have here the single expression for the $\mathfrak{L}$-transforms of the f_α:

(11) $$\mathfrak{L}\{f_\alpha\} = s^{-\alpha} \qquad (\alpha \text{ arbitrary real}).$$

One can easily verify the rule:

(12) $$D^k f_\alpha = f_{\alpha-k} \qquad (k = 1, 2, \dots)$$

for the family of the f_α.

14. Rules of Mapping for the $\mathfrak{L}$-Transformation of Distributions

Throughout this Chapter 14, we shall presume that

$$T \in \mathcal{D}_0', \text{ so that } T = D^k h(t),$$

where $h(t)$ represents a function which complies with the conditions (12.4,5), and we shall use:

$$\mathfrak{L}\{T\} = s^k \mathfrak{L}\{h\} = s^k H(s) = F(s).$$

The Translation of a Distribution

The shift or translation of a function $\varphi(t)$ through a distance of positive or negative length b or, equivalently, the translation of the origin to the position b, may be visualized as the effect caused by an operator τ_b:

(1) $$\tau_b \varphi(t) = \varphi(t - b).$$

The translation of a distribution T through the distance b, thus creating the shifted distribution $\tau_b T$, is defined by means of formula:

(2) $$\langle \tau_b T, \varphi(t) \rangle = \langle T, \tau_{-b} \varphi(t) \rangle = \langle T, \varphi(t + b) \rangle.$$

For the special situation in which T is actually determined by some locally integrable function $f(t)$ we find, by definition (see App. No. 9),

$$\langle \tau_b f(t), \varphi(t) \rangle = \langle f(t), \varphi(t + b) \rangle = \int_{-\infty}^{+\infty} f(t)\, \varphi(t + b)\, dt$$

$$= \int_{-\infty}^{+\infty} f(t - b)\, \varphi(t)\, dt = \langle f(t - b), \varphi(t) \rangle,$$

that is, in fact, $\tau_b f(t) = f(t - b)$. It follows that the above suggested definition of the shift of a distribution is consistent with the conventional shift of functions.

Example: For $T = \delta$ one finds (compare App. No. 17):

$$\langle \tau_b \delta, \varphi(t) \rangle = \langle \delta, \varphi(t + b) \rangle = \varphi(b) = \langle \delta_b, \varphi(t) \rangle,$$

therefore,

(3) $$\tau_b \delta = \delta_b = \delta(t - b).$$

Theorem 14.1. *For every $b > 0$, it follows that*

$$\tau_b T \circ\!\!-\!\!\bullet\, e^{-bs} F(s).$$

Proof: We express $\tau_b T$ by means of $h(t)$; according to (2) we have

$$\langle \tau_b T, \varphi(t) \rangle = \langle D^k h(t), \varphi(t + b) \rangle;$$

hence, by App. No. 13,

$$\langle \tau_b T, \varphi \rangle = (-1)^k \langle h(t), \varphi^{(k)}(t+b) \rangle = (-1)^k \int_0^\infty h(t)\, \varphi^{(k)}(t+b)\, dt$$

$$= (-1)^k \int_b^\infty h(t-b)\, \varphi^{(k)}(t)\, dt = (-1)^k \int_{-\infty}^{+\infty} h(t-b)\, \varphi^{(k)}(t)\, dt,$$

since $h(t-b) = 0$, for $t < b$. Whence

$$\langle \tau_b T, \varphi \rangle = \langle D^k h(t-b), \varphi(t) \rangle,$$

which implies that

$$\tau_b T = D^k h(t-b). \tag{4}$$

A shift of T is equivalent to a like shift of $h(t)$.

Application of (12.6) and Theorem 7.2 yields:

$$\mathfrak{L}\{\tau_b T\} = s^k \mathfrak{L}\{h(t-b)\} = s^k e^{-bs} \mathfrak{L}\{h(t)\} = e^{-bs} \mathfrak{L}\{T\}.$$

The Translation of the Image Function (Analogue of the Damping Theorem)

Theorem 14.2. *The correspondence*

$$e^{-\alpha t} T \circ\!\!-\!\!\bullet F(s+\alpha)$$

holds for any arbitrary complex α.

Proof: We represent $e^{-\alpha t} T$ by means of $h(t)$, invoke App. No. 18 and No. 13, and we find:

$$\langle e^{-\alpha t} T, \varphi(t) \rangle = \langle T, e^{-\alpha t} \varphi(t) \rangle = \langle D^k h(t), e^{-\alpha t} \varphi(t) \rangle$$

$$= (-1)^k \langle h(t), D^k (e^{-\alpha t} \varphi(t)) \rangle$$

$$= (-1)^k \int_0^\infty h(t) \frac{d^k}{dt^k} [e^{-\alpha t} \varphi(t)]\, dt$$

$$= (-1)^k \int_0^\infty h(t) \sum_{\nu=0}^{k} \binom{k}{\nu} (-\alpha)^\nu e^{-\alpha t} \varphi^{(k-\nu)}(t)\, dt^{1})$$

$$= (-1)^k \sum_{\nu=0}^{k} \binom{k}{\nu} (-\alpha)^\nu \int_0^\infty h(t)\, e^{-\alpha t} \varphi^{(k-\nu)}(t)\, dt$$

$$= (-1)^k \sum_{\nu=0}^{k} \binom{k}{\nu} (-\alpha)^\nu (-1)^{k-\nu} \langle D^{k-\nu} [e^{-\alpha t} h(t)], \varphi(t) \rangle.$$

[1] We employ here Leibniz' rule for the differentiation of a product:

$$\frac{d^k}{dt^k} [f(t)\, g(t)] = \sum_{\nu=0}^{k} \binom{k}{\nu} f^{(\nu)}(t)\, g^{(k-\nu)}(t).$$

It follows that

$$(5) \qquad e^{-at}\, T = \sum_{\nu=0}^{k} \binom{k}{\nu} a^{\nu} D^{k-\nu} [e^{-at} h(t)].$$

Application of Theorem 7.7 produces:

$$\mathfrak{L}\{e^{-at}\, T\} = \sum_{\nu=0}^{k} \binom{k}{\nu} a^{\nu} s^{k-\nu} \mathfrak{L}\{e^{-at} h(t)\} = (s+a)^k H(s+a)$$
$$= F(s+a).$$

The Derivative of a Distribution

Theorem 14.3. *For the nth distribution-derivative of T we have the correspondence:*

$$D^n T \circ\!\!-\!\!\bullet\, s^n F(s).$$

Proof: We have

$$T = D^k h(t) \quad \text{and} \quad D^n T = D^{n+k} h(t),$$

hence

$$\mathfrak{L}\{T\} = s^k \mathfrak{L}\{h\} \quad \text{and} \quad \mathfrak{L}\{D^n T\} = s^{n+k} \mathfrak{L}\{h\};$$

consequently,

$$\mathfrak{L}\{D^n T\} = s^n \mathfrak{L}\{T\}.$$

Theorem 14.3 differs from the classical Differentiation Theorem 9.3 in the absence of initial values. Indeed, initial values would not have meaning, since a distribution does not have a specific value at a specified point. Nevertheless, Theorem 14.3 agrees with Theorem 9.3 for the special situation that T is, in fact, a function $f(t)$. Any function $f(t)$ as a distribution in $\mathscr{D}'_+$ is to be assigned the value zero for $t < 0$; hence all limits from the left at $t = 0$, of $f, f', f'', \ldots$ are zero. If $f, f', f'', \ldots$ exist as functions, having, for $t \to +0$, the respective limits $f(0^+)$, $f'(0^+)$, $f''(0^+)$, .., then, by App. No. 20,

$$(6) \qquad D^n f = f^{(n)} + f^{(n-1)}(0^+)\,\delta + \ldots + f(0^+)\,\delta^{(n-1)}.$$

If $\mathfrak{L}\{f^{(n)}\}$ and, consequently, $\mathfrak{L}\{f\}$ too exist in the classical sense, then, by Theorem 12.2, these also exist in the sense of the theory of distributions. The $\mathfrak{L}$-transform of (6) is, by Theorem 14.3, using (13.2):

$$s^n \mathfrak{L}\{f\} = \mathfrak{L}\{f^{(n)}\} + f^{(n-1)}(0^+) + \ldots + f(0^+)\, s^{n-1}.$$

This is the conclusion of Theorem 9.3.

The Differentiation of the Image Function

Theorem 14.4. *For the nth derivative of $F(s) = \mathfrak{L}\{T\}$, we find the correspondence:*

$$F^{(n)}(s) \bullet\!\!-\!\!\circ (-t)^n T.$$

Proof: We verify the above conclusion for $n = 1$; the verification for $n > 1$ follows by iteration.

The representation $F(s) = s^k H(s)$ implies that

$$F'(s) = k\, s^{k-1} H(s) + s^k H'(s).$$

Employing the correspondence $H'(s) \bullet\!\!-\!\!\circ (-t)h(t)$ we find, by (12.6):

$$F'(s) \bullet\!\!-\!\!\circ k\, D^{k-1} h(t) + D^k[-t\, h(t)].$$

By App. No. 19:

$$D^k[t\, h(t)] = \binom{k}{0} t\, D^k h(t) + \binom{k}{1} \cdot 1 \cdot D^{k-1} h(t) + \binom{k}{2} \cdot 0,$$

hence

$$F'(s) \bullet\!\!-\!\!\circ k\, D^{k-1} h(t) - t\, D^k h(t) - k\, D^{k-1} h(t)$$
$$= -t\, D^k h(t) = -t\, T.$$

The Convolution of Distributions

The convolution of two distributions can, in general, be defined only in a quite difficult manner, and may actually fail to exist. Fortunately, a simple definition can be devised for distributions from the space $\mathscr{D}'_0$, in a manner which guarantees the existence of the convolution.

Suppose $T_1 = D^{k_1} h_1(t)$ and $T_2 = D^{k_2} h_2(t)$ are distributions from the space $\mathscr{D}'_0$. We define the convolution of T_1 and T_2 by means of the formula:

(7) $$T_1 * T_2 = D^{k_1+k_2}[h_1(t) * h_2(t)],$$

whereby $h_1 * h_2$ is to be understood as the convolution of functions as explained in Chapter 10. The continuous functions h_1 and h_2 are $\mathfrak{J}_0$-functions. Hence, by Theorem 10.2, $h_1 * h_2$ is also continuous and it assumes, for $t < 0$, the value 0. $\mathfrak{L}\{h_1\}$ and $\mathfrak{L}\{h_2\}$ converge absolutely by hypothesis. Hence, by Theorem 10.1, $\mathfrak{L}\{h_1 * h_2\}$ too converges absolutely. We conclude that $T_1 * T_2$ is a distribution in the space $\mathscr{D}'_0$.

The definition (7) is a consistent extension of the classical definition of the convolution of functions; that is, for the special situation where T_1 and T_2 are generated by means of right-sided, locally integrable functions f_1 and f_2, definition (7) agrees with the classical definition (10.2). To demonstrate this fact, let us form the functions:

$$h_1(t) = f_1 * 1, \quad h_2(t) = f_2 * 1.$$

These are continuous for all t, and vanish for $t < 0$. By Theorem 12.1 we find that

$$T_1 = f_1 = D\, h_1, \quad T_2 = f_2 = D\, h_2,$$

hence, by (7),

$$T_1 * T_2 = D^2(f_1 * 1 * f_2 * 1) = D^2(f_1 * f_2 * 1 * 1).$$

Theorem 12.1 applied twice yields:

$$T_1 * T_2 = f_1 * f_2.$$

In the space $\mathscr{D}'_0$ one can verify the Convolution Theorem.

Theorem 14.5 (Convolution Theorem). *Given two distributions T_1 and T_2 from the space $\mathscr{D}'_0$ with the respective $\mathfrak{L}$-transforms $F_1(s)$ and $F_2(s)$, we conclude that*

$$T_1 * T_2 \circ\!\!-\!\!\bullet\, F_1(s) \cdot F_2(s).$$

Proof: Applying the $\mathfrak{L}$-transformation to formula (7), as specified in (12.6), one obtains:

$$\mathfrak{L}\{T_1 * T_2\} = s^{k_1+k_2}\,\mathfrak{L}\,\{h_1 * h_2\}.$$

By Theorem 10.1,

$$\mathfrak{L}\{h_1 * h_2\} = \mathfrak{L}\{h_1\} \cdot \mathfrak{L}\{h_2\},$$

hence

$$\mathfrak{L}\{T_1 * T_2\} = s^{k_1}\,\mathfrak{L}\{h_1\} \cdot s^{k_2}\,\mathfrak{L}\{h_2\} = \mathfrak{L}\{T_1\} \cdot \mathfrak{L}\{T_2\}.$$

The *commutative* property of the convolution follows from formula (7); it is inherited from the commutative property of $h_1 * h_2$. The *associative* property of (7) is shown by means of the Convolution Theorem (compare p. 47).

The Convolution with the Distribution δ

Formula (12.2) implies that:

$$\delta = D^2\,[u(t) * 1] = D^2\,[1 * 1].$$

For $T = D^k h(t)$ one finds, by (7),

$$T * \delta = D^{k+2}\,[h(t) * 1 * 1] = D^k\, h(t),$$

hence

(8) $$T * \delta = T.$$

The same conclusion is reached by means of the Convolution Theorem:

$$\mathfrak{L}\{T * \delta\} = \mathfrak{L}\{T\} \cdot \mathfrak{L}\{\delta\} = \mathfrak{L}\{T\} \cdot 1 = \mathfrak{L}\{T\}.$$

Invoking the uniqueness of the inverse $\mathfrak{L}$-transformation, one obtains equation (8).

In an algebra where the product is defined by the convolution, the distribution δ plays the rôle of the unit element. In particular, one observes that

(9) $$\delta * \delta = \delta, \quad \delta * \delta * \delta = \delta, \ldots$$

Formula (13.2) implies that

$$\mathfrak{L}\{T * \delta^{(n)}\} = \mathfrak{L}\{T\} \cdot \mathfrak{L}\{\delta^{(n)}\} = \mathfrak{L}\{T\} \cdot s^n = \mathfrak{L}\{D^n T\},$$

whence

$$T * \delta^{(n)} = D^n T. \tag{10}$$

In the space of distributions, one may represent the distribution-derivative by means of a convolution. Consequently, one can write a distribution-derivative equation (the analogue to the differential equation of functions) as an equation of convolutions (compare Chapter 18).

The Distribution-Derivative of the Convolution

Theorem 14.6. *The distribution-derivative of the convolution of two distributions taken from the space $\mathcal{D}'_0$ is obtained by distribution-differentiation of one of the two distributions:*

$$D^m [T_1 * T_2] = (D^m T_1) * T_2 = T_1 * (D^m T_2).$$

Proof: Use of (10) and the associative property of the convolution produces:

$$D^m [T_1 * T_2] = \delta^{(m)} * [T_1 * T_2] = [\delta^{(m)} * T_1] * T_2 = (D^m T_1) * T_2.$$

The same result also follows from (7), the definition of the convolution:

$$(D^m T_1) * T_2 = (D^{m+k_1} h_1) * (D^{k_2} h_2) = D^{m+k_1+k_2} (h_1 * h_2) = D^m [T_1 * T_2].$$

Let us compare this result with Theorem 10.5 for functions, for the special case: $m = 1$. If f'_1 exists for $t > 0$, and $f_1(0^+)$ exists, then one finds, by App. No. 20,

$$D f_1 = f'_1 + f_1(0^+)\,\delta,$$

and consequently:

$$(D f_1) * f_2 = f'_1 * f_2 + f_1(0^+)\,\delta * f_2 = f'_1 * f_2 + f_1(0^+)\,f_2(t).$$

15. The Initial Value Problem of Ordinary Differential Equations with Constant Coefficients

An important application of the Differentiation Theorem 9.1 and the Convolution Theorem 10.1accrues when these are called to aid with the problem of integrating ordinary linear differential equations with constant coefficients in the interval $t \geqq 0$, for specified values of the solution and some of its derivatives at $t = 0$, the initial values (Initial Value Problem). This is a problem which may be solved by a familiar classical technique: First one produces a sufficient number of fundamental solutions of the homogeneous equation, then one constructs the general solution as a linear combination of these; by means of the "variation of the constants" one seeks a special solution of the inhomogeneous equation. Addition of the latter to the former yields the general solution of the inhomogeneous equation. Lastly one must "adjust" the arbitrary coefficients so that the general solution agrees with the specified initial values. Theoretically, this technique seems simple; however it creates great difficulties in practice, particularly when applied to differential equations of higher order. By contrast, we shall observe that the method based upon the $\mathfrak{L}$-transformation provides the solution of such problems with a minimum of technical effort.

The Differential Equation of First Order

We shall develop the method with the simplest case, the initial value problem of the differential equation of first order, thereby exhibiting all essential characteristics of the method. Discussions are deliberately extensive and detailed in order to demonstrate the particularities of the method. Whence, the reader should be well prepared for the problems of differential equations of arbitrary order, which may be dealt with more briefly.

Consider the differential equation

$$y'(t) + c\,y(t) = f(t) \tag{1}$$

with the unknown y. The coefficients of y and of y' are constants; the one of y' is already reduced to one, by division; $f(t)$ on the right hand side designates an arbitrary function which, in physical applications, is generally referred to as *disturbing function* or *excitation*.[1] The value of $y(t)$ at $t = 0$ is specified so that uniqueness of the solution is guaranteed. We seek the solution for $t > 0$; the interval $t < 0$ is not considered.

Let us be more specific; we need to find a function $y(t)$ which satisfies the differential equation (1) for $t > 0$. Obviously, one cannot expect to find such a solution for every conceivable disturbing function $f(t)$. This is demonstrated by the simplest problem: $y'(t) = f(t)$, which does not necessarily have a solution for an arbitrary $f(t)$. The solution $y(t)$ should extend continuously to the value specified at $t = 0$, say y_0. In other words, y_0 is to be the limit of the solution $y(t)$ as t approaches zero through positive values:

$$\lim_{t\to+0} y(t) = y_0, \quad \text{briefly:} \quad y(0+) = y_0. \tag{2}$$

The reader ought to recognize early that it is not the *value* of $y(t)$ at $t = 0$ that is of importance, but the *limit* of $y(t)$ as $t \to +0$; this latter interpretation of the "initial value problem" should be accepted from the very beginning. This precise distinction may seem superfluous for a single ordinary differential equation, because its solution, which we shall find, and the derivatives of the solution

[1] In engineering literature $f(t)$ is often called the input, and $y(t)$ the output.

assume the specified initial values at $t = 0$, and they are *continuous* at $t = 0$; that is, "limit" and "value" coincide. However, this is no longer true, in general, for systems of ordinary differential equations or for partial differential equations.

Textbooks on differential equations usually restrict the discussion to disturbing functions that are continuous in the interval of integration. This restriction is too severe for practical applications which frequently involve disturbing functions with jumps (these are the simplest discontinuities), as for instance $u(t - a)$ with $a > 0$ (compare Chapter 2, Example 2). Thus we assume:

A₁ *The disturbing function $f(t)$ is continuous for $t > 0$, with the possible exception of isolated points of discontinuity a where the function jumps; that is, where the one-sided limits, $f(a^-)$ and $f(a^+)$, exist, however $f(a^-) \neq f(a^+)$.*

From A_1 we immediately conclude that $f(t)$ is bounded and properly integrable in every interval: $0 < T_1 \leqq t \leqq T_2$. Moreover, we require that $f(t)$ must be at least improperly, absolutely integrable at $t = 0$. We may express this (see p. 45) thus:

A₂. *The disturbing function $f(t)$ belongs to the class of $\mathfrak{J}_0$-functions.*

Thus, we do admit disturbing functions like $t^{-1/2}$ which are not considered in most textbooks; functions of this type cannot be avoided, for suppose the differential equation (1) has the innocent solution $2t^{1/2}$; then $y'(t) = t^{-1/2}$, and $f(t) = t^{-1/2} + 2ct^{1/2}$.

The differential equation (1) indicates that at those points where $f(t)$ exhibits jumps, either $y(t)$ or $y'(t)$, or both $y(t)$ and $y'(t)$ must also jump. When admitting jumps for $y(t)$, we cannot possibly expect uniqueness of the solution. This is quickly demonstrated by a simple example: $y'(t) = 0$ with the initial condition $y(0^+) = 1$, for which we have as solution not only $y(t) = 1$, but also every staircase function (a sectionally constant function) which starts at $t = 0$ with the value 1. Hence, we must require:

R₁ *The solution must be continuous for $t > 0$.*

By this requirement, only $y'(t)$ may exhibit jumps at discontinuity points, where the one-sided derivatives from the right y'_+ and from the left y'_- differ in value.[2]

Now we require:

R₂ *The differential equation* (1) *must be satisfied for every $t > 0$, at least from the right and from the left; that is:*

$$y'_- + cy = f(t^-), \quad y'_+ + cy = f(t^+).$$

We then have:

(3) $$y'_+(t) - y'_-(t) = f(t^+) - f(t^-).$$

[2] If, at some point a, $y(t)$ is continuous, and $f(t)$ has one-sided limits from the left and from the right, then by the differential equation $y'(t)$ too has one-sided limits from the left and from the right. This implies, by the Lemma on p. 42, the existence of the one-sided derivatives from the left and from the right, at a.

At those discontinuity points where $f(t)$ exhibits a jump, $y'(t)$ too has a jump of the same magnitude, and $y(t)$ has a kink. By R_2, $y(t)$ satisfies the differential equation (1) strictly, wherever $f(t)$ is continuous.

It remains to be seen whether or not these requirements can be satisfied.

Now we may proceed with the integration of the differential equation. In the previous Chapters we applied the $\mathfrak{L}$-transformation to *functions*. The essential idea of the method presented here is that we apply the $\mathfrak{L}$-transformation to the *Equation* (1); that is, we $\mathfrak{L}$-transform both sides of the Equation (1). Utilizing the linearity property of the $\mathfrak{L}$-operator, we find:

$$\mathfrak{L}\{y'\} + c\,\mathfrak{L}\{y\} = \mathfrak{L}\{f\}. \tag{4}$$

However, equation (4) has meaning only if two additional hypotheses are satisfied.

H$_1$ *The disturbing function $f(t)$ has a $\mathfrak{L}$-transform.*

H$_2$ *The derivative of the solution $y'(t)$ has a $\mathfrak{L}$-transform.*

From H_2 we conclude, by means of Theorem 9.1, that $y(t)$ too has a $\mathfrak{L}$-transform.[3]

Indubitably, there are differential equations that comply with these hypotheses, and others which do not. An example of the former is: $y' + y = 2e^t$ with $y(0+) = 1$, having the solution $y(t) = e^t$; an example for the latter is: $y' + y = 2(t+1)e^{t^2}$ with $y(0+) = 1$, having the solution $y(t) = e^{t^2}$. However, no method can sensibly be developed without some hypotheses that guarantee the applicability of the method, although such hypotheses are often tacitly overlooked. Moreover, we shall later dispense with the hypotheses H_1 and H_2.

We may now exploit the properties of the $\mathfrak{L}$-transformation, in particular the ones described in the Differentiation Theorem 9.1 which enable us to express $\mathfrak{L}\{y'\}$ in terms of $\mathfrak{L}\{y\}$. At those discontinuity points where $f(t)$ exhibits a jump, the derivative y' does not really exist, and we must resort to Theorem 9.2, the generalization of Theorem 9.1. Suppose that between 0 and t we find two discontinuity points t_0 and t_1, where $f(t)$ jumps, then

$$\int_0^{t_0} y'(\tau)\,d\tau = y(t_0^-) - y(0+),$$

$$\int_{t_0}^{t_1} y'(\tau)\,d\tau = y(t_1^-) - y(t_0^+),$$

$$\int_{t_1}^{t} y'(\tau)\,d\tau = y(t^-) - y(t_1^+),$$

defining for the first integral $y'(t_0) = y'_-(t_0)$, for the second integral $y'(t_0) = y'_+(t_0)$ and $y'(t_1) = y'_-(t_1)$, and for the third integral $y'(t_1) = y'_+(t_1)$. The solu-

[3] Observe that H_1 and H_2 are interdependent through the differential equation.

tion $y(t)$ is continuous, thus $y(t_0^-) = y(t_0^+)$, $y(t_1^-) = y(t_1^+)$; hence, by addition:

$$y(t) = y(0^+) + \int_0^t y'(\tau)\, d\tau,$$

and all hypotheses of Theorem 9.2 are satisfied with $y^{(1)} = y'$ for $t \neq t_0, t_1$, irrespective of the manner of definition of $y^{(1)}(t)$ at those discontinuity points t_0 and t_1.

We may apply Theorem 9.2 to $\mathfrak{L}\{y'\}$ in equation (4), thereby using the limit $y(0^+)$, that is the initial value y_0, exactly as explained earlier. Using upper case letters to designate the respective $\mathfrak{L}$-transforms, we write:

(5) $$s\,Y(s) - y_0 + c\,Y(s) = F(s).$$

Conforming with established terminology, we shall call Eq. (1) the **Original Equation,** and Eq. (5) the **Image Equation.**

One observes that the image equation poses a far easier problem than the original equation since it is a linear *algebraic equation.* Moreover, the specified *initial value* y_0 which occurred separately with the original equation is now incorporated in the image equation; it is *automatically considered.*

The image equation can readily be solved:

(6) $$Y(s) = F(s)\,\frac{1}{s+c} + y_0\,\frac{1}{s+c}.$$

Supposing that the differential equation (1), complying with hypothesis H_1, has a solution $y(t)$, which satisfies requirements R_1 and R_2 and hypothesis H_2, this solution must be amongst the original functions of $Y(s)$. To find the sought inverse function, we consult the Table of Transforms at the rear of this book, possibly requiring the help of the "grammatical rules" developed in previous Chapters. For this particular problem we employ the correspondence

$$\frac{1}{s+c} \bullet\!\!-\!\!\circ\; e^{-ct}$$

and the Extended Convolution Theorem 10.4, and we find the original function which corresponds to $Y(s)$; it is:

(7) $$y(t) = f(t) * e^{-ct} + y_0\, e^{-ct},$$

to which an arbitrary nullfunction may be added. However, $y(t)$ is continuous by R_1, and the right hand side of (7) is continuous by Theorem 10.2 and assumption A_2. By Theorem 5.5, there corresponds but one continuous original function to an image function, whence (7) is the only function that need be considered.

We are not certain that (7) is indeed a solution of the stated problem, for it could be that no solution exists which satisfies all requirements and hypotheses; if the problem fails to have a solution, the presented argumentation is unfounded. Consequently, verification of the solution is imperative. In the process of verification we shall employ an important principle, which we shall apply to other functional equations in the sequel. Obviously, we must retain the assumptions A_1 and A_2 regarding $f(t)$, and the requirements R_1 und R_2, for these serve to specify the problem. By contrast, the hypotheses H_1 and H_2 were introduced solely for the purpose of making the $\mathfrak{L}$-transformation applicable to the problem. Were we now able to demonstrate that (7) satisfies the requirements R_1 and R_2, independently of H_1 and H_2, we could disregard these hypotheses, in retrospect. This amounts, in effect, *to finding a function by any means, and then investigating under what most general conditions this function is a solution of the problem.* One may envisage this process in the following manner: one starts in a constrained domain of given and sought functions in which all hypotheses are satisfied; the solution, established in this constrained domain, may then be extended into the largest domain, where it has meaning. Thus, we shall call the basic concept of the technique expounded here the **Principle of Extension.**

It is indeed easy to show that (7) with the assumptions A_1 and A_2 satisfies the requirements R_1 and R_2 as well as the specified initial condition. The convolution integral of (7) tends towards zero for $t \to 0$, since e^{-ct} is bounded in a neighbourhood of zero (compare Theorem 10.3); thus (7) satisfies the initial condition. The continuity of $y(t)$ for $t > 0$ follows by Theorem 10.2. Theorem 10.5 (compare the foot note of this Theorem), when applied to $f_1(t) \equiv e^{-ct}$, $f_2(t) \equiv f(t)$, shows that $y(t)$ has at least one-sided derivatives for $t > 0$, hence:

$$y'_-(t) = -c\,e^{-ct} * f(t) + f(t^-) - c\,y_0\,e^{-ct},$$

$$y'_+(t) = -c\,e^{-ct} * f(t) + f(t^+) - c\,y_0\,e^{-ct},$$

and we find that:

$$y'_- + c\,y = f(t^-), \qquad y'_+ + c\,y = f(t^+).$$

For practical applications, we summarize the results of this lengthy theoretical discussion, and we obtain a brief set of instructions:

When the disturbing function $f(t)$ satisfies the assumptions A_1 and A_2, write under the differential equation

$$y' + c\,y = f(t)$$

its image equation

$$s\,Y - y_0 + c\,Y = F(s);$$

determine the solution of the latter:

$$Y = F(s)\,\frac{1}{s+c} + y_0\,\frac{1}{s+c},$$

and find the corresponding original function

$$y = f * e^{-ct} + y_0\,e^{-ct}.$$

This is the solution of the differential equation; it assumes the specified initial value y_0, and it satisfies the requirements R_1 and R_2.

This method is represented by the

Diagram

Original Space:	differential equation + initial condition	solution
	|	↑
	$\mathfrak{L}$-transformation	$\mathfrak{L}^{-1}$-transformation
	↓	|
Image space:	algebraic equation ⟶	solution

Explanation: Instead of solving the initial value problem directly in the original space, we take a roundabout way and proceed via the image space following the arrows.

The image function $F(s)$ of the disturbing function may or may not exist; it need not be found explicitly. Nevertheless, it might be convenient to actually determine the image function $F(s)$, if this is possible, and to seek the $\mathfrak{L}^{-1}$-transform of $F(s)/(s+c)$ as an entity rather than by means of the Convolution Theorem. Consider, for instance, $f(t) = u(t-a)$ with $a > 0$, which leads to a rather inconvenient convolution integral, since $f(t)$ is defined differently in the respective intervals: $0 \leqq t < a$ and $a < t < \infty$. The differential equation

$$y' + c\,y = u(t-a) \quad \text{with} \quad y(0+) = y_0$$

is $\mathfrak{L}$-transformed into

$$s\,Y - y_0 + c\,Y = \frac{e^{-as}}{s},$$

hence

$$Y = \frac{e^{-as}}{s(s+c)} + \frac{y_0}{s+c}.$$

Using

$$\frac{1}{s(s+c)} = \frac{1}{c}\left(\frac{1}{s} - \frac{1}{s+c}\right) \bullet\!\!-\!\!\circ \frac{1}{c}\left(1 - e^{-ct}\right)$$

and the first Translation Theorem 7.2, we find:

$$\frac{e^{-as}}{s(s+c)} \bullet\!\!-\!\!\circ \left\{\begin{array}{ll} 0 & \text{for } 0 \leqq t < a \\ \frac{1}{c}\left(1 - e^{-c(t-a)}\right) & \text{for } t \geqq a \end{array}\right\} = \frac{1}{c}\left(1 - e^{-c(t-a)}\right) u(t-a),$$

and the solution of the differential equation can be written compactly as

$$y(t) = \frac{1}{c}\left(1 - e^{-c(t-a)}\right) u(t-a) + y_0\,e^{-ct},$$

and explicitly as

$$y(t) = \begin{cases} y_0\,e^{-ct} & \text{for } 0 \leqq t < a \\ \frac{1}{c} - \left(\frac{e^{ac}}{c} - y_0\right) e^{-ct} & \text{for } t \geqq a\,. \end{cases}$$

For the three possibilities $y_0 \lesseqqgtr e^{ac}/c$, we observe entirely different patterns of behaviour of $y(t)$ for $t > a$.

Prior to the extension of the above considerations to differential equations of arbitrary order, we insert a section on the

Partial Fraction Expansion of a Rational Function

In the sequel we shall encounter image functions that are rational functions $g(s)/p(s)$, where $g(s)$ and $p(s)$ represent polynomials in s. We shall show by Theorem 23.2 that every $\mathfrak{L}$-transform of a function must tend towards zero for real-valued $s \to +\infty$. It follows that *the degree of* $g(s)$ *must be lower than the degree of* $p(s)$. This, we shall presume for the remainder of this section. Envisage $g(s)$ and $p(s)$, each expressed as the product of linear factors, common factors already being cancelled; then we recognize the zeros of $p(s)$ as the poles of the rational function $g(s)/p(s)$. Suppose α_μ is a k_μ-fold zero of p, then the main part of the Laurent series of g/p about α_μ is given by

$$\frac{d_{\mu 1}}{s-\alpha_\mu} + \frac{d_{\mu 2}}{(s-\alpha_\mu)^2} + \cdots + \frac{d_{\mu k_\mu}}{(s-\alpha_\mu)^{k_\mu}}.$$

Subtracting from g/p the respective main parts corresponding to the several poles of p: α_μ $(\mu = 1, 2, \ldots, m)$ we obtain a rational function without poles, an entire rational function. For $s \to +\infty$, this difference obviously tends towards zero; hence it is the constant zero. Thus we have established for g/p the *partial fraction expansion:*

$$\frac{g(s)}{p(s)} = \sum_{\mu=1}^{m} \left(\frac{d_{\mu 1}}{s-\alpha_\mu} + \cdots + \frac{d_{\mu k_\mu}}{(s-\alpha_\mu)^{k_\mu}} \right). \tag{8}$$

Next we need to *evaluate the coefficients* d. By the most primitive method one writes the right hand side of (8) as a single fraction, thus reproducing the denominator $p(s)$. Then one equates the corresponding coefficients of powers of s in the numerator to those in $g(s)$. In this manner one obtains a system of linear equations in the $d_{\mu\nu}$. The solution of these equations is troublesome, particularly for large numbers of unknowns. We propose another, simpler method which will be introduced firstly for the special case of simple poles.

a) $p(s)$ *has only simple zeros*

In this case, the partial fraction expansion (8) simplifies to

$$\frac{g(s)}{p(s)} = \sum_{\mu=1}^{n} \frac{d_\mu}{s-\alpha_\mu},$$

with mutually distinct zeros α_μ. Suppose α_ν is one of these zeros. We multiply the equation by $(s - \alpha_\nu)$ and we obtain, for $s \neq \alpha_\nu$,

$$\frac{g(s)(s-\alpha_\nu)}{p(s)} = d_\nu + \sum_{\substack{\mu=1\\ \mu\neq\nu}}^{n} \frac{d_\mu (s-\alpha_\nu)}{s-\alpha_\mu}.$$

Noticing the fact that $p(\alpha_\nu) = 0$, we write the left hand side of the last equation in the form:

$$\frac{g(s)}{\dfrac{p(s) - p(\alpha_\nu)}{s - \alpha_\nu}},$$

which has, for $s \to \alpha_\nu$, the limit $g(\alpha_\nu)/p'(\alpha_\nu)$. Notice that, α_ν being a simple zero of p, we necessarily have $p'(\alpha_\nu) \neq 0$, and the limit has meaning. The limit of the right hand side, for $s \to \alpha_\nu$, is clearly d_ν, hence

$$d_\nu = \frac{g(\alpha_\nu)}{p'(\alpha_\nu)},$$

and the partial fraction expansion can be written thus

$$\frac{g(s)}{p(s)} = \sum_{\mu=1}^{n} \frac{g(\alpha_\mu)}{p'(\alpha_\mu)} \frac{1}{s - \alpha_\mu}. \tag{9}$$

In particular, for $g(s) \equiv 1$, we find:

$$\frac{1}{p(s)} = \sum_{\mu=1}^{n} \frac{1}{p'(\alpha_\mu)} \frac{1}{s - \alpha_\mu}. \tag{10}$$

b) $p(s)$ *has multiple zeros*

Here we extract from (8) those terms which are related to some specific zero α_ν, and call the remainder $h(s)$ which is holomorphic at $s = \alpha_\nu$. Thus we have:

$$\frac{g(s)}{p(s)} = \frac{d_{\nu 1}}{s - \alpha_\nu} + \cdots + \frac{d_{\nu k_\nu}}{(s - \alpha_\nu)^{k_\nu}} + h(s).$$

In a power series with positive exponents, one can express the coefficients by derivatives. We generate positive exponents thus:

$$g(s) \frac{(s - \alpha_\nu)^{k_\nu}}{p(s)} = d_{\nu 1}(s - \alpha_\nu)^{k_\nu - 1} + \cdots + d_{\nu\lambda}(s - \alpha_\nu)^{k_\nu - \lambda} + \cdots + d_{\nu k_\nu}$$
$$+ h(s)(s - \alpha_\nu)^{k_\nu}.$$

Remembering that

$$p(s) = C(s - \alpha_1)^{k_1} \cdots (s - \alpha_m)^{k_m}$$

we realize that the function

$$r_\nu(s) = \frac{p(s)}{(s - \alpha_\nu)^{k_\nu}}$$

is obtained from $p(s)$ simply by deleting the factor $(s - \alpha_\nu)^{k_\nu}$; it is a polynomial that assumes a non-zero value at $s = \alpha_\nu$. We have

$$d_{\nu k_\nu} + \cdots + d_{\nu\lambda}(s - \alpha_\nu)^{k_\nu - \lambda} + \cdots + d_{\nu 1}(s - \alpha_\nu)^{k_\nu - 1} = \frac{g(s)}{r_\nu(s)} - h(s)(s - \alpha_\nu)^{k_\nu},$$

and by the Taylor formula:

$$d_{\nu\lambda} = \frac{1}{(k_\nu - \lambda)!}\left\{\left[\frac{g(s)}{r_\nu(s)} - h(s)\,(s-\alpha_\nu)^{k_\nu}\right]^{(k_\nu-\lambda)}\right\}_{s=\alpha_\nu} \quad (\lambda = 1, \cdots, k_\nu).$$

The $(k_\nu - \lambda)^{th}$ derivative of $h(s)\,(s - \alpha_\nu)^{k_\nu}$ vanishes at $s = \alpha_\nu$ for $k_\nu - \lambda = 0, 1, 2, \ldots, k_\nu - 1$; hence,

$$d_{\nu\lambda} = \frac{1}{(k_\nu - \lambda)!}\left\{\left[\frac{g(s)}{r_\nu(s)}\right]^{(k_\nu-\lambda)}\right\}_{s=\alpha_\nu} \quad (\lambda = 1, \cdots, k_\nu). \tag{11}$$

For simple zeros α_ν, we have $k_\nu = \lambda = 1$, and we find:

$$d_{\nu 1} = \frac{g(\alpha_\nu)}{r_\nu(\alpha_\nu)}.$$

In particular, for $g(s) \equiv 1$, we obtain

$$d_{\nu\lambda} = \frac{1}{(k_\nu - \lambda)!}\left\{\left[\frac{1}{r_\nu(s)}\right]^{(k_\nu-\lambda)}\right\}_{s=\alpha_\nu}. \tag{12}$$

The Differential Equation of Order n

The differential equation

$$y^{(n)} + c_{n-1}\,y^{(n-1)} + \cdots + c_1\,y' + c_0\,y = f(t) \tag{13}$$

is to be integrated in the interval $t \geqq 0$. To permit the determination of a unique solution, the initial values of the solution and of the first n-1 derivatives of the solution (that is, a total of n values) must be specified. Recalling the arguments of p. 70, we state this as follows:

$$y(0+) = y_0,\; y'(0+) = y_0',\; \cdots,\; y^{(n-1)}(0+) = y_0^{(n-1)}. \tag{14}$$

With regard to the disturbing function $f(t)$, we retain the assumptions A_1 and A_2 as stated for $n = 1$. In analogy to the requirements R_1 and R_2 for $n = 1$, we formulate here the requirements for the solution.

R$_1$ *The functions $y(t)$, $y'(t)$, $\ldots$, $y^{(n-1)}(t)$ must be continuous for $t > 0$.*

Hence only $y^{(n)}(t)$ can exhibit jumps at the discontinuity points of $f(t)$, where the one-sided n^{th} derivatives, the one from the left $y_-^{(n)}$ and the one from the right $y_+^{(n)}$, differ.

R$_2$ *The differential equation* (13) *must be satisfied at every $t > 0$, at least from the left and from the right;* that is:

$$y_-^{(n)} + c_{n-1}\,y^{(n-1)} + \cdots + c_1\,y' + c_0\,y = f(t^-),$$

$$y_+^{(n)} + c_{n-1}\,y^{(n-1)} + \cdots + c_1\,y' + c_0\,y = f(t^+).$$

At discontinuity points of $f(t)$, $y^{(n)}(t)$ exhibits a jump of the same magnitude as the one of $f(t)$:

$$y_+^{(n)}(t) - y_-^{(n)}(t) = f(t^+) - f(t^-).$$

At those points where $f(t)$ is continuous, the differential equation (13) is strictly satisfied.

We postulate two hypotheses to make the problem amenable to the method of the $\mathfrak{L}$-transformation.

H$_1$ *The disturbing function $f(t)$ has a $\mathfrak{L}$-transform.*

H$_2$ *The nth derivative of the solution $y^{(n)}(t)$ has a $\mathfrak{L}$-transform.*

From H_2 we conclude, by means of Theorem 9.3, that $y, y', \ldots, y^{(n-1)}$ too have $\mathfrak{L}$-transforms. Similarly, as on p. 72 for $n = 1$, we conclude that a generalized n^{th} derivative may be introduced when $y^{(n)}$ exhibits jumps. Thus we may apply Theorem 9.3 (compare the footnote of the theorem) to the original equation (13), and we find the *image equation*

$$\left.\begin{array}{l} [s^n\, Y - y_0\, s^{n-1} - y_0'\, s^{n-2} - \cdots - y_0^{(n-2)}\, s - y_0^{(n-1)}] \\ + c_{n-1}\, [s^{n-1}\, Y - y_0\, s^{n-2} - y_0'\, s^{n-3} - \cdots - y_0^{(n-2)}] \\ \cdots\cdots\cdots\cdots\cdots\cdots \\ + c_1\, [s\, Y - y_0] \\ + c_0\, Y \end{array}\right\} = F(s).$$

We define the *characteristic polynomial* of the differential equation (13):

$$p(s) = s^n + c_{n-1}\, s^{n-1} + \cdots + c_1\, s + c_0\,; \tag{15}$$

with this, the image equation can be written as follows:

$$\begin{aligned} p(s)\, Y(s) = F(s) &+ y_0\, (s^{n-1} + c_{n-1}\, s^{n-2} + \cdots + c_2\, s + c_1) \\ &+ y_0'\, (s^{n-2} + c_{n-1}\, s^{n-3} + \cdots + c_2) \\ &\cdots\cdots\cdots\cdots\cdots\cdots \\ &+ y_0^{(n-2)}\, (s + c_{n-1}) \\ &+ y_0^{(n-1)}\,. \end{aligned}$$

This is a linear algebraic equation with the solution:[4]

$$
\begin{aligned}
Y(s) = \frac{F(s)}{p(s)} &+ y_0 \frac{s^{n-1} + c_{n-1}s^{n-2} + \cdots + c_2 s + c_1}{p(s)} \\
&+ y_0' \frac{s^{n-2} + c_{n-1}s^{n-3} + \cdots + c_2}{p(s)} \\
&\cdots\cdots\cdots\cdots\cdots\cdots \\
&+ y_0^{(n-2)} \frac{s + c_{n-1}}{p(s)} \\
&+ y_0^{(n-1)} \frac{1}{p(s)}.
\end{aligned}
\tag{16}
$$

Supposing that the initial value problem stated by (13) with (14), complying with hypothesis H_1, has a solution which satisfies hypothesis H_2 and requirements R_1 and R_2, this solution must have the image function (16). Observe that the *initial values* (14) which have to be specified separately with the original equation (13), are now incorporated into the image equation (16). Thus, the initial values are *automatically introduced into the solution.*

Now we need to determine the original function of $Y(s)$. In order to attain greater clarity, we separate the problem into two parts. Firstly we consider in (16) only the first term of the right hand side which involves $F(s)$; that is, we are investigating the special case of zero initial values: the inhomogeneous differential equation with vanishing initial values. Secondly we consider the remaining terms of the right hand side of (16); that is, we are investigating the special case with $F(s) \equiv 0$, or equivalently, $f(t) \equiv 0$: the homogeneous differential equation with arbitrary initial values. We begin with the latter case.

1. *The homogeneous differential equation with arbitrary initial values*

In this special case we know a priori that all hypotheses and the assumptions of our method are satisfied, since the function $f(t) \equiv 0$ trivially satisfies the assumptions A_1 and A_2 and the hypothesis H_2. We know that every solution of the homogeneous differential equation is a linear combination of fundamental solutions of the type $t^l e^{\alpha t}$; all derivatives of these fundamental solutions are continuous and have $\mathfrak{L}$-transforms; thus the requirements R_1 and R_2 and the hypothesis H_2 are satisfied. The solution of the initial value problem certainly exists, and it is produced by our method.

First we investigate the *special case* with the following initial values:

$$y_0 = y_0' = \cdots = y_0^{(n-2)} = 0, \quad y_0^{(n-1)} = 1.$$

[4] The polynomials of the numerators, when read successively, beginning with the last term, are the successive stages of the evaluation procedure of $p(s)$ by Horner's method.

With these initial values, (16) becomes:

$$y(s) = \frac{1}{p(s)}.$$

We define:

$$\frac{1}{p(s)} = G(s);$$

and accordingly we denote the solution $y(t)$ of the differential equation for this special case by $g(t)$. We have

$$g(t) \circ\!\!-\!\!\bullet\, G(s) = \frac{1}{p(s)}, \tag{17}$$

$$g^{(n)} + c_{n-1}\, g^{(n-1)} + \cdots + c_1 g' + c_0 g = 0, \tag{18}$$

$$g(0+) = g'(0+) = \cdots = g^{(n-2)}(0+) = 0,\; g^{(n-1)}(0+) = 1. \tag{19}$$

In order to find $g(t)$, we express $G(s)$ by its partial fraction expansion. For this purpose we use formula (9) if all zeros α_μ of $p(s)$ are simple, formula (8) if $p(s)$ has multiple zeros. For either case we can readily find the corresponding original function (compare Tab. No. 5,24):

$$g(t) = \sum_{\mu=1}^{n} \frac{1}{p'(\alpha_\mu)}\, e^{\alpha_\mu t} \quad (\text{all } \alpha_\mu \text{ simple}), \tag{20}$$

$$g(t) = \sum_{\mu=1}^{m} \left(d_{\mu 1} + \frac{d_{\mu 2}}{1!}\, t + \cdots + \frac{d_{\mu k_\mu}}{(k_\mu - 1)!}\, t^{k_\mu - 1}\right) e^{\alpha_\mu t} \quad (\alpha_\mu\ k_\mu\text{-multiple}), \tag{21}$$

with coefficients $d_{\mu\lambda}$ by formula (12).

Using the explicit solution (20), we obtain, by (19), the remarkable relations:

$$\sum_{\mu=1}^{n} \frac{1}{p'(\alpha_\mu)} = 0,\quad \sum_{\mu=1}^{n} \frac{\alpha_\mu}{p'(\alpha_\mu)} = 0, \ldots, \quad \sum_{\mu=1}^{n} \frac{\alpha_\mu^{n-2}}{p'(\alpha_\mu)} = 0, \tag{22}$$

$$\sum_{\mu=1}^{n} \frac{\alpha_\mu^{n-1}}{p'(\alpha_\mu)} = 1 \quad (\text{all } \alpha_\mu \text{ simple}). \tag{23}$$

The solution of the homogeneous differential equation with *arbitrary* initial values can be obtained by application of the method of partial fractions to the rational functions of (16), using formula (9) or formula (8), and the subsequent $\mathfrak{L}^{-1}$-transformation of the several terms. One can avoid these steps and develop the solution directly by means of the solution $g(t)$ of the above special case. By the Differentiation Theorem 9.3, we find, with (19):

$$\frac{1}{p(s)} \bullet\!\!-\!\!\circ\, g(t), \quad \frac{s}{p(s)} \bullet\!\!-\!\!\circ\, g'(t), \quad \frac{s^2}{p(s)} \bullet\!\!-\!\!\circ\, g''(t), \ldots, \quad \frac{s^{n-1}}{p(s)} \bullet\!\!-\!\!\circ\, g^{(n-1)}(t) \tag{24}$$

$$\frac{s^n}{p(s)} - 1 \bullet\!\!-\!\!\circ\, g^{(n)}(t). \tag{25}$$

Thus we obtain *the solution of the homogeneous equation with specified initial values:*

$$\begin{aligned} y(t) &= y_0 \left[g^{(n-1)}(t) + c_{n-1}\, g^{(n-2)}(t) + \cdots + c_2\, g'(t) + c_1\, g(t)\right] \\ &\quad + y_0' \left[g^{(n-2)}(t) + c_{n-1}\, g^{(n-3)}(t) + \cdots + c_2\, g(t)\right] \\ &\quad \cdots\cdots\cdots\cdots\cdots\cdots\cdots\cdots\cdots \\ &\quad + y_0^{(n-2)} \left[g'(t) + c_{n-1}\, g(t)\right] \\ &\quad + y_0^{(n-1)}\, g(t). \end{aligned} \tag{26}$$

The functions $g'(t), g''(t), \cdots$ have the same basic structure as $g(t)$; that is, they are exponential functions, possibly multiplied by powers of t. They too are solutions of the homogeneous differential equation, having, however, different initial values. By means of (18) and (19), one can verify that $y(t)$, given by (26), satisfies the specified initial values.

2. *The inhomogeneous differential equation with vanishing initial values*

For this case, we have the solution of the image equation, by (16),

$$Y(s) = \frac{1}{p(s)}\, F(s) = G(s)\, F(s). \tag{27}$$

For this, we find by the Convolution Theorem 10.4, using (17), the original function:

$$y(t) = g(t) * f(t) = \int_0^t g(t-\tau)\, f(\tau)\, d\tau. \tag{28}$$

Invoking the principle of extension, we shall demonstrate that $y(t)$, by (28), satisfies the requirements R_1 and R_2, irrespective of the hypotheses H_1 and H_2, provided $f(t)$ satisfies the assumptions A_1 and A_2. First we form the derivatives of $y(t)$, using Theorem 10.5. By (19), we have $g(0+) = g'(0+) = \cdots = g^{(n-2)}(0+) = 0$;

hence $y'(t), y''(t), \cdots, y^{(n-2)}(t)$ exist in the conventional sense for $t > 0$; they are given by

(29) $$y'(t) = g'(t) * f(t), \cdots, y^{(n-1)}(t) = g^{(n-1)}(t) * f(t).$$

However, $y^{(n)}(t)$ exists from the left and from the right; we have, with $g^{(n-1)}(0^+) = 1$:

(30) $$y_-^{(n)}(t) = g^{(n)}(t) * f(t) + f(t^-),$$
$$y_+^{(n)}(t) = g^{(n)}(t) * f(t) + f(t^+).$$

Hence, with (18),

$$y_-^{(n)} + c_{n-1}\,y^{(n-1)} + \cdots + c_1 y' + c_0 y$$
$$= [g^{(n)} + c_{n-1}\,g^{(n-1)} + \cdots + c_1 g' + c_0 g] * f + f(t^-) = f(t^-),$$
$$y_+^{(n)} + c_{n-1}\,y^{(n-1)} + \cdots + c_1 y' + c_0 y = f(t^+).$$

Moreover, $g, g', \cdots, g^{(n-1)}$ are bounded in every finite interval so that, by Theorem 10.3,

(31) $$y(0^+) = y'(0^+) = \cdots = y^{(n-1)}(0^+) = 0.$$

We conclude that formula (28) always produces *the solution of the inhomogeneous differential equation with vanishing initial values, irrespective of the method of derivation employed.*

Having thus resolved all questions theoretically, one may, without hesitation, apply the method of the $\mathfrak{L}$-transformation in every special case, feeling secure that the solution of the differential equation with the specified initial values will be obtained. This method excels the classical method by one further characteristic; that is, when using the $\mathfrak{L}$-transformation we immediately find the solution with the prescribed initial values. By contrast, when using the classical method, one must adjust the "general solution" to fit the specified initial values. This latter process involves considerable labour, particularly for higher values of n, since n simultaneous linear equations in n unknowns must be solved. The problem with vanishing initial values is most frequently encountered in practical applications. It implies no simplifications in the course of the classical method; however, it does afford a particularly simple form of solution when the $\mathfrak{L}$-transformation is employed.

In practical applications involving specific numerical data, one does not use the above presented general formulae for the solution, instead *one executes the method through the several steps:* derivation of the image equation, partial fraction expansion of the solution, $\mathfrak{L}^{-1}$-transformation into the original space. In this way one can make use of all possible simplifications. In particular, it is frequently possible to avoid the convolution representation of the solution (27) of the inhomogeneous differential equation. This is done by explicit evaluation of $F(s)/p(s)$, followed by the $\mathfrak{L}^{-1}$-transformation of the function $F(s)/p(s)$ as an entity.[5]

The Transfer Function

The solution in the image space: $Y(s) = G(s) \cdot F(s)$ has an intriguingly simple form. It is for this reason that in technical applications involving differential equations, one attempts to extract information regarding the solution from the image function. In technical literature the function $G(s)$ is called the *transfer function* of the system which is described by the differential equation, since $G(s)$ "transfers" the input function $F(s)$ into the output function $Y(s)$.[6]

The function $g(t)$ in (28) is, in mathematical terminology, the *Green's function* of the initial value problem. In technical literature it is called the *weighting function*, since in formula (28) for the solution, every value $f(\tau)$ of the excitation appears associated with the weighting factor $g(t - \tau)$, which depends upon the time interval $t - \tau$ between the action of the excitation f at time τ and the observation of y at time t.

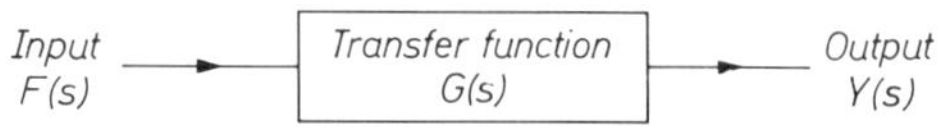

Figure 7

The interrelation between input, output and transfer function can be clearly presented in graphical form by a *block diagram* as shown in Fig. 7: $F(s)$ enters the box, $Y(s)$ leaves the box, the box is inscribed with $G(s)$.

Such block diagrams prove particularly useful when several systems are connected in series or in some other manner, so that the output of one system serves

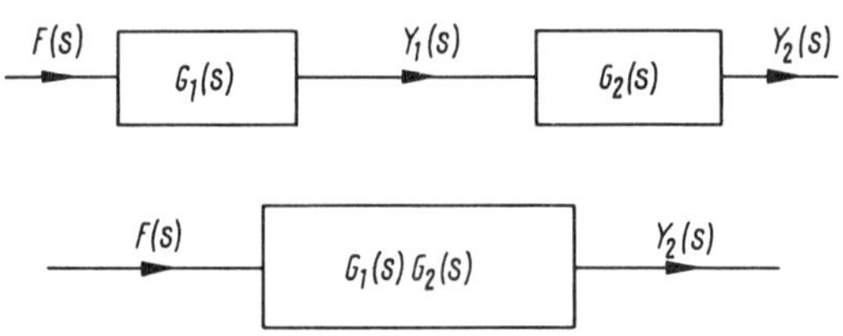

Figure 8

[5] A number of completely evaluated solutions of differential equations of second to fifth order can be found in the author's book: *Guide to the applications of the Laplace and Z-Transforms. Second edition, Van Nostrand Reinhold Company, London 1971.*

[6] In engineering, the term input is used for both $f(t)$ and $F(s)$, and the term output for both $y(t)$ and $Y(s)$.

as the input of another system. In this way, the interrelation presented in Fig. 8 is readily understood. For we have

$$Y_1(s) = G_1(s)\, F(s), \quad Y_2(s) = G_2(s)\, Y_1(s),$$

and, consequently,

$$Y_2(s) = G_1(s)\, G_2(s)\, F(s).$$

From this we can conclude that the series connection of two systems, $G_1(s)$ and $G_2(s)$ respectively, behaves equivalently to a single system having the transfer function $G_1(s)\, G_2(s)$. As a typical, practical example of such a combination consider: some force excites a mechanical system; the mechanical motion of this system affects a second mechanical, or an electrical system.

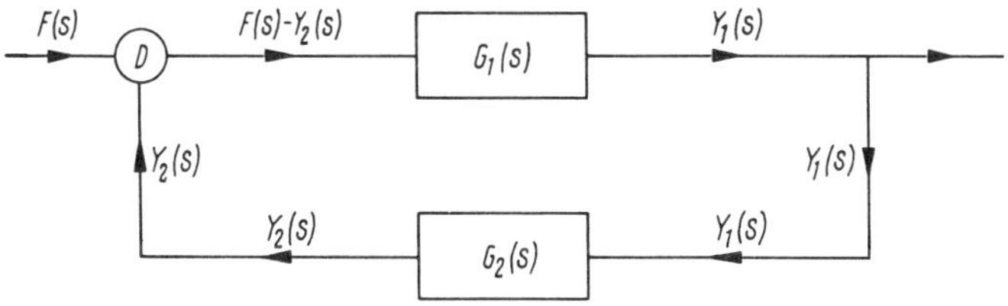

Figure 9

Feedback is another connection of systems that can easily be explained by a block diagram; it is shown in Fig. 9: the function $Y_1(s)$ leaves the box which is designated by $G_1(s)$; $Y_1(s)$ enters the box marked $G_2(s)$ and is altered into the function $Y_2(s)$; $Y_2(s)$ and another function $F(s)$ are combined in the instrument marked D in which the difference $F(s) - Y_2(s)$ is formed;[7] this difference enters the box marked $G_1(s)$ as input function. According to Fig. 9 we have the equations

$$Y_1(s) = G_1(s)\,[F(s) - Y_2(s)], \quad Y_2(s) = G_2(s)\, Y_1(s).$$

Of interest in feedback systems is the relationship between $F(s)$ and $Y_1(s)$. It is obtained by elimination of $Y_2(s)$:

$$Y_1(s) = \frac{G_1(s)}{1 + G_1(s)\, G_2(s)}\, F(s) = \frac{1}{\dfrac{1}{G_1(s)} + G_2(s)}\, F(s).$$

The transfer functions $G_1(s)$ and $G_2(s)$ are the reciprocal expressions of the characteristic polynomials, $p_1(s)$ and $p_2(s)$, of the differential equations that describe the behaviour of the respective boxes. Hence,

$$Y_1(s) = \frac{1}{p_1(s) + \dfrac{1}{p_2(s)}}\, F(s) = \frac{p_2(s)}{p_1(s)\, p_2(s) + 1}\, F(s).$$

[7] In reality, one forms $f(t) - y_2(t)$; however, we always interchange lower case symbols with upper case symbols.

The degree of $p_1(s)\,p_2(s)+1$ is higher than the degree of $p_2(s)$; consequently, we can find the partial fraction expansion of the factor of $F(s)$, and its original function. With this, we write the output $y_1(t)$ as a convolution. In case $F(s)$ is a rational function, as, for instance, for $f(t) \equiv u(t)$ with $F(s) \equiv 1/s$, one can inversely transform $Y_1(s)$ directly as an entity.

16. The Ordinary Differential Equation, specifying Initial Values for Derivatives of Arbitrary Order, and Boundary Values

The application of the $\mathfrak{L}$-transformation in Chapter 15 presupposed the knowledge of the values of the function and its first $(n-1)$ derivatives at $t=0$. However, one could encounter some initial value problem with n specified values at $t=0$ for derivatives of arbitrary order. For instance, for same third order differential equation one might specify the initial values[1] $y(0)$, $y^{\mathrm{III}}(0)$, $y^{\mathrm{IV}}(0)$. In this case, we would solve the problem as if $y(0)$, $y'(0)$, and $y''(0)$ were given. Then we would form the higher derivatives, $y^{\mathrm{III}}(t)$ and $y^{\mathrm{IV}}(t)$. For $t=0$, we would obtain two linear equations in the unknowns $y'(0)$ and $y''(0)$. Having solved these equations, we can write the complete solution $y(t)$.

An analogous process is suggested to find the solutions of *boundary value problems*; that is, when the n values are specified not at one point, but instead at two or more points. If, for instance, for some third order differential equation, the values are specified at 0, and at $l>0$, say $y(0)$, $y(l)$ and $y'(l)$, then we derive the solution with $y(0)$, $y'(0)$, and $y''(0)$, form the values $y(l)$ and $y'(l)$ and obtain two equations in the unknowns $y'(0)$, $y''(0)$.

The same procedure can be employed when linear combinations of boundary values are specified instead of individual boundary values.

To demonstrate the proposed technique, let us consider a special *boundary value problem*, the solution of which will be needed later on (see pp. 282, 294). We are given the equation

$$y''-\alpha^2 y=f(t) \qquad (\alpha \neq 0, \text{ complex})$$

with continuous $f(t)$, and the boundary values $y(0)$ and $y(l)$. Proceeding as if $y'(0)$ were given instead of the specified $y(l)$, we find the image equation:

$$s^2\,Y-y(0)\,s-y'(0)-\alpha^2\,Y=F(s)$$

which has the solution:

$$Y(s)=\frac{F(s)}{s^2-\alpha^2}+y(0)\,\frac{s}{s^2-\alpha^2}+y'(0)\,\frac{1}{s^2-\alpha^2}\,.$$

---

[1] For brevity, we shall write $y(0)$ instead of $y(0^+)$, etc, and $y(l)$ instead of $y(l^-)$, etc.

Hence (compare Tab. No. 15,20),

$$y(t) = \frac{1}{a} f(t) * \sinh a t + y(0) \cosh a t + \frac{1}{a} y'(0) \sinh a t. \tag{1}$$

To gain better insight, let us separate this problem into two parts:

I. $f(t) \equiv 0$; $y(0)$ and $y(l)$ arbitrary.

Substituting into (1) $f(t) \equiv 0$, we find at $t = l$:

$$y(l) = y(0) \cosh a l + \frac{1}{a} y'(0) \sinh a l;$$

this implies that

$$\frac{1}{\alpha} y'(0) = \frac{y(l) - y(0) \cosh \alpha l}{\sinh \alpha l}.$$

It follows that

$$y(t) = y(0) \cosh a t + (y(l) - y(0) \cosh a l) \frac{\sinh a t}{\sinh a l}$$

$$= y(0) \frac{\sinh a(l-t)}{\sinh a l} + y(l) \frac{\sinh a t}{\sinh a l}. \tag{2}$$

II. $f(t) \not\equiv 0$; $y(0) = y(l) = 0$.

Substituting $y(0) = 0$ into (1), we find:

$$y(t) = \frac{1}{a} f(t) * \sinh a t + \frac{1}{a} y'(0) \sinh a t,$$

and, at $t = l$:

$$0 = y(l) = \frac{1}{a} \int_0^l f(\tau) \sinh a(l - \tau)\, d\tau + \frac{1}{a} y'(0) \sinh a l,$$

which implies that

$$\frac{1}{a} y'(0) = -\frac{1}{a \sinh a l} \int_0^l f(\tau) \sinh a(l - \tau)\, d\tau;$$

hence,

$$y(t) = \frac{1}{a} \int_0^t f(\tau) \sinh a(t - \tau)\, d\tau - \frac{1}{a} \frac{\sinh a t}{\sinh a l} \int_0^l f(\tau) \sinh a(l - \tau)\, d\tau. \tag{3}$$

The solution (3) can be improved in the following manner. First we split the interval of integration from 0 to l into two sections, one from 0 to t, the other from t to l. Then we combine the two integrals from 0 to t into one, with the common denominator $\sinh \alpha l$; for this integral we have, in the integrand, the following expression:

$$\sinh a(t-\tau)\sinh a\,l - \sinh a(l-\tau)\sinh a\,t$$

$$= (\sinh a\,t \cosh a\,\tau - \cosh a\,t \sinh a\,\tau)\sinh a\,l$$

$$- (\sinh a\,l \cosh a\,\tau - \cosh a\,l \sinh a\,\tau)\sinh a\,t$$

$$= -\sinh a\,\tau\,(\sinh a\,l \cosh a\,t - \cosh a\,l \sinh a\,t) = -\sinh a\,\tau \sinh a(l-t).$$

Thus, (3) can be modified into:

$$y(t) = -\frac{1}{a}\frac{\sinh a(l-t)}{\sinh a\,l}\int_0^t f(\tau)\sinh a\,\tau\,d\tau - \frac{1}{a}\frac{\sinh a\,t}{\sinh a\,l}\int_t^l f(\tau)\sinh a\,(l-\tau)\,d\tau.$$

Introducing the following "Green's Function".

(4) $$\gamma(t,\tau;a) = \begin{cases} -\dfrac{1}{a}\,\dfrac{\sinh a\,\tau \sinh a(l-t)}{\sinh a\,l} & \text{for } 0 \leqq \tau \leqq t \\[2ex] -\dfrac{1}{a}\,\dfrac{\sinh a\,t \sinh a(l-\tau)}{\sinh a\,l} & \text{for } t \leqq \tau \leqq l, \end{cases}$$

we obtain the simplified representation of the solution:

(5) $$y(t) = \int_0^l \gamma(t,\tau;a)\,f(\tau)\,d\tau.$$

The general solution of the boundary value problem is obtained by superposition of (2) with (5).

Obviously, the solution has meaning only if the denominator $\sinh \alpha l \neq 0$. The values $\alpha^2 \neq 0$, for which $\sinh \alpha l = 0$, that is

$$a^2 = -n^2\left(\frac{\pi}{l}\right)^2 \quad (n = 1, 2, \cdots)$$

are the *characteristic values (eigenvalues)* of the boundary value problem. The homogeneous problem: $f(t) \equiv 0$, $y(0) = y(l) = 0$, has the non-trivial solutions $\sin n\,(\pi/l)\,t$ (the *characteristic solutions, eigensolutions*), provided α^2 is a characteristic value; otherwise it has only the solution $y(t) \equiv 0$.

The Unbounded Interval

Elsewhere we shall need the special case with $l = \infty$. Naturally, we must interpret the boundary value $y(\infty)$ as $\lim_{t\to\infty} y(t)$. The function (1) represents the totality of all solutions of the differential equation in $0 \leqq t < \infty$; hence it must contain the solution of the boundary value problem for $l = \infty$. We presume now, that

$$\alpha^2 \text{ is not negative real-valued},$$

in order to exclude the characteristic values. Let α designate the root of α^2 with positive real part. Replacing, in (1), the hyperbolic functions by their exponential representations and collecting respectively the terms with e^{at}, and those with e^{-at}, we obtain:

$$y(t) = \left\{ \frac{1}{2a}\int_0^t e^{-a\tau} f(\tau)\, d\tau + \frac{y(0)}{2} + \frac{y'(0)}{2a} \right\} e^{at} \tag{6}$$
$$+ \left\{ -\frac{1}{2a}\int_0^t e^{a\tau} f(\tau)\, d\tau + \frac{y(0)}{2} - \frac{y'(0)}{2a} \right\} e^{-at},$$

and we observe that, in general, $y(t)$ does not have a limit as $t \to \infty$. This indicates the need for some assumption regarding the behaviour of $f(t)$, as $t \to \infty$, so that $y(\infty)$ does exist, and $y'(0)$ may be evaluated. We shall discover that a sufficient assumption is the existence of $\lim_{t\to\infty} f(t) = f(\infty)$. Beyond that, we shall demonstrate that $y(\infty)$ cannot be specified arbitrarily; it must be $y(\infty) = -f(\infty)/\alpha^2$. First we verify the

Lemma. *If $f(t)$ is continuous for $t \geqq 0$, and the $\lim_{t\to\infty} f(t) = f(\infty)$ exists, then we have, for $\Re\alpha > 0$:*

$$\left.\begin{array}{l} \displaystyle\int_0^t e^{-a(t-\tau)} f(\tau)\, d\tau \\[2ex] \displaystyle\int_t^\infty e^{-a(\tau-t)} f(\tau)\, d\tau \end{array}\right\} \to \frac{f(\infty)}{a} \quad \text{as} \quad t \to \infty.$$

Proof: We have

$$\frac{f(\infty)}{a} = \int_{-\infty}^t e^{-a(t-\tau)} f(\infty)\, d\tau;$$

hence,

$$\int_0^t e^{-a(t-\tau)} f(\tau)\, d\tau - \frac{f(\infty)}{a} = \int_0^t e^{-a(t-\tau)} [f(\tau) - f(\infty)]\, d\tau - f(\infty)\int_{-\infty}^0 e^{-a(t-\tau)}\, d\tau.$$

The difference $f(\tau) - f(\infty)$ is continuous for $\tau \geqq 0$; it approaches zero as $\tau \to \infty$; hence, it is bounded:

$$|\, f(\tau) - f(\infty)\, | < M \quad \text{for } \tau \geqq 0,$$

and for every $\varepsilon > 0$, there exists a $T > 0$, so that

$$|f(\tau) - f(\infty)| < \varepsilon \quad \text{for } \tau \geqq T.$$

Consequently, for $t > T$,

$$\left| \int_0^t e^{-\alpha(t-\tau)} f(\tau)\,d\tau - \frac{f(\infty)}{\alpha} \right| \leqq M \int_0^T e^{-\Re\alpha(t-\tau)}\,d\tau + \varepsilon \int_T^t e^{-\Re\alpha(t-\tau)}\,d\tau + \left|\frac{f(\infty)}{\alpha}\right| e^{-\Re\alpha t}$$

$$\leqq \frac{M\,e^{\Re\alpha T}}{\Re\alpha} e^{-\Re\alpha t} + \frac{\varepsilon}{\Re\alpha} + \left|\frac{f(\infty)}{\Re\alpha}\right| e^{-\Re\alpha t}.$$

For all sufficiently large t, the established bound is $< 3\varepsilon/\Re\alpha$. This verifies the first conclusion. For the second conclusion, we use the equation

$$\frac{f(\infty)}{\alpha} = \int_t^\infty e^{-\alpha(\tau-t)} f(\infty)\,d\tau,$$

and we find, for $t > T$,

$$\left| \int_t^\infty e^{-\alpha(\tau-t)} f(\tau)\,d\tau - \frac{f(\infty)}{\alpha} \right| = \left| \int_t^\infty e^{-\alpha(\tau-t)} [f(\tau) - f(\infty)]\,d\tau \right|$$

$$\leqq \varepsilon \int_t^\infty e^{-\Re\alpha(\tau-t)}\,d\tau = \frac{\varepsilon}{\Re\alpha},$$

thus establishing the second conclusion.

We return to the presentation (6) of the solution $y(t)$. We assume that $f(t)$ is continuous for $t \geqq 0$, and that $f(\infty)$ exists; then

$$\int_0^\infty e^{-\alpha t} f(t)\,dt = F(\alpha) \quad (\Re\alpha > 0)$$

converges. Therefore,we can re-write the first integral of (6) in the following form:

$$F(\alpha) - \int_t^\infty e^{-\alpha\tau} f(\tau)\,d\tau.$$

Applying the above Lemma to the modified expression, we recognize that each of the two integrals, multiplied with the respective exponential function, does have a limit. It follows that $y(t)$ too has a limit $y(\infty)$ as $t \to \infty$ if and only if

(7) $$\frac{F(\alpha)}{\alpha} + y(0) + \frac{y'(0)}{\alpha} = 0.$$

Presuming this condition, one finds:

$$y(\infty) = -\frac{f(\infty)}{\alpha^2}. \tag{8}$$

This is the only possible boundary value at $t = \infty$. Using (7), we evaluate $y'(0)$ and substitute its value into (6). In this manner we obtain the solution of the boundary value problem in the interval $(0, \infty)$; it is:

$$y(t) = y(0)\, e^{-\alpha t} - \frac{1}{2\alpha} e^{-\alpha t} \int_0^t e^{\alpha\tau} f(\tau)\, d\tau - \frac{1}{2\alpha} e^{\alpha t} \int_t^\infty e^{-\alpha\tau} f(\tau)\, d\tau$$

$$+ \frac{1}{2\alpha} e^{-\alpha t} \int_0^\infty e^{-\alpha\tau} f(\tau)\, d\tau$$

$$= y(0)\, e^{-\alpha t} - \frac{1}{2\alpha} e^{-\alpha t} \int_0^t \left(e^{\alpha\tau} - e^{-\alpha\tau}\right) f(\tau)\, d\tau - \frac{1}{2\alpha}\left(e^{\alpha t} - e^{-\alpha t}\right) \int_t^\infty e^{-\alpha\tau} f(\tau)\, d\tau.$$

Using the "Green's Function":

$$\gamma_\infty(t, \tau; \alpha) = \begin{cases} -\dfrac{1}{\alpha} e^{-\alpha t} \sinh \alpha\tau & \text{for } 0 \leqq \tau \leqq t \\ -\dfrac{1}{\alpha} e^{-\alpha\tau} \sinh \alpha t & \text{for } t \leqq \tau < \infty \end{cases} \tag{9}$$

one can write the solution compactly:

$$y(t) = y(0)\, e^{-\alpha t} + \int_0^\infty \gamma_\infty(t, \tau; \alpha)\, f(\tau)\, d\tau. \tag{10}$$

Theorem 16.1. *Given the boundary value problem in the infinite interval:*

$$\left| \begin{array}{l} y'' - \alpha^2 y = f(t) \quad (\alpha^2 \textit{ complex, not negative real-valued and } \neq 0;\ \Re\alpha > 0), \\ y(0) \textit{ and } y(\infty) \textit{ are specified;} \end{array} \right.$$

if the function $f(t)$ is continuous for $t \geqq 0$, and $\lim_{t\to\infty} f(t) = f(\infty)$ does exist, then the boundary value problem has a solution if and only if $y(\infty) = -f(\infty)/\alpha^2$. The solution is given by (10).

Consider the limits of the following expressions, as $l \to \infty$,

$$\frac{\sinh \alpha(l - t)}{\sinh \alpha l} = \frac{e^{\alpha(l-t)} - e^{-\alpha(l-t)}}{e^{\alpha l} - e^{-\alpha l}} = \frac{e^{-\alpha t} - e^{-\alpha(2l-t)}}{1 - e^{-2\alpha l}} \to e^{-\alpha t}$$

and

$$\frac{\sinh \alpha t}{\sinh \alpha l} = \frac{e^{\alpha t} - e^{-\alpha t}}{e^{\alpha l} - e^{-\alpha l}} \to 0.$$

One could produce, with these limits, the same solution from (2) and (5). In (2) the term with $y(l)$ would vanish and the remaining terms of (2) and (5) would assume the form of expression (10). This approach is objectionable, for we cannot be certain a priori that the solution of the limiting problem equals the limit of the solution. Moreover, in this manner one could neither explain why the boundary value at $t = \infty$ does not appear in the solution, nor indicate what boundary value the solution assumes at $t = \infty$. Both questions are completely answered by the derivation presented here under the hypothesis that $f(\infty)$ exists. Naturally, this need not be the only productive hypothesis. For instance, one might investigate the results that can be obtained with the hypothesis that

$$\int_0^\infty |f(t)|\, dt, \quad \text{or} \quad \int_0^\infty |f(t)|^2\, dt$$

exists.

17. The Solutions of the Differential Equation for Specific Excitations

When the differential equation (15.13) describes some physical system, then the solution $y(t)$ of the homogeneous equation represents the action of the system that is left undisturbed, starting from an initial situation which is defined by the initial values $y_0, y_0', \cdots, y_0^{(n-1)}$. This solution is a linear combination of functions of the type $t^k e^{at}$; it is easy to survey. Therefore we disregard this part of the problem here, and we presume that

(1) $$y_0 = y_0' = \ldots = y_0^{(n-1)} = 0;$$

that is, the system is assumed to be initially at rest. Thus we find the corresponding solution in the image space:

(2) $$Y(s) = G(s)\, F(s)$$

and, consequently, in the original space:

(3) $$y(t) = g(t) * f(t).$$

Occasionally we shall call the input function $f(t)$ the *excitation*, and we shall extend this designation to the corresponding image function $F(s)$; also, we shall designate the output function $y(t)$, and its image function $Y(s)$, as the *response*. In physics and in engineering, special test functions are often employed as excitations to gain insight into the behaviour of the physical system. In this Chapter, we shall study the response of a system to two specific excitations: the unit step function, and the complex oscillations.

1. The Step Response

When using the unit step function $u(t)$ as excitation, we obtain the *step response* (formerly sometimes called the *transient function*) $y_u(t)$; its image function will be designated by $Y_u(s)$. For this function we have, by (15.27, 28),

(4) $$Y_u(s) = G(s)\frac{1}{s} = \frac{1}{s\,p(s)},$$

and

(5) $$y_u(t) = g(t) * 1 = \int_0^t g(\tau)\,d\tau,$$

since $\mathfrak{L}\{u(t)\} = \mathfrak{L}\{1\} = 1/s$.

One could evaluate the convolution using the expressions (15.21) or (15.22) for $g(t)$; it is simpler, however, to first express the rational function $Y_u(s)$ by its partial fraction expansion, and then to transform the latter into the original space. General formulae are quite complicated and of questionable practical value when $p(s)$ has multiple roots. For actual problems of this type one can write the particular partial fraction expressions by (15.8) and (15.12) without difficulties. However, for most practical problems, $p(s)$ has only simple, non-zero (the latter is true for $c_0 \neq 0$) roots. In this case, $s\,p(s)$ *has only simple roots*, and a general formula can easily be written, for we have

$$[s\,p(s)]' = p(s) + s\,p'(s) = \begin{cases} p(0) & \text{for } s = 0 \\ \alpha_\mu\,p'(\alpha_\mu) & \text{for } s = \alpha_\mu\,. \end{cases}$$

Replacing $p(s)$ by $s\,p(s)$ in formula (15.10), we produce:

$$Y_u(s) = \frac{1}{s\,p(s)} = \frac{1}{p(0)}\,\frac{1}{s} + \sum_{\mu=1}^{n} \frac{1}{\alpha_\mu p'(\alpha_\mu)}\,\frac{1}{s-\alpha_\mu}$$

which has the original function:

(6) $$y_u(t) = \frac{1}{p(0)} + \sum_{\mu=1}^{n} \frac{e^{\alpha_\mu t}}{\alpha_\mu p'(\alpha_\mu)}\,.$$

Formula (6) indicates that the system, which is initially at rest, responds to the excitation by the unit step function with the following combination: the jump $u(t)/p(0) = u(t)/c_0$ superimposed with an aggregate of the proper oscillations $e^{\alpha_\mu t}$. The latter diminishes in magnitude towards zero, for increasing t, provided we have for all roots $\Re\,\alpha_\mu < 0$; this aggregate of oscillations represents the transient behaviour of the system between the initial rest position and the steady state $1/p(0) = 1/c_0$, for large values of t.

The relationship between the step response $y_u(t)$ and the function $g(t)$ is, by (5),

$$y_u'(t) = g(t). \tag{7}$$

That is, as soon as $y_u(t)$ is determined by theoretical or by experimental means, one can find the response of the system to any arbitrary input function $f(t)$, for we have, by (3) and (7),

$$y(t) = y_u'(t) * f(t). \tag{8}$$

Invoking Theorem 10.5, we can write the relationship (8) as follows:

$$y(t) = \frac{d}{dt}[y_u * f], \tag{9}$$

since $y_u(0) = 0$. The name *Duhamel's formula* is associated with both formulae (8) and (9).

2. Sinusoidal Excitations. The Frequency Response

In consideration of the remarks of Chapter 1, we prefer the use of the complex oscillations $f(t) = e^{i\omega t}$ to the real oscillations $\cos\omega t$ and $\sin\omega t$ as excitations. When we want to know the response of the system to the real oscillations, we simply use the real part or the imaginary part of the complex oscillation response. This convenient property stems from the following facts. Although the coefficients $c_0, c_1, \ldots, c_n$ may, in principle, be complex valued (compare Chapter 15), they are real valued whenever the differential equation describes a physical system. Hence the coefficients of the polynomial $p(s)$ are real valued, and the complex roots of $p(s)$ occur in conjugate pairs: $\alpha_\nu, \bar{\alpha}_\nu$. Accordingly, in (15.20) we have conjugate pairs of terms

$$\frac{1}{p'(\alpha_\nu)} e^{\alpha_\nu t} \quad \text{and} \quad \frac{1}{p'(\bar{\alpha}_\nu)} e^{\bar{\alpha}_\nu t},$$

which combine to form real valued expressions, and $g(t)$ is real valued. The same argumentation can be applied to (15.21). The above conclusions follow directly since $g(t)$ is the derivative of the step response, a real valued function.

We shall use the notation $y_\omega(t)$ to designate the response of the system to the excitation $f(t) \equiv e^{i\omega t}$, and the symbol $Y_\omega(s)$ for the corresponding image function. The image function is given by:

$$Y_\omega(s) = \frac{G(s)}{s - i\omega},$$

since $\mathfrak{L}\{e^{i\omega t}\} = 1/(s - i\omega)$, and the original function by:

$$y_\omega(t) = g(t) * e^{i\omega t} = e^{i\omega t}\int_0^t e^{-i\omega\tau} g(\tau)\, d\tau. \tag{10}$$

The function $g(t)$ is the superposition of the proper oscillations (characteristic oscillations) of the system: $e^{\alpha_\mu t}$, possibly multiplied by powers of t (compare (15.20) and (15.21)).[1] These functions grow without bound for increasing t, if either $\Re\alpha_\mu > 0$, or $\Re\alpha_\mu = 0$ for a multiple root, such that $e^{\alpha_\mu t}$ is multiplied by powers of t. Physical argumentation indicates that the amplitude of any proper oscillation can grow beyond every finite bound only if the system possesses inner sources of energy. For *passive systems* without inner sources of energy, we must have $\Re\alpha_\mu \leqq 0$ for all roots of $p(s)$, and roots with $\Re\alpha_\mu = 0$ must be simple roots of the characteristic polynomial. For these latter roots, which must be of the form iy_μ, the proper oscillations have constant amplitudes. Resorting once more to physical argumentation, we recognize that these oscillations with constant amplitudes are possible only for systems without internal energy losses. For passive systems which dissipate energy through mechanical friction, Ohmic resistance, or the like, one can predict: $\Re\alpha_\mu < 0$ for all roots of the characteristic polynomial $p(s)$.

For such a system one can readily determine the *steady state* of $y_\omega(t)$, as $t \to \infty$. If

$$0 > \Re\alpha_1 \geqq \Re\alpha_2 \geqq \cdots \geqq \Re\alpha_m,$$

then obviously $G(s) = \mathfrak{L}\{g(t)\}$ converges for $\Re s > \Re\alpha_1$; it certainly converges on the imaginary axis. That is

$$G(i\omega) = \int_0^\infty e^{-i\omega t}\, g(t)\, dt$$

exists, and we can write (10) as follows:

$$y_\omega(t) = e^{i\omega t}\left(\int_0^\infty e^{-i\omega\tau}\, g(\tau)\, d\tau - \int_t^\infty e^{-i\omega\tau}\, g(\tau)\, d\tau\right)$$

(11)
$$= G(i\omega)\, e^{i\omega t} - e^{i\omega t}\int_t^\infty e^{-i\omega\tau}\, g(\tau)\, d\tau.$$

The last term tends towards zero when $t \to \infty$. It follows that $y_\omega(t)$ differs from

(12)
$$\tilde{y}_\omega(t) = G(i\omega)\, e^{i\omega t}$$

by an arbitrarily small amount, for sufficiently large values of t. Hence $\tilde{y}_\omega(t)$ does represent the steady state.

One could express $Y_\omega(s)$ by its partial fraction expansion. When $p(s)$ has only simple roots, all different from $i\omega$, then we find:

$$Y_\omega(s) = \frac{1}{(s-i\omega)\,p(s)} = \frac{G(i\omega)}{s-i\omega} + \sum_{\mu=1}^{n} \frac{1}{(\alpha_\mu - i\omega)\,p'(\alpha_\mu)}\,\frac{1}{s-\alpha_\mu},$$

[1] It is customary to call the solutions of the homogeneous equation (proper solutions, characteristic solutions, eigensolutions) characteristic oscillations, proper oscillations, or eigenoscillations of the system, although for real valued α_μ the functions are not oscillatory but aperiodic.

and in the original space:

$$y_\omega(t) = G(i\,\omega)\, e^{i\omega t} + \sum_{\mu=1}^{n} \frac{e^{\alpha_\mu t}}{(\alpha_\mu - i\,\omega)\, p'(\alpha_\mu)}\,.$$

The previous expression (11) for $y_\omega(t)$ has more general usefulness than the last one, since (11) is valid if, in the representation of the image function: $Y_\omega(s) = G(s)/(s - i\omega)$, *the function* $G(s)$ *is any arbitrary* $\mathfrak{L}$*-transform which converges on the imaginary axis.*

So far we have presumed that the initial values of $y_\omega(t)$ vanish. Observe that the steady state is independent of the initial values, for the superimposed solution (15.26) of the homogeneous equation which accounts for the contribution by the initial values tends towards zero, since $\Re\alpha_\mu < 0$.

In general, $G(i\omega)$ is a complex valued function:

$$G(i\,\omega) = |G(i\,\omega)|\; e^{i\psi(\omega)}. \tag{13}$$

Thus,for the steady state we have:

$$\tilde{y}_\omega(t) = |G(i\,\omega)|\; e^{i(\omega t + \psi(\omega))}; \tag{14}$$

it is an *oscillation having the frequency* ω *of the excitation;* its *amplitude* is determined by $|G(i\omega)|$, and its *initial phase* by $\psi(\omega)$.

The function $G(i\omega)e^{i\omega t}$ describes the steady state of the system in response to the excitation $e^{i\omega t}$; it is a function of the independent variable ω, the frequency of the excitation. We call $G(i\omega)e^{i\omega t}$ the *frequency response* of the system; the modulus $|G(i\omega)|$ and the initial phase $\psi(\omega)$ are the *frequency characteristics* of the system. These are of outstanding importance in engineering; for the steady state, they tell how much the output is amplified or diminished and how much the phase of the output is shifted, when compared to the input $e^{i\omega t}$.

A *special advantage* lies in the fact that for oscillatory excitations one can determine the steady state of the system exclusively from G in the image space, without transformation into the original space.

There is a *simple relationship that connects the frequency response to the step response.* We have, by (4),

$$G(i\,\omega) = i\,\omega\, Y_u(i\,\omega) = [s\,\mathfrak{L}\{y_u\}]_{s=i\omega} = \mathfrak{L}\{y_u'\}_{s=i\omega}\,, \tag{15}$$

a remarkable formula of considerable technical interest, for it permits the determination of the frequency response, when the step response is known. It is more important that, inversely, the step response may be found, when the frequency response is given.[2] For this purpose we recall the previous remarks (see p. 94)

[2] The following remarks are included here for systematic reasons; they will be fully appreciated only after the study of Chapter 24.

concerning the real-valued coefficients of differential equations which describe realistic physical systems; accordingly, $g(t)$ is real-valued. The transform $\mathfrak{L}\{g(t)\}$ converges for $\Re s > \Re \alpha_1$ and consequently for $\Re s > 0$. Hence, by the Integration Theorem 8.1,

$$\mathfrak{L}\{y_u(t)\} = \mathfrak{L}\{g(t) * 1\} = \frac{G(s)}{s} \quad \text{for } \Re s > 0.$$

We now need the Complex Inversion Formula of the $\mathfrak{L}$-transformation which will be verified in Chapter 24. The hypotheses of Theorem 24.3 are clearly satisfied; hence,[3] for $x > 0$,

$$\frac{1}{2\pi i} \int_{x-i\infty}^{x+i\infty} e^{ts} \frac{G(s)}{s} \, ds = \begin{cases} y_u(t) & \text{for } t > 0 \\ 0 & \text{for } t < 0. \end{cases}$$

We alter the integral so that the path of integration can be moved to the imaginary axis. This is not possible for the integral in its present form, since $G(s)/s$ has a singularity at $s = 0$. Thus we modify the integral by subtracting the equation (compare (24.19))

$$\frac{1}{2\pi i} \int_{x-i\infty}^{x+i\infty} e^{ts} \frac{G(0)}{s} \, ds = \begin{cases} G(0) & \text{for } t > 0 \\ 0 & \text{for } t < 0, \end{cases}$$

and we obtain

$$\frac{1}{2\pi i} \int_{x-i\infty}^{x+i\infty} e^{ts} \frac{G(s) - G(0)}{s} \, ds = \begin{cases} y_u(t) - G(0) & \text{for } t > 0 \\ 0 & \text{for } t < 0. \end{cases}$$

At $s = 0$, we assign to the function $[G(s) - G(0)]/s$ its limit, $G'(0)$. Thus, by Riemann's theorem,[4] this function becomes analytic at $s = 0$ and, consequently, at the entire imaginary axis and, moreover, in a strip beyond the imaginary axis. This function tends uniformly towards zero in $0 \leqq \Re s \leqq x$ when $s \to \infty$, and, following the process of the proof on p. 159 one may move the path of integration onto the imaginary axis, where $s = iy$; thus one obtains:

$$(16) \qquad \frac{1}{2\pi} \int_{-\infty}^{+\infty} e^{it\omega} \frac{G(i\omega) - G(0)}{i\omega} \, d\omega = \begin{cases} y_u(t) - G(0) & \text{for } t > 0 \\ 0 & \text{for } t < 0. \end{cases}$$

In Eq. (16) we already express $y_u(t)$ by $G(i\omega)$. However this expression may be

---

[3] The symbol *V.P.* before the integral is not needed in this case.

[4] If the function $\varphi(s)$ is analytic and bounded in $0 < |s - s_0| < \varrho$, then it has a limit l as $s \to s_0$; when defining $\varphi(s_0) = l$, the so completed function is analytic at s_0. In particular, we may invoke this theorem if we already know the existence of $\lim_{s \to s_0} \varphi(s)$.

further simplified. Consider the integration over the interval $(-\infty, 0)$, and modify the integral as follows:

$$\int_{-\infty}^{0} e^{it\omega} \frac{G(i\omega) - G(0)}{i\omega} d\omega = \int_{0}^{\infty} e^{-it\omega} \frac{G(-i\omega) - G(0)}{-i\omega} d\omega .$$

The function $g(t)$ is real-valued, hence $G(0)$ is a real number and

$$G(-i\omega) = \overline{G(i\omega)},$$

thus

$$e^{-it\omega} \frac{G(-i\omega) - G(0)}{-i\omega} = \overline{e^{it\omega} \frac{G(i\omega) - G(0)}{i\omega}} .$$

Combining the two integrals, the one over the interval $(-\infty, 0)$ as modified, and the other one over the interval $(0, \infty)$, one obtains as the new integrand the sum of the integrand plus its conjugate; it is twice the real part of the integrand. Thus (16) is changed into

$$\frac{1}{\pi} \int_{0}^{\infty} \Re \left\{ e^{it\omega} \frac{G(i\omega) - G(0)}{i\omega} \right\} d\omega = \begin{cases} y_u(t) - G(0) & \text{for } t > 0 \\ 0 & \text{for } t < 0 . \end{cases}$$

For the integrand we find, when using (13),

$$\Re \frac{|G(i\omega)| e^{i(t\omega + \psi)} - G(0) e^{it\omega}}{i\omega} = \frac{|G(i\omega)| \sin(t\omega + \psi) - G(0) \sin t\omega}{\omega} ;$$

hence,

$$\frac{1}{\pi} \int_{0}^{\infty} \frac{1}{\omega} \{ |G(i\omega)| \sin(t\omega + \psi) - G(0) \sin t\omega \} d\omega = \begin{cases} y_u(t) - G(0) & \text{for } t > 0 \\ 0 & \text{for } t < 0 . \end{cases}$$

The integral of the second part of the integrand exists; it is

$$\int_{0}^{\infty} \frac{\sin t\omega}{\omega} d\omega = \begin{cases} \displaystyle\int_{0}^{\infty} \frac{\sin u}{u} du = \frac{\pi}{2} & \text{for } t > 0 \\ \displaystyle -\int_{0}^{\infty} \frac{\sin u}{u} du = -\frac{\pi}{2} & \text{for } t < 0 . \end{cases}$$

It follows that the integral of the first part of the integrand also exists. Thus, we obtain:

$$\text{(17)} \qquad \frac{1}{\pi} \int_{0}^{\infty} \frac{|G(i\omega)|}{\omega} \sin(t\omega + \psi) d\omega = \begin{cases} y_u(t) - \frac{G(0)}{2} & \text{for } t > 0 \\ -\frac{G(0)}{2} & \text{for } t < 0 . \end{cases}$$

From the first line of formula (17) we summarize the following conclusions.

Theorem 17.1. *A differential equation with real-valued coefficients is given. For* $\Re\alpha_\mu < 0$, *we can determine the step response from the frequency characteristics* $|G(i\omega)|$ *and* $\psi(\omega)$, *by means of the formula*

$$y_u(t) = \frac{G(0)}{2} + \frac{1}{\pi}\int_0^\infty \frac{|G(i\omega)|}{\omega} \sin\big(t\omega + \psi(\omega)\big)\, d\omega \quad (t > 0)\,. \tag{18}$$

The frequency characteristics can be obtained from a plot of the frequency function $G(i\omega)$ in polar coordinates. When plotting $G(i\omega)$ in rectangular coordinates, we use the components[5] $U(\omega)$ and $V(\omega)$:

$$G(i\,\omega) = U(\omega) + i\,V(\omega)\,. \tag{19}$$

Attention is called to the fact that $y_u(t)$ may actually be evaluated by the use of only one of these components. We have

$$\sin(t\omega + \psi) = \sin t\omega\,\cos\psi + \cos t\omega\,\sin\psi\,,$$

$$|G(i\,\omega)|\sin(t\omega + \psi) = U(\omega)\sin t\omega + V(\omega)\cos t\omega\,.$$

We write the first line of (17) for $t > 0$, and the second line of (17) for $-t$, using the same t,

$$\frac{1}{\pi}\int_0^\infty \frac{1}{\omega}\,(U(\omega)\sin t\omega + V(\omega)\cos t\omega)\,d\omega = y_u(t) - \frac{G(0)}{2}\,,$$

$$\frac{1}{\pi}\int_0^\infty \frac{1}{\omega}\,\big(-U(\omega)\sin t\omega + V(\omega)\cos t\omega\big)\,d\omega = -\frac{G(0)}{2}\,.$$

The desired result follows by subtraction and addition, with $G(0) = U(0)$:

Theorem 17.2. *A differential equation with real-valued coefficients is given. For* $\Re\alpha_\mu < 0$, *we can determine the step response by the use of only one of the components of the frequency function* ($t > 0$): *when the real component is given, then we employ the formula*

$$y_u(t) = \frac{2}{\pi}\int_0^\infty \frac{U(\omega)}{\omega}\sin t\omega\, d\omega\,; \tag{20}$$

when the imaginary component is available, then we use the formula

$$y_u(t) = \frac{2}{\pi}\int_0^\infty \frac{V(\omega)}{\omega}\cos t\omega\, d\omega + U(0)\,. \tag{21}$$

---

[5] These quantities are actually measured when investigating practical problems.

For many practical engineering problems the determination of the frequency response is easier than the determination of the step response; it is for this reason that the presented formulae are important. Special numerical techniques have been developed for the practical evaluation of the integrals (18), (20), and (21); these techniques benefit from the fact that ω occurs in the denominator of the integrand which thus becomes negligibly small beyond some value of the dummy variable of integration.

We must especially warn against the use of the concept of the frequency response in such cases where not all $\Re\alpha_\mu$ are negative. To demonstrate the danger, consider the following technically important problem, for which the condition $\Re\alpha_\mu < 0$ is not satisfied for all roots, and which shows a behaviour entirely different from that outlined above. The differential equation of second order

$$y'' + c_1 y' + c_0 y = f(t)$$

describes the physical behaviour of a *linear mechanical oscillator* or of an *electric RCL circuit*, or the likes. The roots of the characteristic polynomial $p(s) = 0$ are given by:

$$\alpha_1 = -\frac{c_1}{2} + \sqrt{\frac{c_1^2}{4} - c_0}, \quad \alpha_2 = -\frac{c_1}{2} - \sqrt{\frac{c_1^2}{4} - c_0}.$$

For $c_1^2/4 - c_0 \geqq 0$ we have real roots α_1 and α_2; the corresponding characteristic solutions $e^{\alpha_1 t}$ and $e^{\alpha_2 t}$, or $e^{\alpha t}$ and $te^{\alpha t}$ for $\alpha_1 = \alpha_2 = \alpha$, are also real-valued, aperiodic functions. For $c_1^2/4 - c_0 < 0$, both α_1 and α_2 are complex-valued, and the characteristic solutions are oscillatory. In this case we have $c_0 > (c_1/2)^2$, and we may define real valued δ and ω, so that

$$c_1 = 2\,\delta, \quad c_0 = \delta^2 + \omega^2.$$

Thus the characteristic solutions may be written as follows:

$$e^{-\delta t}\, e^{i\omega t} \quad \text{and} \quad e^{-\delta t}\, e^{-i\omega t};$$

that is, they represent oscillations with frequency ω which are damped for $\delta > 0$. Writing the second order differential equation in the convenient form

$$y'' + 2\,\delta\, y' + (\delta^2 + \omega^2)\, y = f(t),$$

we can directly read the proper frequency of the system ω, and the damping coefficient δ. Thus, when oscillations are expected, then it is advisable to write the differential equation in this particular form at the outset.

Now we shall investigate the special case which is encountered during the *tuning of a receiver*, that is when the input function $f(t)$ is an oscillation with the same frequency and damping as the characteristic oscillations of the system:

$$f(t) = E_0\, e^{i\varphi}\, e^{-\delta t}\, e^{i\omega t},$$

where E_0 and φ determine the initial amplitude and the initial phase of the excitation. For this case we have the differential equation:

$$(22) \qquad y'' + 2\,\delta\, y' + (\delta^2 + \omega^2)\, y = E_0\, e^{i\varphi}\, e^{-\delta t}\, e^{i\omega t},$$

which has the image equation

$$s^2\, Y - y_0\, s - y_0' + 2\,\delta\,(s\, y - y_0) + (\delta^2 + \omega^2)\, Y = E_0\, e^{i\varphi} \frac{1}{s + \delta - i\,\omega}$$

and the solution

$$(23) \qquad Y = E_0\, e^{i\varphi} \frac{1}{(s + \delta - i\,\omega)\,[(s+\delta)^2 + \omega^2]} + y_0 \frac{s + 2\,\delta}{(s+\delta)^2 + \omega^2} + y_0' \frac{1}{(s+\delta)^2 + \omega^2}.$$

Firstly we investigate *undamped waves*, with $\delta = 0$. In this case, the input is of the form const. $e^{i\omega t}$, like the excitations used in the study of the frequency response. However, the roots α_1 and α_2 lack a real negative part; they are located on the imaginary axis: $\alpha_1 = i\,\omega$, $\alpha_2 = -i\,\omega$. Consequently, we cannot apply the above conclusion concerning the frequency response; instead, we must directly evaluate $y(t)$ to obtain the sought information regarding the steady state of $y(t)$ for $t \to \infty$, For $\delta = 0$, we have

$$(24) \qquad Y = E_0\, e^{i\varphi} \frac{1}{(s - i\,\omega)\,(s^2 + \omega^2)} + y_0 \frac{s}{s^2 + \omega^2} + y_0' \frac{1}{s^2 + \omega^2}.$$

The denominator of the first fraction of (24) is $(s - i\,\omega)^2\,(s + i\,\omega)$; its partial fraction expansion has the form:

$$\frac{1}{(s - i\,\omega)\,(s^2 + \omega^2)} = \frac{a}{(s - i\,\omega)^2} + \frac{b}{s - i\,\omega} + \frac{c}{s + i\,\omega}.$$

Determining the several coefficients as explained on p. 78, we obtain:

$$\frac{1}{(s - i\,\omega)\,(s^2 + \omega^2)} = -\frac{i}{2\,\omega} \frac{1}{(s - i\,\omega)^2} + \frac{1}{4\,\omega^2} \frac{1}{s - i\,\omega} - \frac{1}{4\,\omega^2} \frac{1}{(s + i\,\omega)}$$

$$= \frac{i}{2\,\omega}\left(-\frac{1}{(s - i\,\omega)^2} + \frac{1}{s^2 + \omega^2}\right).$$

The original functions of the second and the third fraction of (24) can be determined directly (see Tab. Nos. 8, 14, 19), hence we have the original function of (24):

$$(25) \qquad y(t) = E_0\, e^{i\varphi} \frac{i}{2\,\omega}\left(-t\, e^{i\omega t} + \frac{1}{\omega} \sin \omega t\right) + y_0 \cos \omega t + y_0' \frac{1}{\omega} \sin \omega t.$$

Formula (25) shows that the excitation with frequency ω causes in the system, not only an oscillation with frequency ω having constant amplitude, but also another oscillation, the amplitude of which increases proportionally with time t. Moreover, the characteristic oscillations $\cos\omega t$ and $\sin\omega t$ do not diminish in amplitude with time. Obviously, the earlier presented concept of frequency response is not applicable here.

Next we consider *damped waves* with $\delta > 0$. In the first term and the third term of (24), we replace s by $(s + \delta)$, to obtain the respective terms of (23); according to Theorem 7.7, this substitution is paired with the multiplication by $e^{-\delta t}$ of the original functions. The second term of (23) may be written as follows:

$$y_0 \left(\frac{s+\delta}{(s+\delta)^2 + \omega^2} + \frac{\delta}{(s+\delta)^2 + \omega^2} \right);$$

its original function is easily found. Thus we have, altogether,

$$\begin{aligned} y(t) = {} & E_0\, e^{i\varphi} \frac{i}{2\omega}\, e^{-\delta t} \left(-t\, e^{i\omega t} + \frac{1}{\omega} \sin\omega t \right) \\ & + y_0\, e^{-\delta t} \left(\cos\omega t + \frac{\delta}{\omega} \sin\omega t \right) + y_0' \frac{1}{\omega} e^{-\delta t} \sin\omega t . \end{aligned} \tag{26}$$

For growing t, the damping factor $e^{-\delta t}$ dominates also the term with the factor t, and $y(t)$ tends towards zero, like the excitation $f(t)$.

The physical oscillations of the system are described by the real part of (26); it is:

$$\begin{aligned} y(t) = {} & E_0 \frac{1}{2\omega}\, e^{-\delta t} \left(t \sin(\omega t + \varphi) - \frac{\sin\varphi}{\omega} \sin\omega t \right) \\ & + y_0\, e^{-\delta t} \left(\cos\omega t + \frac{\delta}{\omega} \sin\omega t \right) + y_0' \frac{1}{\omega} e^{-\delta t} \sin\omega t . \end{aligned} \tag{27}$$

Again we emphasize the advantages gained by the use of the complex oscillations $e^{i\omega t}$ instead of the real oscillations $\sin\omega t$ or $\cos\omega t$. When using the oscillation $\sin\omega t$ in the case $\delta = 0$, we would find in the denominator of the first fraction of (24) $(s^2 + \omega^2)$ instead of $(s - i\omega)$; the denominator would be a fourth order expression; its partial fraction evaluation would be more complicated.

The response of the system to a third test function, the impulse input, will be presented in Chapter 18.

18. The Ordinary Linear Differential Equation in the Space of Distributions

In the space of distributions, the derivative must be replaced by the distribution-derivative and, consequently, also differential equations by "distribution-derivative equations." In the latter, the given and the sought quantities are distributions. To emphasize the analogy to the case of functions, we shall employ here for the designation of distributions lower case letters like $f, y, \ldots$ (which are usually reserved for functions) instead of the letters $T, U, \cdots$. A distribution-derivative equation with constant coefficients has the form:

(1) $$D^n y + c_{n-1} D^{n-1} y + \cdots + c_1 D y + c_0 y = f.$$

Earlier, we investigated differential equations in the interval $t \geqq 0$; accordingly we shall specify here that y and f belong to the space $\mathscr{D}'_+$. Moreover, we intend to employ the $\mathfrak{L}$-transformation; thus for f and y, we must specify the more restricted subspace $\mathscr{D}'_0$ (compare Chapter 12). For the situation where f and y represent function-distributions which are generated by locally integrable functions, we must not overlook the requirement that these functions have the value zero for $t < 0$.

By (14.8, 10), the distribution-derivative equation (1) can also be written as a convolution equation:

(2) $$(\delta^{(n)} + c_{n-1}\, \delta^{(n-1)} + \cdots + c_1\, \delta' + c_0\, \delta) * y = f.$$

Upon development of an algebra which has the convolution as product operation, one could attack (2) by algebraic methods. When employing the $\mathfrak{L}$-transformation, it is immaterial whether one starts with Eq. (1) or with Eq. (2). For if we borrow the usual notations for functions,

$$\mathfrak{L}\{y\} = Y(s), \quad \mathfrak{L}\{f\} = F(s),$$

then the application of either the Distribution-Derivative Theorem 14.3 to Eq. (1) or of the Convolution Theorem 14.5 to Eq. (2) will yield the same image equation:

(3) $$(s^n + c_{n-1} s^{n-1} + \cdots + c_1 s + c_0)\, Y(s) = F(s).$$

We can write the solution of (3) with the functions $p(s)$ and $G(s)$ introduced in Chapter 15:

(4) $$Y(s) = \frac{1}{p(s)} F(s) = G(s)\, F(s).$$

Corresponding to $G(s)$ we have the weighting function $g(t)$ as the original function in the space of functions (compare p. 84), and, by Theorem 12.2, also in the space

$\mathscr{D}'_0$ of distributions. However, we must here extend the definition of $g(t)$ assigning to it the value zero for $t < 0$, and consider it as a distribution. Consequently, we have:

$$g(0^-) = g'(0^-) = \cdots = g^{(n-1)}(0^-) = 0$$

and, by (15.19),

$$(5) \qquad g(0^+) = g'(0^+) = \cdots = g^{(n-2)}(0^+) = 0, \text{ however } g^{(n-1)}(0^+) = 1;$$

hence, by App. No. 20,

$$(6) \qquad Dg = g', \ldots, D^{n-1}g = g^{(n-1)}, \text{ however, } D^n g = g^{(n)} + \delta.$$

Obviously, g is in $\mathscr{D}'_0$. The original function of $Y(s)$ is, by Theorem 14.5,

$$(7) \qquad y = g * f.$$

y is the convolution of two distributions of $\mathscr{D}'_0$, hence it is a distribution in $\mathscr{D}'_0$.

In the case where $f = D^k h(t)$, whereby $h(t)$ complies with both conditions (12.4, 5), one finds, by (14.7),

$$y = D^k [g * h].$$

We verify that the distribution (7) satisfies the Eq. (1). Because of (6), we have

$$\begin{aligned} p(D)g &= D^n g + c_{n-1} D^{n-1} g + \cdots + c_1 Dg + c_0 g \\ &= g^{(n)} + c_{n-1} g^{(n-1)} + \cdots + c_1 g' + c_0 g + \delta; \end{aligned}$$

hence, by (15.18),

$$(8) \qquad p(D)g = \delta.$$

Theorem 14.6 produces:

$$p(D)[g * f] = [p(D)g] * f = \delta * f = f.$$

The method of the $\mathfrak{L}$-transformation encompasses every solution of (1) in $\mathscr{D}'_0$; it follows that the distribution (7) is, indeed, the only solution in $\mathscr{D}'_0$.

We shall consider a few special excitations, represented by distributions.

The Impulse Response

For many physical systems, important conclusions regarding the characteristics of the system may be drawn from its response to a strong, shock-like impulse excitation. Such a "shock" is mathematically represented by means of the distribution δ, which derived its name "impulse" from this very interpretation. The solution of (1) for the excitation $f \equiv \delta$ is known in mathematics as the elementary solution, in physics as the *impulse response*. We shall designate this solution by

y_δ, and its $\mathfrak{L}$-transform by $Y_\delta(s)$. We find, by Eqs. (4) and (7), using $\mathfrak{L}\{\delta\} = 1$,

$$(9) \qquad Y_\delta(s) = G(s), \quad y_\delta = g(t),$$

whereby $g(t)$ is to be interpreted as a distribution. Eq. (8) provides the verification of this solution. Whilst $g(t)$, when considered as a function (the weighting function) satisfies, by (15.18), the differential equation with 0 at the right hand side, $g(t)$, when considered as a distribution (the impulse response), satisfies the distribution-derivative equation with the right hand side δ, due to the property $D^k g = g^{(k)} + \delta$. In technical literature one often encounters the designation impulse response for the weighting function; this designation is incorrect, considering the distinction between the impulse response y_δ and the weighting function g which is shown by the relationship

$$(10) \qquad D^n y_\delta = g^{(n)} + \delta.$$

The distribution δ is "equal to zero" (compare App. No. 15) for $t > 0$; consequently, this confusion is not dangerous as long as one is interested merely in the interval $t > 0$, and not in the point $t = 0$. Thus, one can employ empirical methods in practical problems to find the weighting function through the response of the physical system to the shock excitation.

Response to the Excitation $\delta^{(m)}$

In physics, one also considers pairs of shocks, the shocks having opposite signs, the second one immediately following the first one. This excitation is mathematically represented by δ'.[1] Indeed, this development is extended further to excitations which are mathematically represented by $\delta^{(m)}$. We shall designate the corresponding responses by y_m, and the $\mathfrak{L}$-transforms by $Y_m(s)$. Using $\mathfrak{L}\{\delta^{(m)}\} = s^m$ we find, by Theorem 14.3,

$$(11) \qquad Y_m(s) = s^m G(s), \quad y_m = D^m g(t).$$

Assigning to $g(t)$ the value zero for $t < 0$, we find for all derivatives, as $t \to -0$, the limits 0; hence, by App. No. 20,

$$D^m g(t) = [g^{(m)}(t)] + g^{(m-1)}(0+)\,\delta + g^{(m-2)}(0+)\,\delta' + \cdots + g(0+)\,\delta^{(m-1)},$$

where we explicitly employ the rectangular brackets to emphasize the interpretation of $g^{(m)}(t)$ as a distribution (compare App. No. 9).

We must distinguish two cases:

a) $m < n$. According to (5),

$$g^{(m-1)}(0+) = g^{(m-2)}(0+) = \cdots = g(0+) = 0,$$

---

[1] One may attempt to approximately visualize the impulse δ by means of a narrow, bell-shaped curve with steeply sloping flanks; seeking the derivative of this approximation, one first obtains large positive values which are followed by large negative values.

hence

(12) $$y_m = [g^{(m)}(t)].$$

b) $m \geqq n$. According to (5),

(13) $$y_m = [g^{(m)}(t)] + g^{(m-1)}(0+)\,\delta + g^{(m-2)}(0+)\,\delta' + \dots + g^{(n)}(0+)\,\delta^{(m-n-1)} + \delta^{(m-n)}.$$

The Response to Excitation by a Pseudofunction

Suppose the excitation is described by

$$f = \mathrm{Pf}\,[t^{-\lambda}\,u(t)] \quad (\lambda > -1, \text{ not an integer}).$$

By App. No. 22d), we find explicitly

$$\mathrm{Pf}\,[t^{-\lambda}\,u(t)] = \frac{(-1)^m}{(\lambda-1)\dots(\lambda-m)}\,D^m\,[t^{-\lambda+m}\,u(t)],$$

whereby m is presumed to be an integer, $m > \lambda - 1$. We impose the condition $m < \lambda < m + 1$, that is $m = [\lambda]$, thus producing a determinate value m. It follows that $-1 < -\lambda + m < 0$.[2]

The response to this excitation is given by[3]

(14) $$y = \frac{(-1)^m}{(\lambda-1)\dots(\lambda-m)}\,\{D^m\,[t^{-\lambda+m}\,u(t)]\} * [g(t)\,u(t)].$$

This can be rewritten, by Theorem 14.6, as follows:

(15) $$y = \frac{(-1)^m}{(\lambda-1)\dots(\lambda-m)}\,D^m\,\{[t^{-\lambda+m}\,u(t)] * [g(t)\,u(t)]\}$$

(in the parentheses we have the convolution of two integrable functions), or

(16) $$y = \frac{(-1)^m}{(\lambda-1)\dots(\lambda-m)}\,[t^{-\lambda+m}\,u(t)] * D^m\,[g(t)\,u(t)].$$

In Equ. (16) one can express $D^m g(t)$ by (12) or (13). One obtains: for $m = [\lambda] < n$,

$$y = \frac{(-1)^m}{(\lambda-1)\dots(\lambda-m)}\,[t^{-\lambda+m}\,u(t)] * [g^{(m)}(t)\,u(t)]$$

---

[2] The derivatives of $t^{-\lambda+m}$ are not integrable functions; hence one cannot replace D^m by d^m/dt^m, which latter would simply produce the function $t^{-\lambda}$. Moreover, the formula App. No. 20 cannot be employed here, since the derivatives fail to have limits as $t \to +0$. One must avoid the erroneous interpretation of (14) as the solution of a differential equation with the non-integrable excitation $t^{-\lambda}$. Both the given f and the unknown y are distributions and they have meaning only as functionals over the space $\mathcal{D}$ (compare App. No. 5,6).

[3] We write here, more precisely, $g(t)\,u(t)$ instead of $g(t)$, since $t^{-\lambda+m}$ too has been properly augmented by $u(t)$.

(this is the convolution of two integrable functions, that is a function or, more precisely, a function-distribution);
for $m = [\lambda] \geqq n$,

$$y = \frac{(-1)^m}{(\lambda-1)\dots(\lambda-m)} [t^{-\lambda+m}\, u(t)] * [g^{(m)}(t)\, u(t) + g^{(m-1)}(0+)\,\delta + \dots + g^{(n)}(0+)\,\delta^{(m-n-1)} + \delta^{(m-n)}]$$

$$= \frac{(-1)^m}{(\lambda-1)\dots(\lambda-m)} \{[t^{-\lambda+m}\, u(t)] * [g^{(m)}(t)\, u(t)] + g^{(m-1)}(0+)\, t^{-\lambda+m}\, u(t) + g^{(m-2)}(0+)\, D[t^{-\lambda+m}\, u(t)] + \dots + g^{(n)}(0+)\, D^{m-n-1}\, [t^{-\lambda+m}\, u(t)] + D^{m-n}\, [t^{-\lambda+m}\, u(t)]\}.$$

For the special case $m = n$ distribution-derivatives of $t^{-\lambda+m} u(t)$ do not occur, and y_m is a function-distribution,

The same conclusions can be deduced by means of the $\mathfrak{L}$-transform. By (13.7),

$$\mathfrak{L}\{\mathrm{Pf}\,[t^{-\lambda}\, u(t)]\} = \frac{\Gamma(-\lambda+1)}{s^{-\lambda+1}};$$

consequently, by (4),

$$Y(s) = \Gamma(-\lambda+1)\, s^{\lambda-1}\, G(s).$$

For $m < \lambda < m + 1$, one can rewrite the last expression thus:

$$Y(s) = \Gamma(-\lambda+1)\, s^m\, [s^{\lambda-m-1}\, G(s)].$$

The function $s^{\lambda-m-1}$ is the $\mathfrak{L}$-transform of the function $t^{-\lambda+m}/\Gamma(-\lambda+m+1)$, since $-1 < \lambda - m - 1 < 0$; thus one finds, by Theorem 14.3,

$$y(t) = \frac{\Gamma(-\lambda+1)}{\Gamma(-\lambda+m+1)} D^m\, [t^{-\lambda+m} * g(t)].$$

This result agrees with the representation (15).

Excitations with integer-valued λ were excluded in the above development; responses to excitations of this type can be constructed by means of App. No. 22c).

A New Interpretation of the Concept Initial Value

A distribution is a functional which does not ascribe values to specified points of the t-axis. Whence the term initial values of a solution is meaningless in connection with distribution-derivative equations. However, the distributions include the classical functions, and the distribution-derivative equations embrace the differential equations. Thus, we are led to the question: What meaning has the solution (7), obtained in the space of distributions, for the special case that both f and y are classical functions? Inspection shows that the solution (7) of the distribution-derivative equation corresponds to the solution (15.8) of the differential equation for *vanishing initial values*. The fact that (7) yields precisely this solution and not the general solution apparently results from the imposed conditions: the solution of the distribution-derivative equation was not to be selected from the space $\mathcal{D}'$ of all distributions, but from the subspace $\mathcal{D}'_0$ of those distributions which, in the interpretation of App. No. 10, are zero in the interval $t < 0$.

This seemingly incidental remark, nevertheless, suggests a new understanding of the concept "initial value". In the space of functions, we understood the initial value of the function $y(t)$ as the limit $y(0^+)$ which is obtained when approaching $t = 0$ from the *right*; this limit is exclusively determined by the values of the function $y(t)$ for $t > 0$. (When deriving the Differentiation Theorem of the $\mathfrak{L}$-transformation, we had to interpret the initial value in exactly this manner.) In the frame of distribution theory, the fact that (7) represents the solution for vanishing initial values reflects the condition that y is defined to be zero for $t < 0$; hence the initial value of y is to be understood as the limit $y(0^-)$, from the *left*; it is determined by the values of y for $t < 0$.

Actually, this interpretation of the initial value as the limit from the left is quite natural; indeed, it harmonizes with physical concepts. Suppose that the differential equation describes a physical process, with t representing the time variable; then we understand the initial value as the state of the solution at the begin of the process. However, this state is clearly the result of the past of this physical variable, that is, it is determined by the values of this variable for negative t. The concept "initial value" in the physical sciences can be understood only when the past, that is the interval $t < 0$, has been included in our considerations. This occurs naturally for distributions which, without exception, are defined on the entire t-axis.

Thus, the term initial value affords two interpretations: The initial values y_0, y'_0, $\cdots$ can either be interpreted as the limits from the right: $y(0^+)$, $y'(0^+)$, $\cdots$, or else as the limits from the left: $y(0^-)$, $y'(0^-)$, $\cdots$. For the problem represented by a single ordinary differential equation, this ambiguity is not quoted in the usual representation. Indeed, for this case the distinction is without consequence; for if we interpret the initial values as limits from the left and, nevertheless, substitute these into formula (15.26), we shall create a $y(t)$ which has the same initial values from the right; this can easily be confirmed by verification (compare the remark on p. 82).

The above argumentation indicates that for a process which can be described by a single ordinary differential equation, the future for $t > 0$ joins the past with $t < 0$ continuously at $t = 0$, whatever may have been the state in the past. This conclusion seems trivial; however, it is not really self-evident. When studying systems of ordinary differential equations, of which the single ordinary differential equation is but a special case, one may encounter solutions which fail to exhibit the above presented continuity. The above considerations are presented here to prepare the reader for situations of such character which, indeed, can be managed only through the theory of distributions.

19. The Normal System of Simultaneous Differential Equations

We found in the $\mathfrak{L}$-transformation a superior tool for the solution of the initial value problem involving a single differential equation of nth order, when compared with the classical method. This was due to the fact that with the latter one has to adapt the derived solution to the specified initial values, while with the former this is accomplished automatically in the process of solution. In the course of adapting the solution in the classical method, one has to solve a system of n simultaneous linear equations in n unknowns – a time-consuming task, particularly for $n > 3$. This favourable characteristic of the Laplace transformation is particularly appreciated when solving initial value problems that involve systems of N simultaneous linear differential equations in N unknown functions. Indeed, the Laplace transformation provides the only practical method of solution of such problems for $N > 2$, requiring a tolerable amount of calculation.

When considering a system of N *differential equations* of order n, one could, in principle, encounter in every equation with each of the N unknown functions a differential operator of order n. Thus, one finds in the αth equation for $y_\beta(t)$ the expression:

$$c_n^{\alpha\beta}\, y_\beta^{(n)} + c_{n-1}^{\alpha\beta}\, y_\beta^{(n-1)} + \cdots + c_1^{\alpha\beta}\, y_\beta' + c_0^{\alpha\beta}\, y_\beta .$$

Some of the coefficients c may, of course, be zero; indeed, $y_\beta(t)$ may be missing entirely in the α-th equation, or only derivatives of $y_\beta(t)$ of order less than n may occur. To devise a more readable presentation of the equations, we introduce the polynomial

$$c_n^{\alpha\beta}\, s^n + c_{n-1}^{\alpha\beta}\, s^{n-1} + \cdots + c_1^{\alpha\beta}\, s + c_0^{\alpha\beta} = p_{\alpha\beta}(s) . \tag{1}$$

With this we can symbolically write the differential expression as follows:

$$c_n^{\alpha\beta}\, y_\beta^{(n)} + \cdots + c_1^{\alpha\beta}\, y_\beta' + c_0^{\alpha\beta}\, y_\beta = p_{\alpha\beta}\left(\frac{d}{dt}\right) y_\beta ,$$

and the system of equations assumes the form

$$\left\{\begin{array}{l} p_{11}\left(\frac{d}{dt}\right) y_1 + p_{12}\left(\frac{d}{dt}\right) y_2 + \cdots + p_{1N}\left(\frac{d}{dt}\right) y_N = f_1(t) \\ \cdots\cdots\cdots\cdots\cdots\cdots\cdots\cdots\cdots\cdots\cdots \\ p_{N1}\left(\frac{d}{dt}\right) y_1 + p_{N2}\left(\frac{d}{dt}\right) y_2 + \cdots + p_{NN}\left(\frac{d}{dt}\right) y_N = f_N(t). \end{array}\right. \tag{2}$$

System (2) looks like a system of algebraic equations; that suggests the use of matrix calculus. Indeed, by this method one may reduce the process of solution to a scheme similar to that of a single differential equation.[1] We shall avoid this

---

[1] Compare G. Doetsch: Handbuch der Laplace-Transformation, Vol. II, p. 311, Birkhäuser Verlag, Basel und Stuttgart 1955, revised edition 1972.

method here to make the presentation easier to understand. Moreover, to limit the extent of the formulae, we shall set $n = 2$; thus restricting ourselves to the case that is most frequently encountered in practical problems. For this case we can eliminate one of the indices when setting

$$p_{\alpha\beta}(s) = a_{\alpha\beta}\, s^2 + b_{\alpha\beta}\, s + c_{\alpha\beta}\,,$$

instead of (1). Based upon hypotheses analogous to those required for a single differential equation, one can write, for (2), the system of image equations

$$(3)\qquad \begin{cases} p_{11}(s)Y_1 + \cdots + p_{1N}(s)Y_N \\ \qquad = F_1(s) + \sum\limits_{\nu=1}^{N} \{(a_{1\nu}\, s + b_{1\nu})\; y_\nu(0^+) + a_{1\nu}\, y'_\nu\,(0^+)\} \\ \cdots\cdots\cdots\cdots\cdots\cdots\cdots\cdots\cdots\cdots\cdots\cdots \\ p_{N1}(s)Y_1 + \cdots + p_{NN}(s)\,Y_N \\ \qquad = F_N(s) + \sum\limits_{\nu=1}^{N} \{(a_{N\nu}\, s + b_{N\nu})\; y_\nu(0^+) + a_{N\nu} y'_\nu\,(0^+)\}, \end{cases}$$

a system of N algebraic equations in the image functions $Y_1(s), \cdots, Y_N(s)$. The solution of this system depends upon the behaviour of the determinant of the $p_{\alpha\beta}(s)$[2]

$$\Delta(s) = \det \| p_{\alpha\beta}(s) \| \cdot$$

Each of the $p_{\alpha\beta}(s)$ is a polynomial in s of, at most, second degree; hence $\Delta(s)$ is a polynomial in s of, at most, degree $2N$. We propose now, for the remainder of this Chapter, a very important hypothesis: *the degree of* $\Delta(s)$ *is exactly* $2N$*;* in which case we shall call the system *normal*; otherwise, if the degree of $\Delta(s)$ is less than $2N$, we shall call the system anomalous. The anomalous case will be investigated later on. In $\Delta(s)$, we replace each $p_{\alpha\beta}(s)$ by its explicit expression $a_{\alpha\beta}s^2 + b_{\alpha\beta}s + c_{\alpha\beta}$, and we resolve $\Delta(s)$ into a sum of determinants so that each of these in each column has only one of the terms $a_{\alpha\beta}s^2$, $b_{\alpha\beta}s$, or $c_{\alpha\beta}$. The determinant which has s in its highest power, that is s^{2N}, is the one which is formed of all the terms $a_{\alpha\beta}s^2$. *Consequently, we have a normal system if and only if*

$$(4)\qquad A = \det \| a_{\alpha\beta} \| \neq 0 \text{ (accordingly, } \textit{for the general case:}\ \det \| c_n^{\alpha\beta} \| \neq 0).$$

The polynomial $\Delta(s)$ has $2N$ zeros, some of which might be repeated. In the half-plane to the right of the root with the largest real component, we have $\Delta(s) \neq 0$, hence the system (3) affords a unique solution $Y_1, \cdots, Y_N$. Repeating the process

---

[2] We employ here the following notation for matrices: $\| p_{\alpha\beta} \|$; that is, we write the general element of the matrix between vertical double bars. For the determinant of the matrix, we write: $\det \| p_{\alpha\beta} \|$.

that we employed for the solution of one single differential equation, we shall attempt the solution in two separate steps: Firstly, we shall presume vanishing functions $F_1, \cdots, F_N$, and necessarily also $f_1, \cdots, f_N$ (the homogeneous system of differential equations); the initial values being arbitrary. Secondly, we shall admit arbitrary excitations $f_1, \cdots, f_N$ (the inhomogeneous system), while assuming zero initial values. The general solution can then be constructed by superposition of the solutions obtained for these two separate problems.

1. The Normal Homogeneous System, for Arbitrary Initial Values

In the general theory of differential equations, it is known and easy to verify that the solutions of the homogeneous system must be sums of exponential functions, possibily multiplied by powers. Thus we know a priori that the hypotheses of our method are satisfied, and the method will produce the correct solutions. We shall firstly treat a *special case*, specifying the initial values by the following equations, where k is some fixed integer, $1 \leq k \leq N$:

(5)
$$\begin{cases} y_1(0^+) = y_2(0^+) = \cdots = y_N(0^+) = 0; \\ \sum_{\nu=1}^{N} a_{1\nu}\, y_\nu'(0^+) = \cdots = \sum_{\nu=1}^{N} a_{k-1\,\nu}\, y_\nu'(0^+) = 0, \\ \sum_{\nu=1}^{N} a_{k\nu}\, y_\nu'(0^+) = 1, \\ \sum_{\nu=1}^{N} a_{k+1\,\nu}\, y_\nu'(0^+) = \cdots = \sum_{\nu=1}^{N} a_{N\nu}\, y_\nu'(0^+) = 0. \end{cases}$$

Although system (5) does not specify the $y'_\nu(0^+)$ explicitly, we could, by (4), calculate these values uniquely, for we have an inhomogeneous system of equations with the non-zero determinant A. With these initial values we have, for (3), when setting $F_1 = \cdots = F_N = 0$, the following expressions:

(6)
$$\begin{cases} p_{11}(s)\, Y_1 + \cdots + p_{1N}(s)\, Y_N = 0 \\ \cdots\cdots\cdots\cdots\cdots\cdots\cdots\cdots \\ p_{k1}(s)\, Y_1 + \cdots + p_{kN}(s)\, Y_N = 1 \\ \cdots\cdots\cdots\cdots\cdots\cdots\cdots\cdots \\ p_{N1}(s) Y_1 + \cdots + p_{NN}(s) Y_N = 0. \end{cases}$$

For each fixed $k = 1, \cdots, N$ we obtain a set of solutions $Y_1, \cdots, Y_N$. To emphasize the dependency of these solutions upon the choice of k, we shall write, for the kth set of solutions, $G_{k1}, \cdots, G_{kN}$. The evaluation of these solutions follows simply by Cramer's rule: We designate the co-factor of $p_{\alpha\beta}(s)$ by $\Delta_{\alpha\beta}(s)$; it is the determinant of that submatrix of $\Delta(s)$, which is obtained by deletion of the αth row and the βth column, multiplied by $(-1)^{\alpha+\beta}$. We then have:

(7)
$$G_{kl}(s) = \frac{\Delta_{kl}(s)}{\Delta(s)} \quad (k = 1, \cdots, N;\ l = 1, \cdots, N).$$

The co-factor $\Delta_{\alpha\beta}(s)$ is a polynomial in s of, at most, degree $2(N-1)$; thus, the degree of the numerator is always smaller than the degree of the denominator of (7), and the G_{kl} is a Laplace transform; the original function $g_{kl}(t)$ may be obtained by the method of partial fraction expansion:

$$g_{kl}(t) \circ\!\!-\!\!\bullet\, G_{kl}(s) = \frac{\Delta_{kl}(s)}{\Delta(s)} \quad (k = 1, \cdots, N;\ l = 1, \cdots, N). \tag{8}$$

The N^2 functions may be arranged in the pattern of a matrix; for any fixed k, they are solutions of the homogeneous system of equations; by (5), they satisfy the specifications:

$$g_{kl}(0) = 0 \quad (k, l = 1, \cdots, N), \tag{9}$$

$$\sum_{\nu=1}^{N} a_{i\nu}\, g'_{k\nu}(0) = \begin{cases} 0 & \text{for } i \neq k \\ 1 & \text{for } i = k \end{cases} \quad (i, k = 1, \cdots, N). \tag{10}$$

Now, we can develop the solution of the homogeneous system of equations for arbitrary initial values. Cramer's rule provides, for (3):

$$Y_l(s) = \sum_{k=1}^{N} \frac{\Delta_{kl}(s)}{\Delta(s)} \cdot \sum_{\nu=1}^{N} \{(a_{k\nu}\, s + b_{k\nu})\ y_\nu(0+) + a_{k\nu} y'_\nu\ (0+)\} \quad (l = 1, \cdots, N).$$

We have, by (8),

$$\frac{\Delta_{kl}(s)}{\Delta(s)} \bullet\!\!-\!\!\circ\, g_{kl}(t)$$

and, by the Differentiation Theorem 9.1, because of (9),

$$s\, \frac{\Delta_{kl}(s)}{\Delta(s)} \bullet\!\!-\!\!\circ\, g'_{kl}(t);$$

hence

$$\begin{aligned} y_l(t) &= \sum_{\nu=1}^{N} y_\nu(0+) \sum_{k=1}^{N} \{a_{k\nu}\, g'_{kl}(t) + b_{k\nu}\, g_{kl}(t)\} \\ &\quad + \sum_{\nu=1}^{N} y'_\nu\, (0+) \sum_{k=1}^{N} a_{k\nu}\, g_{kl}(t) \quad (l = 1, \cdots, N). \end{aligned} \tag{11}$$

The functions (11) are solutions of the homogeneous case since the existence of the $\mathfrak{L}$-transforms of the $y_l(t)$ is assured; using the relations (9) and (10), one can easily verify that they assume the specified initial values.

2. The Normal Inhomogeneous System with Vanishing Initial Values

For vanishing initial values, we are left on the right hand sides of (3) with the functions $F_1, \cdots, F_N$. Thus we have, by Cramer's rule,[3]

$$Y_l(s) = \sum_{k=1}^{N} \frac{\Delta_{kl}(s)}{\Delta(s)} F_k(s)$$

and consequently, because of (8),

$$y_l(t) = \sum_{k=1}^{N} g_{kl}(t) * f_k(t) \qquad (l = 1, \cdots, N). \tag{12}$$

We now demonstrate that these are correct solutions irrespective of the required hypothesis regarding the existence of the $\mathfrak{L}$-transforms of the functions involved.[4] We have, by Theorem 10.5, considering (9),

$$y_l'(t) = \sum_{k=1}^{N} g_{kl}'(t) * f_k(t), \tag{13}$$

and

$$y_l''(t) = \sum_{k=1}^{N} \{g_{kl}''(t) * f_k(t) + g_{kl}'(0) f_k(t)\}. \tag{14}$$

For simplicity sake, we presume continuous functions $f_k(t)$. Should some of these functions have points of jump, then at these points we would have to form the respective derivatives from the right and from the left, in a manner analogous to that of (15.30). The functions $g_{kl}(t)$ are aggregates of exponential functions and powers; these are bounded in a neighbourhood of the origin; hence, by (12) and (13), all $y_l(0^+)$ and all $y_{el}'(0^+)$ vanish. Moreover, we have, by (12), (13), and (14),

$$\begin{aligned} p_{il}\left(\frac{d}{dt}\right) y_l(t) = {} & a_{il} \sum_{k=1}^{N} g_{kl}''(t) * f_k(t) + a_{il} \sum_{k=1}^{N} g_{kl}'(0) f_k(t) \\ & + b_{il} \sum_{k=1}^{N} g_{kl}'(t) * f_k(t) \\ & + c_{il} \sum_{k=1}^{N} g_{kl}(t) * f_k(t) \\ = {} & \sum_{k=1}^{N} \left\{p_{il}\left(\frac{d}{dt}\right) g_{kl}(t)\right\} * f_k(t) + a_{il} \sum_{k=1}^{N} g_{kl}'(0) f_k(t), \end{aligned}$$

[3] Here, the determinant $\Delta(s)$ assumes the rôle corresponding to that of the characteristic polynomial $p(s)$ of a *single* differential equation.

[4] This verification is not superfluous. This fact will become clear during the discussion of the anomalous system, when we shall discover that the solutions of the anomalous system generally produce initial values (from the right) different from the specified ones.

whence:

$$\sum_{l=1}^{N} p_{il}\left(\frac{d}{dt}\right) y_l(t) = \sum_{k=1}^{N} f_k(t) * \sum_{l=1}^{N} p_{il}\left(\frac{d}{dt}\right) g_{kl}(t) + \sum_{k=1}^{N} f_k(t) \sum_{l=1}^{N} a_{il}\, g'_{kl}(0),$$

for any fixed $i = 1, 2, \cdots, N$. For any fixed k, the $g_{kl}(t)$ represent solutions of the homogeneous system. It follows that

$$\sum_{l=1}^{N} p_{il}\left(\frac{d}{dt}\right) g_{kl}(t) = 0 \quad (k = 1, \cdots, N);$$

thus the first sum vanishes. Also, we have, by (10),

$$\sum_{l=1}^{N} a_{il}\, g'_{kl}(0) = \begin{cases} 0 & \text{for } k \neq i \\ 1 & \text{for } k = i, \end{cases}$$

and we are left with:

$$\sum_{l=1}^{N} p_{il}\left(\frac{d}{dt}\right) y_l(t) = f_i(t) \quad (i = 1, \cdots, N).$$

It follows that the $Y_l(t)$ do satisfy the inhomogeneous system.

Having shown, by theoretical considerations, that the method of the $\mathfrak{L}$-transformation produces the solution of a normal system of differential equations for specified initial values, we may make practical use of this technique in every specific case, without further concern. One should not, however, use the derived and presented formulae; instead one is advised to execute the several indicated steps, in this manner benefitting from simplifications that occur in practical problems.[5]

Advantages of the Method of the $\mathfrak{L}$-Transformation

1. Only one system of linear algebraic equations in N unknowns, that of the N image equations in the $Y_l(s)$, need be solved.

2. When employing the classical method, one solves firstly the homogeneous system by proposing the solutions as exponential functions; the solution of the inhomogenous system is then obtained by the technique of variation of the constants. By contrast, the method of the $\mathfrak{L}$-transformation produces the solution of the inhomogeneous system with vanishing initial values directly, and independently of the homogeneous system.

3. The initial values are incorporated in the system of image equations; in this manner they are automatically considered. To this we compare the efforts expended by the classical method: It firstly produces the general solution which must

[5] When solving practical problems, one encounters normal systems less frequently than anomalous systems. A completely executed example of a normal system of three differential equations with three unknown functions which represents a physical problem can be found on p. 79 of the book cited in the footnote on p. 84. Also, compare the explicitly solved normal system in Chapter 21.

subsequently be adapted to the specific initial values, a process which necessitates the solution of another system of linear equations. In physics and in engineering, one encounters most offen problems with vanishing initial values; for this case, an appreciable simplification results when using the method of the $\mathfrak{L}$-transformation, no reduction of labour results for the classical method.

4. An outstanding advantage of the method of the $\mathfrak{L}$-transformation arises from its facility to evaluate any individual unknown function with no knowledge of all others. The classical method offers no such possibility for specified initial values. This is an important qualification, for one is, in fact, most often interested in but one particular unknown, while the others have importance only in the construction of the system of equations.

20. The Anomalous System of Simultaneous Differential Equations, with Initial Conditions which can be fulfilled

The normal system of N differential equations of order n has been characterized by the specification of a non-vanishing determinant of the matrix of the coefficients of the highest derivatives, the nth derivatives: det $\| c_n^{\alpha\beta} \| \neq 0$.

As a consequence of this specification we observed the following three properties:

1. When solving the image equations, we encountered the *determinant* $\Delta(s)$; it is a polynomial in s of the *highest possible degree* nN. All its co-factors $\Delta_{\alpha\beta}(s)$ are necessarily polynomials in s of lower degree, and the quotients

$$G_{kl}(s) = \frac{\Delta_{kl}(s)}{\Delta(s)}$$

are necessarily rational functions, the numerator polynomials of which have degrees lower than that of the denominator polynomial. Thus, we had conventional functions $g_{kl}(t)$ in the original space.

2. Contributions to the solutions due to the excitations are composed of products of the form: $G_{kl}(s)\, F_k(s)$. Corresponding to these, in the original space, we have the convolutions $g_{kl}(t) * f_k(t)$ which are *continuous* functions irrespective of possible discontinuities of the input functions $f_k(t)$. This indicates that the output functions do *not repeat* possible *jumps* of the excitations; such discontinuities are repeated only in the nth derivatives of the output functions (compare (15.30)). Any physical system which is described by a normal system exhibits a smoothing character.

3. In connection with Eqs. (19.11) and 19.12) we observed that each of the N derived solutions satisfies its n specified initial conditions. These initial conditions were incorporated in the image equations during the $\mathfrak{L}$-transformation of the original equations, according to the Differentiation Theorem, as limits from the *right*; they are assumed by the original solutions *in this sense*.

We may summarize our observations: the solutions of a normal system behave exactly in the same manner as the solution of a single differential equation. Indeed, the latter always represents, when visualized as a system with $N = 1$, the normal case, since the determinant of the matrix of the coefficients of the nth derivatives degenerates to the one coefficient $c_n \neq 0$.

The anomalous system is characterized by

$$\det \| c_n^{\alpha\beta} \| = 0.$$

We anticipate the conclusions of the subsequent investigation of the anomalous system, and enumerate its characteristic properties which are quite different from those presented for the normal system.

1. The determinant $\Delta(s)$ is a polynomial in s of degree $< nN$. Consequently, it may happen that some co-factors $\Delta_{\alpha\beta}(s)$ are polynomials in s of degrees equal to, or higher than, that of $\Delta(s)$. Thus the corresponding rational functions $G_{kl}(s)$ may have numerator polynomials of degrees equal to, or larger than, the degree of the denominator polynomial. Accordingly, in the original space we do *not* have *conventional functions*; instead, we encounter distributions.

2. For the same reason, to the products $G_{kl}(s)\, F_k(s)$, there correspond not only convolutions of the excitations $f_k(t)$ with certain functions, but the functions $f_k(t)$ *themselves* and, possibly, also derivatives of these. If, as for the normal system, we admit continuous input functions with isolated points of discontinuity, then we are faced with another difficulty, since the output functions may fail to have the required derivatives in the classical sense which are needed to satisfy the differential equations.

3. In mathematics it is customary to require that the specified initial values be assumed as limits from the *right*. If we follow this convention, then we cannot arbitrarily specify the initial values; instead these initial values, together with the initial values of the excitations have to satisfy certain *relations*. When specifying initial values such that these relations are not satisfied, we have an initial value problem which *has no solution in the mathematical sense*. Should we nevertheless proceed, and produce solutions in the usual manner, using the specified initial values, we would generate solutions which have limits from the right different from the ones specified.

The fact that *for physical problems*, in general, arbitrary initial values are given which need not comply with these relations, and that, nevertheless, a physical process ensues, that is, solutions do exist, forces us to review the concept of the initial value problem. A possibility of avoiding the inconsistencies is shown by the deliberations on p. 108: We need to include into our considerations the *past*, that is the time interval $t < 0$, and interpret the initial values as limits as $t \to 0$ from the *left*. The limits from the left and the limits from the right need no longer agree at $t = 0$. Having thus solutions which are not differentiable at $t = 0$, we encounter another incentive to consider *distributions*, an interpretation which was already suggested by the difficulties observed in remark 2 above.

The classical method of solution lacks lucidity, for it requires the solution of the homogeneous system for the construction of the solution of the inhomogeneous system; it is for this reason that no attention was formerly given to the facts that are explained in the above remarks 1, 2, and 3. It was only the method the $\mathfrak{L}$-transformation which made it possible to investigate these problems, for it permits the transformation of the transcendental differential equations into algebraic image quations.

Before entering the detailed discussion of the anomalous system, a remark concerning the *number of initial values* is in order. We use the term "system of differential equations of order n" to imply that the highest occuring derivatives are of order n. This by no means precludes the possibility that in some of the equations the highest derivatives of some of the unknown functions might have order less than n Should the n-th derivative of some function $y_\beta(t)$ be missing from all equations, then we must, of course, specify fewer initial values for that function y_β. In this case, the system is certainly anomalous, since all elements of the βth column of the matrix $||c_n{}^{\alpha\beta}||$ are zero (compare p. 109); hence, its determinant has value zero.

First, we want to explain the background of the *relations between the initial values*, in the sense of limits from the right, which were mentioned in the above remark 3. Consider the matrix of the coefficients of the nth derivatives of an anomalous system of N differential equations of order n: $||c_n{}^{\alpha\beta}||$. The determinant of this matrix is zero, hence the rank r_n of this matrix is less than N. Therefore, one can completely eliminate the derivatives $y_\mu{}^{(n)}$ from $(N - r_n)$ equations and thus obtain r_n equations which are truly of order n, that is, they contain, at least, one derivative $y_\mu{}^{(n)}$, while the remaining $(N - r_n)$ equations are, at most, of order $n - 1$. Next, we inspect the matrix of the coefficients of the $y_\mu{}^{(n-1)}$ in these $(N - r_n)$ equations which has $(N - r_n)$ rows and N columns, and we determine from which of these equations the $y_\mu{}^{(n-1)}$ can be completely eliminated. In this manner, one obtains a group of r_{n-1} equations which are truly of order $n - 1$, while the remaining $(N - r_n - r_{n-1})$ equations are, at most, of order $n - 2$. Proceeding thus, one can reduce the given system of equations to an equivalent system of equations which is composed of r_n equations of order n, r_{n-1} equations of order $n - 1, \cdots,$ and r_0 equations of order 0 which are ordinary, algebraic equations.

We write one of the r_{n-1} equations of order $n - 1$, using for the coefficients and the excitation functions the same, unaltered symbols, to avoid the introduction of new symbols:

$$[c_{n-1}^{\alpha 1}\, y_1^{(n-1)} + \cdots + c_1^{\alpha 1}\, y_1' + c_0^{\alpha 1}\, y_1] + [c_{n-1}^{\alpha 2}\, y_2^{(n-1)} + \cdots + c_1^{\alpha 2}\, y_2' + c_0^{\alpha 2}\, y_2]$$
$$+ \cdots = f_\alpha(t)\,.$$

If the functions $y_1, y'_1, \cdots, y_1{}^{(n-1)}; y_2, y'_2, \cdots, y_2{}^{(n-1)}; \cdots$ are to have limits as $t \to +0$, then $f_\alpha(t)$ must have a limit as $t \to +0$, and the following relation must be valid:

$$[c_{n-1}^{\alpha 1}\, y_1^{(n-1)}(0+) + \cdots + c_0^{\alpha 1}\, y_1(0+)] + [c_{n-1}^{\alpha 2}\, y_2^{(n-1)}(0+) + \cdots + c_0^{\alpha 2}\, y_2(0+)]$$
$$+ \cdots = f_\alpha(0+)\,.$$

Each of the given r_{n-1} equations of order $n - 1$ will provide one such relation; we obtain r_{n-1} relations of the above type.

Next, we consider one of the r_{n-2} equations of order $n - 2$:

$$(1)\qquad \left[c_{n-2}^{a1}\, y_1^{(n-2)} + \cdots + c_0^{a1}\, y_1\right] + \left[c_{n-2}^{a2}\, y_2^{(n-2)} + \cdots + c_0^{a2}\, y_2\right] + \cdots = f_a(t)\,.$$

The above considerations may be repeated and we obtain the r_{n-2} relations:

$$\left[c_{n-2}^{a1}\, y_1^{(n-2)}(0+) + \cdots + c_0^{a1}\, y_1(0+)\right] + \left[c_{n-2}^{a2}\, y_2^{(n-2)}(0+) + \cdots + c_0^{a2}\, y_2(0+)\right] + \cdots = f_a(0+)\,.$$

The left hand side of (1) can be differentiated once and the derivatives thus formed are to have limits as $t \to +0$; hence $f'_a(t)$ must exist, and must have a limit as $t \to +0$. In this manner we obtain r_{n-2} additional relations, of the form

$$\left[c_{n-2}^{a1}\, y_1^{(n-1)}(0+) + \cdots + c_0^{a1}\, y_1'(0+)\right] + \left[c_{n-2}^{a2}\, y_2^{(n-1)}(0+) + \cdots + c_0^{a2}\, y_2'(0+)\right] + \cdots = f_a'(0+)\,.$$

Proceeding in this manner, we produce a total of

$$r_{n-1} + 2\, r_{n-2} + \cdots + n\, r_0$$

relations which necessarily must be satisfied if solutions of the system are to exist which assume the specified initial values, in the sense of limits from the right. We call these relations the *compatibility conditions* of the anomalous system; initial values that comply with these compatibility conditions will be called *attainable* initial values.

For the remainder of this Chapter we presume that the initial values exist as *limits from the right*, and that these *comply with the compatibility conditions*. We shall show that under these conditions the anomalous system has, in general, solutions which assume these initial values. That is, the compatibility conditions are not only *necessary* but also *sufficient* for the adoption of the specified initial values. The investigation of the anomalous system of N differential equations of order n would entail great difficulties with notation. Therefore, we shall present the method using as a model the system of two differential equations of first order in two unknown functions. In this manner we gain insight into the process of solution for a system of N equations of first order in N unknown functions. This model actually embraces the most general case, since any system of order n can be replaced by a system of first order.

To substantiate this claim, we start with *one* equation of order n:

$$c_n\, y^{(n)} + c_{n-1}\, y^{(n-1)} + \cdots + c_1\, y' + c_0\, y = f(t)\,.$$

We define:

(2) $$y' = \eta_1,\ y'' = \eta_2, \ldots, y^{(n-1)} = \eta_{n-1},$$

and, with these, we alter the differential equation into:

(3) $$c_n \eta'_{n-1} + c_{n-1}\eta_{n-1} + \cdots + c_1\eta_1 + c_0 y = f(t),$$

together with the $(n-1)$ equations

(4) $$y' = \eta_1,\ \eta'_1 = \eta_2, \ldots, \eta'_{n-2} = \eta_{n-1}.$$

We have a system of $1 + (n-1) = n$ equations of the first order with the unknowns y, η_1, ..., η_{n-1} instead of the one, given equation of order n. The n initial values of the unknown $y: y(0), y'(0), \ldots, y^{(n-1)}(0)$ become the initial values of the n new unknowns: $y(0)$, $\eta_1(0)$, ..., $\eta_{n-1}(0)$.

For a *system of N equations* of order n with N unknown functions, we proceed similarly by defining a new function for each derivative up to the $(n-1)^{th}$ derivative of each function as shown in (2) for one function. With these substitutions, the N equations of order n are altered into N equations of the first order, similar to (3), together with $(n-1)$ equations for each of the N unknowns similar to (4), that is a total of $N(n-1)$ equations. Thus, we obtain alltogether $N + N(n-1) = Nn$ equations of the first order for Nn unknown functions. A rather involved calculation demonstrates that the determinant of the coefficients of the highest derivatives has the same value in the new system as it has in the old system; consequently, a normal system becomes a normal system, an anomalous system an anomalous one.

Consider the system

(5) $$\begin{cases} a_{11} y'_1 + b_{11} y_1 + a_{12} y'_2 + b_{12} y_2 = f_1(t) \\ a_{21} y'_1 + b_{21} y_1 + a_{22} y'_2 + b_{22} y_2 = f_2(t). \end{cases}$$

We presume the existence of the limits of the excitations as $t \to +0$: $f_1(0^+)$, $f_2(0^+)$, deferring further specifications concerning these functions. Let the system be *anomalous*, that is:

(6) $$A = \begin{vmatrix} a_{11} & a_{12} \\ a_{21} & a_{22} \end{vmatrix} = 0.$$

Initial values y_1^0 and y_2^0 of y_1 and y_2 are given as limits as $t \to +0$:

(7) $$y_1(0^+) = y_1^0, \quad y_2(0^+) = y_2^0.$$

Because of (6), we can eliminate y'_1 and y'_2 from the Eqs. (5). In order that the Eqs. (5) represent a system of differential equations, at least one of the coefficients a_{ik} must have a non-zero value; without loss of generality, let $a_{11} \neq 0$. To accomplish the attempted elimination, multiply the first equation by a_{21} and the second equation by a_{11}, and then subtract the first from the second. With

(8) $$\begin{vmatrix} a_{11} & b_{11} \\ a_{21} & b_{21} \end{vmatrix} = B, \qquad \begin{vmatrix} a_{11} & b_{12} \\ a_{21} & b_{22} \end{vmatrix} = C$$

we can write the result compactly thus:

$$B\,y_1(t) + C\,y_2(t) = -a_{21}\,f_1(t) + a_{11}\,f_2(t). \tag{9}$$

The *compatibility condition* is obtained from (9), by the limiting process $t \to +0$:

$$B\,y_1^0 + C\,y_2^0 = -a_{21}\,f_1(0+) + a_{11}\,f_2(0+). \tag{10}$$

If not only the determinant A but also the determinants B and C are each zero, then we must conclude that the coefficients of the second equation of (5) are fixed multiples of the coefficients of the first equation of (5). In this case, either the second equation is equivalent to the first if f_2 too is the same fixed multiple of f_1, or else the equations would contradict one another. Hence, B and C cannot both be zero.

Suppose that $B \neq 0$ and $C \neq 0$. In this case we can arbitrarily select only one of the two initial values, y_1^0 or y_2^0, the other is then implicitely determined through (10). For the case that $B = 0$ and $C \neq 0$, we can arbitrarily select y_1^0, while y_2^0 is determined through (10). (In this case y_2 follows from (9) trivially without integration.) Similar considerations apply to the case: $B \neq 0$ and $C = 0$. In any case, one can freely specify but one of the initial values, the other one is determined by this choice through the compatibility condition (10).

We apply the $\mathfrak{L}$-transformation to the system (5). When applying the Differentiation Theorem, we need the limits $y_1(0+)$ and $y_2(0+)$. We presume that we can use the specified values y_1^0 and y_2^0 for these limits, provided they comply with the compatibility condition (10). Thus we produce the image equations:

$$\begin{aligned}(a_{11}\,s + b_{11})\,Y_1 + (a_{12}\,s + b_{12})\,Y_2 &= F_1(s) + a_{11}\,y_1^0 + a_{12}\,y_2^0\\ (a_{21}\,s + b_{21})\,Y_1 + (a_{22}\,s + b_{22})\,Y_2 &= F_2(s) + a_{21}\,y_1^0 + a_{22}\,y_2^0.\end{aligned} \tag{11}$$

When calculating the determinant $\Delta(s)$ of the system of algebraic equations (11), we find, for s^2, the coefficient $A = 0$. With equations (6) and (8) we introduced short notations for three determinants of the matrix of coefficients of (5); here, we introduce the remaining three determinants:

$$\begin{vmatrix} b_{11} & a_{12} \\ b_{21} & a_{22} \end{vmatrix} = D, \quad \begin{vmatrix} b_{11} & b_{12} \\ b_{21} & b_{22} \end{vmatrix} = E, \quad \begin{vmatrix} a_{12} & b_{12} \\ a_{22} & b_{22} \end{vmatrix} = G. \tag{12}$$

Using the notation of (12), we find:

$$\Delta(s) = (C + D)\,s + E. \tag{13}$$

The solution of (11) yields:

$$\Delta(s)\,Y_1(s) = \begin{vmatrix} F_1(s) & a_{12}\,s + b_{12} \\ F_2(s) & a_{22}\,s + b_{22} \end{vmatrix} + \begin{vmatrix} a_{11}\,y_1^0 + a_{12}\,y_2^0 & a_{12}\,s + b_{12} \\ a_{21}\,y_1^0 + a_{22}\,y_2^0 & a_{22}\,s + b_{22} \end{vmatrix}$$

or, since $A = 0$:

$$\begin{aligned} \Delta(s)\, Y_1(s) &= F_1(s)\,(a_{22}\,s + b_{22}) - F_2(s)\,(a_{12}\,s + b_{12}) + C\,y_1^0 + G\,y_2^0 \\ \Delta(s)\, Y_2(s) &= -F_1(s)\,(a_{21}\,s + b_{21}) + F_2(s)\,(a_{11}\,s + b_{11}) - B\,y_1^0 + D\,y_2^0. \end{aligned} \tag{14}$$

For brevity we set

$$C + D = H.$$

We defer the special case $H = 0$ (compare the remark 1 at the end of this Chapter), and we presume here:

$$H \neq 0, \textit{ that is } \Delta(s) \textit{ is a linear function.}$$

By (14), we have:

$$Y_1(s) = F_1(s)\,\frac{a_{22}s + b_{22}}{Hs + E} - F_2(s)\,\frac{a_{12}s + b_{12}}{Hs + E} + \frac{C}{Hs + E}\,y_1^0 + \frac{G}{Hs + E}\,y_2^0. \tag{15}$$

The first term can be modified thus:[1]

$$F_1(s)\,\frac{a_{22}s + b_{22}}{Hs + E} = F_1(s)\left(\frac{a_{22}}{H} + \frac{K/H}{s + E/H}\right), \text{with } K = b_{22} - a_{22}\,\frac{E}{H}. \tag{16}$$

The corresponding original function is:

$$\frac{a_{22}}{H}\,f_1(t) + f_1(t) * \frac{K}{H}\,e^{-Et/H}.$$

The second term of (15) may be similarly modified. Altogether, for the original function of (15), we find:

$$\begin{aligned} y_1(t) = {} & \frac{a_{22}}{H}\,f_1(t) + \frac{K}{H}\,f_1(t) * e^{-Et/H} - \frac{a_{12}}{H}\,f_2(t) - \frac{L}{H}\,f_2(t) * e^{-Et/H} \\ & + y_1^0\,\frac{C}{H}\,e^{-Et/H} + y_2^0\,\frac{G}{H}\,e^{-Et/H}, \end{aligned} \tag{17}$$

with

$$K = b_{22} - a_{22}\,\frac{E}{H}, \quad L = b_{12} - a_{12}\,\frac{E}{H}. \tag{18}$$

Similarly, one finds:[2]

$$\begin{aligned} y_2(t) = {} & -\frac{a_{21}}{H}\,f_1(t) - \frac{M}{H}\,f_1(t) * e^{-Et/H} + \frac{a_{11}}{H}\,f_2(t) + \frac{N}{H}\,f_2(t) * e^{-Et/H} \\ & - y_1^0\,\frac{B}{H}\,e^{-Et/H} + y_2^0\,\frac{D}{H}\,e^{-Et/H}, \end{aligned} \tag{19}$$

[1] To the factor of $F_1(s)$, there corresponds a distribution for $a_{22} \neq 0$. The introduction of a distribution is avoided through the above modification.

[2] One can verify that the functions (17) and (19) satisfy the relation (9).

with

(20) $$M = b_{21} - a_{21}\frac{E}{H}, \qquad N = b_{11} - a_{11}\frac{E}{H}.$$

When comparing the solution of the normal system with the solution of the anomalous system, one observes the new fact that in the solution of the anomalous system certainly *at least one of the excitations* f_1 and f_2 *occurs* not only in the convolution integrals but also *as isolated term*, since at least one of the coefficients a_{11}, a_{12}, a_{21}, a_{22} must have non-zero value. In order that, at some point, the functions y_1 and y_2 strictly satisfy the differential equations (5), the functions f_1 and/or f_2 must be differentiable at that point. We shall demonstrate that under this hypothesis, the functions (17) and (19) do satisfy the differential equations. Later on, in Chapter 22, we shall show that when using the theory of distributions, we may omit the requirement of differentiability of the excitations.

To verify the solution at some point, we form the derivatives of y_1 and y_2, using Theorem 10.5 for the differentiation of the convolutions (f_1 and f_2 are differentiable and therefore continuous at the point t); for instance,

$$\frac{d}{dt}\left[f_1(t) * e^{-Et/H}\right] = -\frac{E}{H} f_1(t) * e^{-Et/H} + f_1(t).$$

We substitute the functions y_1, y'_1, y_2, y'_2 into the left hand side of the first differential equation of (5), and we collect the terms which contain the same expression as, for instance, f_1, f'_1, f_2, ... In this manner we find:

(21)
$$\begin{aligned}
\frac{1}{H}\Big\{&(a_{11}K + b_{11}a_{22} - a_{12}M - b_{12}a_{21})\, f_1(t)\\
&+ (a_{11}a_{22} - a_{12}a_{21})\, f'_1(t)\\
&+ (-a_{11}L - b_{11}a_{12} + a_{12}N + b_{12}a_{11})\, f_2(t)\\
&+ (-a_{11}a_{12} + a_{12}a_{11})\, f'_2(t)\\
&+ \left(a_{11}K\left(-\frac{E}{H}\right) + b_{11}K - a_{12}M\left(-\frac{E}{H}\right) - b_{12}M\right) f_1(t) * e^{-Et/H}\\
&+ \left(-a_{11}L\left(-\frac{E}{H}\right) - b_{11}L + a_{12}N\left(-\frac{E}{H}\right) + b_{12}N\right) f_2(t) * e^{-Et/H}\\
&+ \left(a_{11}C\left(-\frac{E}{H}\right) + b_{11}C - a_{12}B\left(-\frac{E}{H}\right) - b_{12}B\right) y_1^0\, e^{-Et/H}\\
&+ \left(a_{11}G\left(-\frac{E}{H}\right) + b_{11}G + a_{12}D\left(-\frac{E}{H}\right) + b_{12}D\right) y_2^0\, e^{-Et/H}\Big\}.
\end{aligned}$$

We need to show that this expression equals $f_1(t)$, the right hand side of the differential equation. We immediately see that the term with $f'_1(t)$ and the term with $f'_2(t)$ vanish, since $A = 0$,, a fact which must be remembered in the sequel. Upon substitution of the values for K and M, the factor of $f_1(t)$ becomes:

$$\frac{1}{H}\left\{a_{11}\left(b_{22}-a_{22}\frac{E}{H}\right)+b_{11}a_{22}-a_{12}\left(b_{21}-a_{21}\frac{E}{H}\right)-b_{12}a_{21}\right\}$$
$$=\frac{1}{H^2}\left\{(a_{11}b_{22}-b_{12}a_{21}+b_{11}a_{22}-a_{12}b_{21})\,H-(a_{11}a_{22}-a_{12}a_{21})\,E\right\}$$
$$=\frac{1}{H^2}\left\{(C+D)\,H+0\cdot E\right\}=1.$$

Similarly, upon substitution of the values for L and N, the coefficient of $f_2(t)$ becomes:

$$\frac{1}{H}\left\{-a_{11}\left(b_{12}-a_{12}\frac{E}{H}\right)-b_{11}a_{12}+a_{12}\left(b_{11}-a_{11}\frac{E}{H}\right)+b_{12}a_{11}\right\}$$
$$=\frac{1}{H^2}\left\{(-a_{11}b_{12}-a_{12}b_{11}+a_{12}b_{11}+a_{11}b_{12})\,H+(a_{11}a_{12}-a_{11}a_{12})\,E\right\}=0.$$

The factor of $f_1(t) * e^{-Et/H}$ is equal to

$$\frac{1}{H^2}\left\{-a_{11}\left(b_{22}-a_{22}\frac{E}{H}\right)E+b_{11}\left(b_{22}-a_{22}\frac{E}{H}\right)H+a_{12}\left(b_{21}-a_{21}\frac{E}{H}\right)E\right.$$
$$\left.-b_{12}\left(b_{21}-a_{21}\frac{E}{H}\right)H\right\}$$
$$=\frac{1}{H^2}\left\{-(a_{11}b_{22}-a_{21}b_{12}+b_{11}a_{22}-b_{21}a_{12})\,E+(b_{11}b_{22}-b_{12}b_{21})\,H\right.$$
$$\left.+(a_{11}a_{22}-a_{12}a_{21})\frac{E}{H}E\right\}$$
$$=\frac{1}{H^2}\left\{-(C+D)\,E+E\,H+0\right\}=\frac{1}{H^2}(-H\,E+E\,H)=0.$$

For the factor of $y_1^0e^{-Et/H}$, we find, with $C=H-D$,

$$\frac{1}{H^2}\left\{(-a_{11}H+a_{11}D+a_{12}B)\,E+(b_{11}C-b_{12}B)\,H\right\}.$$

The relations

$$(22)\qquad a_{11}D+a_{12}B=b_{11}A=0,\quad b_{11}C-b_{12}B=a_{11}E$$

can be verified by explicit expression of the determinants, and for the previous expression we obtain:

$$\frac{1}{H^2}(-a_{11}HE+a_{11}EH)=0.$$

By a similar process one can show that both the factor of $f_2(t)*e^{-Et/H}$ and the factor of $y_2^0e^{-Et/H}$ vanish. Indeed, in the sum (21), we are left with $f_1(t)$.

The fact that (17) and (19) satisfy the second differential equation of (5) can be established by steps similar to those above. Thus, we have verified that these functions are solutions, independent of the tacitly employed hypothesis that the given functions and the sought functions have $\mathfrak{L}$-transforms (compare the principle of extension, p. 74).

We have yet to determine what *initial values* the functions y_1 and y_2 assume in case the compatibility condition (10) is satisfied. We have, by (17),

$$(23)\qquad H\,y_1(0+)=a_{22}f_1(0+)-a_{12}f_2(0+)+y_1^0C+y_2^0G.$$

The values y_1^0 and y_2^0 should satisfy the condition:

$$0 = a_{21} f_1(0^+) - a_{11} f_2(0^+) + y_1^0 B + y_2^0 C. \tag{24}$$

Forming a_{11} times Eq. (23) minus a_{12} times Eq. (24), we eliminate both $f_2(0^+)$ and $f_1(0^+)$, the latter since $A = 0$. The remainder becomes, upon replacement of C by $H - D$,

$$a_{11} H y_1(0^+) = y_1^0 (a_{11} H - a_{11} D - a_{12} B) + y_2^0 (a_{11} G - a_{12} C).$$

We now use (22) and the additional relation

$$a_{11} G - a_{12} C = - b_{12} A = 0, \tag{25}$$

and we find:

$$a_{11} H y_1(0^+) = y_1^0 a_{11} H.$$

This shows that $y_1(0^+) = y_1^0$, for we have supposed that $H \neq 0$ and we have assumed at the outset that $a_{11} \neq 0$.

Eq. (19) implies that

$$H y_2(0^+) = - a_{21} f_1(0^+) + a_{11} f_2(0^+) - y_1^0 B + y_2^0 D. \tag{26}$$

Adding functions (26) and (24), one finds:

$$H y_2(0^+) = y_2^0 (C + D) = y_2^0 H,$$

that is, $y_2(0^+) = y_2^0$.

We summarize the conclusions:

Theorem 20.1. *If:* 1. *the system (5) of differential equations is anomalous, that is* $A = 0$*;* 2. $C + D \neq 0$*;* 3. *the excitations* $f_1(t)$ *and* $f_2(t)$ *have the respective limits* $f_1(0^+)$ *and* $f_2(0^+)$ *as* $t \to +0$*;* 4. *the specified initial values* y_1^0 *and* y_2^0 *satisfy the compatibility condition (10); then the functions (17) and (19) which are obtained by means of the* $\mathfrak{L}$*-transformation satisfy the system (5) at those points where* $f_1(t)$ *and* $f_2(t)$ *are differentiable; the functions (17) and (19) assume the specified initial values as limits as* $t \to +0$*, from the right.*

The system is satisfied for all $t > 0$, only if the excitations are differentiable for all $t > 0$; continuity of the excitations, which was sufficient in the normal case, does not suffice here.

Remarks: 1. For the special case that $H = C + D = 0$ and $E \neq 0$, the image function $Y_1(s)$, by (15), has the form:

$$Y_1(s) = \frac{1}{E} \{F_1(s)(a_{22} s + b_{22}) - F_2(s)(a_{12} s + b_{12}) + C y_1^0 + G y_2^0\}.$$

The corresponding original function is, in general, not a conventional function, and theory of distributions is needed for the treatment of this case (compare Chapter 22).

2. The determinant $\Delta(s)$ of an anomalous system of N differential equations of order n is a polynomial of, at most, degree $nN - 1$; the cofactors $\Delta_{\alpha\beta}(s)$ introduced in Chapter 19 are polynomials in s having degree $\leq n(N-1)$. Consequently, numerous possibilities ought to be considered and discussed when treating the "general case". Hence, such a presentation is practically excluded. However, problems encountered in applications with numerically specified coefficients (most of which are usually zero) can be treated in a perspicuous manner, following the pattern set by the solution of the above example.

21. The Normal System in the Space of Distributions

In Chapter 20, we established the following conclusion: If, for some *anomalous* system of differential equations, one seeks the solution in the realm of classical analysis, then one must firstly request that certain hypotheses regarding the differentiability of the *excitation functions* are satisfied, and secondly recognize the fact that the *initial conditions* cannot arbitrarily be specified; instead these have to comply with certain compatibility conditions. However, when investigating problems in physics and in engineering one often encounters excitation functions of general nature, and arbitrary initial conditions. No doubt, physical processes exist for such situations and we are faced with the problem of devising a mathematical description of these. This is, indeed, possible by means of the theory of distributions.

In order to present an elucidating transition to the new point of view, we employ the system (20.5) and begin the development with the *solution of the normal case.*

Using the set of initial values (20.7), which may now be completely arbitrary, we find for the equations (20.5) the system of image equations (20.11). The determinant of the latter is, using the notation of Chapter 20,

$$\Delta(s) = A\, s^2 + (C + D)\, s + E, \tag{1}$$

whereby now $A \neq 0$, since the system is presumed to be normal.

The solutions of the image equations are:

$$\begin{aligned} Y_1(s) &= \frac{1}{\Delta(s)} \begin{vmatrix} F_1(s) + a_{11} y_1^0 + a_{12} y_2^0 & a_{12} s + b_{12} \\ F_2(s) + a_{21} y_1^0 + a_{22} y_2^0 & a_{22} s + b_{22} \end{vmatrix} \\ Y_2(s) &= \frac{1}{\Delta(s)} \begin{vmatrix} a_{11} s + b_{11} & F_1(s) + a_{11} y_1^0 + a_{12} y_2^0 \\ a_{21} s + b_{21} & F_2(s) + a_{21} y_1^0 + a_{22} y_2^0 \end{vmatrix} \end{aligned} \tag{2}$$

Recalling the notations of Chapter 19 we write:

$$
(3)\qquad
\begin{aligned}
\frac{a_{22}s+b_{22}}{\Delta(s)} &= G_{11}(s), & -\frac{a_{12}s+b_{12}}{\Delta(s)} &= G_{21}(s),\\
-\frac{a_{21}s+b_{21}}{\Delta(s)} &= G_{12}(s), & \frac{a_{11}s+b_{11}}{\Delta(s)} &= G_{22}(s),
\end{aligned}
$$

and we obtain:

$$
(4)\qquad
\begin{aligned}
Y_1(s) &= F_1(s)\,G_{11}(s) + F_2(s)\,G_{21}(s) + (a_{11}y_1^0 + a_{12}y_2^0)\,G_{11}(s)\\
&\quad + (a_{21}y_1^0 + a_{22}y_2^0)\,G_{21}(s)\\
Y_2(s) &= F_1(s)\,G_{12}(s) + F_2(s)\,G_{22}(s) + (a_{11}y_1^0 + a_{12}y_2^0)\,G_{12}(s)\\
&\quad + (a_{21}y_1^0 + a_{22}y_2^0)\,G_{22}(s).
\end{aligned}
$$

The degree of the polynomial $\Delta(s)$ is higher than the degrees of the polynomials in the numerators; it follows that, corresponding to the $G_{kl}(s)$, one finds classical functions $g_{kl}(t)$ in the original space. Thus, for the functions (4), one finds in the original space:

$$
(5)\qquad
\begin{aligned}
y_1(t) &= f_1(t) * g_{11}(t) + f_2(t) * g_{21}(t) + (a_{11}y_1^0 + a_{12}y_2^0)\,g_{11}(t)\\
&\quad + (a_{21}y_1^0 + a_{22}y_2^0)\,g_{21}(t)\\
y_2(t) &= f_1(t) * g_{12}(t) + f_2(t) * g_{22}(t) + (a_{11}y_1^0 + a_{12}y_2^0)\,g_{12}(t)\\
&\quad + (a_{21}y_1^0 + a_{22}y_2^0)\,g_{22}(t).
\end{aligned}
$$

The development of Chapter 19 indicates that (5) produces differentiable solutions of the system (20.5) having the initial values (20.7), provided the excitation functions $f_1(t)$ and $f_2(t)$ are continuous. In particular, for a homogeneous system, that is for $f_1 \equiv f_2 \equiv 0$, we have:

$$
\begin{aligned}
&y_1(t) = g_{11}(t),\ y_2(t) = g_{12}(t), \text{ when } a_{11}y_1^0 + a_{12}y_2^0 = 1,\ a_{21}y_1^0 + a_{22}y_2^0 = 0;\\
&y_1(t) = g_{21}(t),\ y_2(t) = g_{22}(t), \text{ when } a_{11}y_1^0 + a_{12}y_2^0 = 0,\ a_{21}y_1^0 + a_{22}y_2^0 = 1.
\end{aligned}
$$

We express this thus: The functions g_{k1}, g_{k2} $(k = 1, 2)$ are solutions of the homogenous system, that is

$$
(6)\qquad
\begin{aligned}
a_{11}g_{k1}' + b_{11}g_{k1} + a_{12}g_{k2}' + b_{12}g_{k2} &= 0\\
a_{21}g_{k1}' + b_{21}g_{k1} + a_{22}g_{k2}' + b_{22}g_{k2} &= 0
\end{aligned}
\qquad (k=1,2).
$$

For their initial values we have the equations:

$$
(7)\qquad a_{i1}g_{k1}(0+) + a_{i2}g_{k2}(0+) = \begin{cases} 0 & \text{for } i \neq k\\ 1 & \text{for } i = k \end{cases} \qquad (i,k=1,2).
$$

From this, one finds the initial values:

$$
(8)\qquad g_{11}(0+) = \frac{a_{22}}{A},\quad g_{12}(0+) = -\frac{a_{21}}{A},\quad g_{21}(0+) = -\frac{a_{12}}{A},\quad g_{22}(0+) = \frac{a_{11}}{A}.
$$

We now investigate

The Normal System in the Frame of the Distribution Theory.

A distribution is defined over the entire real line. To comply with the restriction imposed earlier, confining us to the interval $0 \leqq t < \infty$, we have to restrict the discussion to those distributions which have their support in the right half-line. Moreover, we intend to employ the Laplace transformation, which necessitates the further restriction of the available distributions to those which have $\mathfrak{L}$-transforms. In short, the given and the sought quantities must belong to the space $\mathcal{D}'_0$ (compare Chapter 12). A distribution does not have a value at some specified point; hence, it is not meaningful to specify initial values. One can speak of an initial value only for the special case when the distribution is, in fact, determined by a function. The derivatives must be replaced by distribution-derivatives. Therefore, we shall investigate the following *system of distribution-derivative equations:*

$$\begin{aligned} &a_{11}\, D\, y_1 + b_{11}\, y_1 + a_{12}\, D\, y_2 + b_{12}\, y_2 = f_1 \\ &a_{21}\, D\, y_1 + b_{21}\, y_1 + a_{22}\, D\, y_2 + b_{22}\, y_2 = f_2 \end{aligned} \quad (A \neq 0), \qquad y_1, y_2, f_1, f_2 \in \mathcal{D}'_0. \tag{9}$$

Application of the $\mathfrak{L}$-transformation yields, by Theorem 14.3, the image equations:

$$\begin{aligned} (a_{11}\, s + b_{11})\, Y_1(s) + (a_{12}\, s + b_{12})\, Y_2(s) &= F_1(s) \\ (a_{21}\, s + b_{21})\, Y_1(s) + (a_{22}\, s + b_{22})\, Y_2(s) &= F_2(s); \end{aligned} \tag{10}$$

these agree formally with (20.11), provided in the latter we set $y_1^0 = y_2^0 = 0$. Whence we find the solutions formally from (4), using the same conditions:

$$\begin{aligned} Y_1(s) &= F_1(s)\, G_{12}(s) + F_2(s)\, G_{21}(s) \\ Y_2(s) &= F_1(s)\, G_{12}(s) + F_2(s)\, G_{22}(s). \end{aligned} \tag{11}$$

We have as originals for the $G_{kl}(s)$ the above mentioned functions $g_{kl}(t)$ which, however, are now visualized as distributions. Corresponding to (11) we have, by Theorem 14.5, the following originals from $\mathcal{D}'_0$:

$$\begin{aligned} y_1 &= f_1 * g_{11}(t) + f_2 * g_{21}(t) \\ y_2 &= f_1 * g_{12}(t) + f_2 * g_{22}(t). \end{aligned} \tag{12}$$

For arbitrary distributions f_1 and f_2 from $\mathcal{D}'_0$, these are solutions of (9) which too belong to $\mathcal{D}'_0$.

It is interesting and revealing to investigate in what way the distributions (12) satisfy the distribution-derivative equations (9). According to Theorem 14.6, we find:

$$D y_1 = f_1 * (D g_{11}) + f_2 * (D g_{21}).$$

As a distribution of $\mathcal{D}'_0$, g_{11} is to be defined by zero for $t < 0$; hence, by (8), at $t = 0$, it has the jump:

$$g_{11}(0+) - g_{11}(0-) = \frac{a_{22}}{A} - 0.$$

g_{11} has for $t > 0$ and for $t < 0$ a derivative in the classical sense, hence, by App. No. 20,

$$Dg_{11} = g_{11}' + \frac{a_{22}}{A}\,\delta.$$

Similarly, we find

$$Dg_{21} = g_{21}' - \frac{a_{12}}{A}\,\delta.$$

It follows that

$$Dy_1 = f_1 * g_{11}' + \frac{a_{22}}{A}\,f_1 * \delta + f_2 * g_{21}' - \frac{a_{12}}{A}\,f_2 * \delta$$

or, because of (14.8),

$$Dy_1 = f_1 * g_{11}' + \frac{a_{22}}{A}\,f_1 + f_2 * g_{21}' - \frac{a_{12}}{A}\,f_2. \tag{13}$$

Using the same technique one derives:

$$Dy_2 = f_1 * g_{12}' - \frac{a_{21}}{A}\,f_1 + f_2 * g_{22}' + \frac{a_{11}}{A}\,f_2. \tag{14}$$

Thus, for the left hand side of the first equation of the system (9), we find:

$$\begin{aligned} &a_{11}\left\{f_1 * g_{11}' + \frac{a_{22}}{A}\,f_1 + f_2 * g_{21}' - \frac{a_{12}}{A}\,f_2\right\}\\ &+ b_{11}\left\{f_1 * g_{11} + f_2 * g_{21}\right\}\\ &+ a_{12}\left\{f_1 * g_{12}' - \frac{a_{21}}{A}\,f_1 + f_2 * g_{22}' + \frac{a_{11}}{A}\,f_2\right\}\\ &+ b_{12}\left\{f_1 * g_{12} + f_2 * g_{22}\right\}. \end{aligned}$$

The sums of the convolutions involving f_1 and f_2 vanish according to (6); the factor of f_1 is given by:

$$\frac{1}{A}\,(a_{11}a_{22} - a_{12}a_{21}) = 1,$$

the factor of f_2 is zero. Thus, only f_1 remains, and the first equation of (9) is verified. The same process is employed to verify the second equation of (9).

Attention is called to the fact that Dy_1, given by (13), and Dy_2, given by (14), have the *same form as dy_1/dt and dy_2/dt in the classical case* for continuous f_1 and f_2, which is shown by Theorem 10.5. Here, in contrast, we need no hypothesis regarding f_1 and f_2 other than that both belong to $\mathscr{D}'_0$.

For the special case that $f_1 \equiv \delta$ and that $f_2 \equiv 0$, we have $y_1 = g_{11}(t)$ and $y_2 = g_{12}(t)$, the functions interpreted as distributions; therefore, at $t = 0$, they have the respective jumps of heights $g_{11}(0^+)$ and $g_{12}(0^+)$. The *"weighting functions"* $g_{ik}(t)$, when considered as distributions, assume the rôle of *"impulse responses"* (compare (18.9)).

Applying the result (12) to the *special case* for which f_1 and f_2, and consequently, also y_1 and y_2 are function-distributions, that is *locally integrable functions*, and comparing the result with that obtained in the realm of classical functions, (5), one recognizes two facts:

1. Upon replacing the concept of derivative by distribution-derivative, the solution (12), which agrees with the first part of solution (5), satisfies the system of equations irrespective of the restrictions that f_1 and f_2 be continuous. The

suggested interpretation of the functions as distributions and the derivatives as distribution-derivatives offers the opportunity to admit *non-continuous excitation functions* which are often encountered in physical applications. We merely require that the excitations are locally integrable.

2. The Eqs. (12) yield only those solutions of the system of differential equations which have *zero initial values*. In the discussion of one single differential equation towards the end of Chapter 18, we already presented the rational explanation of this characteristic. The initial values are not to be interpreted as limits from the right: $y_1(0^+), y_2(0^+)$, but instead as *limits from the left:* $y_1(0^-), y_2(0^-)$, an interpretation which agrees well with physical intuition, for they are the result of the influences of the past. Functions, when considered as distributions, are defined on the entire real axis; therefore, the past is always incorporated. Functions which belong to the space $\mathcal{D}'_0$ are, by definition, zero for $t < 0$; hence, at $t = 0$, the limits from the left of these (and their derivatives) equal zero. Clearly, other initial values are excluded.

We are faced with the facts: The replacement of functions by distributions from the space $\mathcal{D}'_0$, and the generalization of the derivative to the distribution-derivative is required if one wants to remove the restriction to continuous excitation functions. Yet this replacement precludes the consideration of functions with arbitrary initial values.

Fortunately, the interpretation of the initial values as limits from the left actually provides the possibility of completing the intended generalization to also include solutions with arbitrary initial values. In the frame of the classical theory we presumed that[1] y has a derivative for $t > 0$, and a limit $y(0^+)$. Within the new theory, we consider y as a distribution in $\mathcal{D}'_0$ which is defined to be zero for $t < 0$, and y' is replaced by Dy. According to App. No. 20 we have:

$$\begin{aligned} Dy &= y' + [y(0^+) - y(0^-)]\,\delta \\ &= y' + y(0^+)\,\delta \qquad\qquad \text{since} \quad y(0^-) = 0. \end{aligned}$$

If we want $y(0^-)$ to have, instead of zero, some arbitrary value, then it is not sufficient to replace y' by Dy; instead, y' must be replaced by $Dy - y(0^-)\,\delta$, in order to obtain: $Dy - y(0^-)\,\delta = y' + [y(0^+) - y(0^-)]\,\delta$. The distribution δ is zero (in the sense of App. No. 10) for both $t < 0$ and $t > 0$; hence, except at $t = 0$, nothing has been altered, only the behaviour of an arbitrary y, at $t = 0$, is properly described from the point of view of the theory of distributions.

The application of the $\mathfrak{L}$-transformation now does not yield $\mathfrak{L}\{Dy\} = s\,Y(s)$, instead one finds, by Eq. (13.1),

$$\mathfrak{L}\,\{Dy - y(0^-)\,\delta\} = s\,Y(s) - y(0^-).$$

When dealing with a *differential equation of higher order*, according to App. No. 20, we must not replace $y^{(k)}$ by $D^k y$, but instead by the completed expression:

$$D^k y - y(0^-)\,\delta^{(k-1)} - y'(0^-)\,\delta^{(k-2)} - \cdots - y^{(k-1)}(0^-)\,\delta$$

[1] It suffices to study one solution; we shall call it y.

or by the expression

$$D^k y - y_0\,\delta^{(k-1)} - y_0'\,\delta^{(k-2)} - \cdots - y_0^{(k-1)}\,\delta, \tag{15}$$

since in our present interpretation $y(0^-)$, $y'(0^-)$, ... represent the specified initial values which we commonly designate by y_0, y'_0, ...

Applying the $\mathfrak{L}$-transformation to (15), one obtains, by formula (13.2),

$$s^k\,Y(s) - y_0\,s^{k-1} - y_0'\,s^{k-2} - \cdots - y_0^{(k-1)}, \tag{16}$$

which is *identical to the expression obtained by the $\mathfrak{L}$-transformation in the classical sense.*

From the above argumentation it follows that to the special system (20.5) by the described completion one obtains formally the same image equations (20.11) containing the initial values, and the same solutions (5) as obtained in the space of functions. However, in the new interpretation, the *excitation functions* may be *arbitrary, locally integrable functions,* provided we interpret the solutions as distributions and, accordingly, form distribution-derivatives instead of derivatives.

We want to verify that the solutions (5), when considered as distributions in $\mathscr{D}'_0$, *satisfy the completed distribution-derivative equations,* and that the specified initial values y_1^0 and y_2^0, which have been introduced here as limits from the left at $t = 0$, are in fact also the *limits from the right* at $t = 0$. This implies that, for the normal case, the "future" state ($t > 0$) joins the "past" state ($t < 0$) continuously at $t = 0$. To begin with, the limits $y_1(0^+)$ and $y_2(0^+)$ are the same as for the classical case, for the convolution integrals involving the excitations tend, for $t \to 0$, towards zero, even for merely locally integrable f_1 and f_2, since the $g_{ik}(t)$ are bounded in a neighbourhood of $t = 0$. The terms which depend upon the initial values tend, exactly as before, towards y_1^0 and y_2^0 respectively, a fact which we want to retain for our subsequent considerations:

Using the abbreviations

$$a_{11}\,y_1^0 + a_{12}\,y_2^0 = R_1, \qquad a_{21}\,y_1^0 + a_{22}\,y_2^0 = R_2, \tag{17}$$

we can write

$$R_1\,g_{11}(0^+) + R_2\,g_{21}(0^+) = y_1^0, \qquad R_1\,g_{12}(0^+) + R_2\,g_{22}(0^+) = y_2^0. \tag{18}$$

The first distribution-derivative equation of the system (9) completed by δ-terms can be written, using $y_1(0^-) = y_1^0$ and $y_2(0^-) = y_2^0$, thus

$$a_{11}(Dy_1 - y_1^0\,\delta) + b_{11}\,y_1 + a_{12}(Dy_2 - y_2^0\,\delta) + b_{12}\,y_2 = f_1. \tag{19}$$

For the solutions (5), when considered as distributions in $\mathscr{D}'_0$, because of

$$D g_{ik}(t) = g'_{ik}(t) + g_{ik}(0^+)\,\delta,$$

with the abbreviations (17), we find the equations:

$$Dy_1 = D\{f_1 * g_{11} + f_2 * g_{21}\} + R_1(g'_{11}(t) + g_{11}(0^+)\,\delta) + R_2(g'_{21}(t) + g_{21}(0^+)\,\delta)$$
$$Dy_2 = D\{f_1 * g_{12} + f_2 * g_{22}\} + R_1(g'_{12}(t) + g_{12}(0^+)\,\delta) + R_2(g'_{22}(t) + g_{22}(0^+)\,\delta).$$

For the purpose of verification we must substitute these expressions, and the representation (5) of y_1 and y_2 into (19). In connection with formulae (13) and (14), we have shown that

$$a_{11} D\{f_1 * g_{11} + f_2 * g_{21}\} + b_{11}\{f_1 * g_{11} + f_2 * g_{21}\}$$
$$+ a_{12} D\{f_1 * g_{12} + f_2 * g_{22}\} + b_{12}\{f_1 * g_{12} + f_2 * g_{22}\} = f_1.$$

Therefore, we must only show that the following expression equals zero:

$$
\begin{aligned}
&a_{11}\{R_1(g'_{11}(t)+g_{11}(0+)\,\delta)+R_2(g'_{21}(t)+g_{21}(0+)\,\delta)-y_1^0\,\delta\}\\
&+b_{11}\{R_1\,g_{11}(t)+R_2\,g_{21}(t)\}\\
&+a_{12}\{R_1(g'_{12}(t)+g_{12}(0+)\,\delta)+R_2(g'_{22}(t)+g_{22}(0+)\,\delta)-y_2^0\,\delta\}\\
&+b_{12}\{R_1\,g_{12}(t)+R_2\,g_{22}(t)\}\\
&=R_1\{a_{11}\,g'_{11}(t)+b_{11}\,g_{11}(t)+a_{12}\,g'_{12}(t)+b_{12}\,g_{12}(t)\}\\
&+R_2\{a_{11}\,g'_{21}(t)+b_{11}\,g_{21}(t)+a_{12}\,g'_{22}(t)+b_{12}\,g_{22}(t)\}\\
&+a_{11}\{R_1\,g_{11}(0+)+R_2\,g_{21}(0+)-y_1^0\}\,\delta\\
&+a_{12}\{R_1\,g_{12}(0+)+R_2\,g_{22}(0+)-y_2^0\}\,\delta.
\end{aligned}
$$

Indeed, by (6), the first two lines of the last expression vanish, and, by (18), the last two also.

The second completed distribution-derivative equation of the system (9) is treated similarly.

Remark: One can trace the presented line of thought in the reverse order. In order to ascertain that the solutions found in the space of distributions in the case that they are functions assume arbitrary initial values, one must take care that the $\mathfrak{L}$-transform of the distribution-derivative $D^k y$ has the form (16). However, this implies that, earlier, $D^k y$ had to be replaced by the completed expression (15). It follows, by App. No. 20, that the $y, y', \ldots$ have limits from the left, at $t = 0$, which are not equal to zero, as customary for functions from $\mathcal{D}'_0$, but equal to $y_0, y'_0, \ldots$.

22. The Anomalous System with Arbitrary Initial Values, in the Space of Distributions

When considering a *normal system*, the interpretation, within the frame of the theory of distributions, of the initial values as *limits from the left* does not conflict with the interpretation of these initial values as *limits from the right*, as interpreted within the frame of the classical theory, since the limits from the right of the discovered solutions agree with the specified initial values which are understood as the limits from the left. This observation might lead to the impression that ultimately the distinction between limits from the left and limits from the right is irrelevant. An entirely different situation is encountered when dealing with *anomalous systems!* For these, the interpretation of the initial values as limits from the left is of fundamental importance. Indeed, this interpretation alone enables us to mathematically describe numerous physical processes which defy treatment within the frame of the classical theory. Here, the application of the theory of distributions is mandatory.

We now consider the model system (20.5) in the space of distributions, that is in the form (21.9), however here for the anomalous case, that is for

$$A = 0.$$

Initial values are not specified for the present for we are dealing with distributions.

Application of the $\mathfrak{L}$-transformation to (21.9) produces the image equations (21.10). The latter agree formally with (20.11), provided we set $y_1^0 = y_2^0 = 0$; hence the solutions of the latter, that is (20.14), for zero initial values are the solutions of the present problem:

$$\begin{aligned} Y_1(s) &= \quad F_1(s)\frac{a_{22}s + b_{22}}{\Delta(s)} - F_2(s)\frac{a_{12}s + b_{12}}{\Delta(s)} \\ Y_2(s) &= -F_1(s)\frac{a_{21}s + b_{21}}{\Delta(s)} + F_2(s)\frac{a_{11}s + b_{11}}{\Delta(s)}, \end{aligned} \tag{1}$$

whereby according to (20.13), using the notation of Chapter 20,

$$\Delta(s) = (C + D)\, s + E = H\, s + E. \tag{2}$$

First we shall consider the special case $H \neq 0$ treated in Chapter 20, then the respective cases $H = 0$, $E \neq 0$ and $H = 0$, $E = 0$.

1. $H \neq 0$; *that is, $\Delta(s)$ is a linear function*

For the construction of the originals of the functions (1), we can simply employ the expressions (20.17, 19), setting $y_1^0 = y_2^0 = 0$:

$$\begin{aligned} y_1 &= \quad \frac{a_{22}}{H} f_1 + \frac{K}{H} f_1 * e^{-Et/H} - \frac{a_{12}}{H} f_2 - \frac{L}{H} f_2 * e^{-Et/H} \\ y_2 &= -\frac{a_{21}}{H} f_1 - \frac{M}{H} f_1 * e^{-Et/H} + \frac{a_{11}}{H} f_2 + \frac{N}{H} f_2 * e^{-Et/H}, \end{aligned} \tag{3}$$

whereby now f_1 and f_2 and, consequently, also y_1 and y_2 are distributions in $\mathcal{D}'_0$.

If, in particular, $f_1 \equiv \delta$, $f_2 \equiv 0$, and $f_1 \equiv 0$, $f_2 \equiv \delta$, respectively, then one obtains the following pairs of "*impulse responses*":

$$y_1 = \quad \frac{a_{22}}{H}\delta + \frac{K}{H} e^{-Et/H}, \quad y_2 = -\frac{a_{21}}{H}\delta - \frac{M}{H} e^{-Et/H},$$

and

$$y_1 = -\frac{a_{12}}{H}\delta - \frac{L}{H} e^{-Et/H}, \quad y_2 = \quad \frac{a_{11}}{H}\delta + \frac{N}{H} e^{-Et/H}.$$

Comparing these with the impulse responses of the normal systems (see p. 128), we discover here not only function-distributions but also impulses.

When applying formulae (3) to the special case that f_1 and f_2 and, consequently, also y_1 and y_2 designate function-distributions, that is *locally integrable functions* with limits as $t \to +0$, we construct solutions which, formally, agree with (20.17, 19), for zero initial values. However, for the present solutions we need not require f_1 and f_2 to be differentiable, provided we replace differentiation by distribution-differentiation. That is, for instance, jumps of the excitation functions are now permissible; such jumps will then recur in the solutions, since the excitations f_1 and f_2 themselves appear in these solutions. (In engineering one speaks of solutions or outputs capable of jumps).

When the compatibility condition (20.10) is not satisfied, then the solutions have, in general, not zero initial values from the right. However, distributions in $\mathcal{D}'_0$ are zero for $t < 0$; hence, the *initial values from the left* are, necessarily, zero.

Interpreting the specified initial values as limits from the left as done in the normal case, the functions (3) are the solutions which correspond to zero initial values. This interpretation enables us to speak of solutions also in the case that the compatibility condition (20.10) is not satisfied and the solutions cannot, for $t \to 0$ from the right, tend towards zero. In this case, the limits of the solutions from the right and from the left are not equal at $t = 0$. This situation can be reasonable explained from the physical point of view: The limits $y_1(0^-)$ and $y_2(0^-)$ which are to be identified with the *given state* of the physical system at $t = 0$ are the result of the (unknown) excitations which acted upon the system before $t = 0$; the limits $y_1(0^+)$ and $y_2(0^+)$, however, result from the excitations f_1 und f_2 which are effective *after* $t = 0$ and which need not be connected with the excitations which were effective before $t = 0$. Thus, when also considering the past, one is able to provide meaningful mathematical descriptions of certain physical processes which pose an unsolvable problem from the classical point of view.

To generate solutions having *arbitrary initial values*, we account for the behaviour of these solutions at $t = 0$ in the very manner which we employed successfully with the normal system: We complete the distribution-derivative equations by δ-terms; that is, we start with the following equations (compare p. 129):

$$\text{(4)} \quad \begin{aligned} a_{11}(Dy_1 - y_1^0\,\delta) + b_{11}\,y_1 + a_{12}(Dy_2 - y_2^0\,\delta) + b_{12}\,y_2 &= f_1 \\ a_{21}(Dy_1 - y_1^0\,\delta) + b_{21}\,y_1 + a_{22}(Dy_2 - y_2^0\,\delta) + b_{22}\,y_2 &= f_2 \end{aligned}$$

rather than with the equations (21.9). Applying the $\mathfrak{L}$-transformation to (4), one obtains the same image equations (20.11) and, therefore, also the same solutions (20.17, 19) that were produced in the classical case,[1] whereby the initial values are now understood as limits from the left which are *not subject to compatibility conditions*. Obviously, it follows that, in general, the limits from the right and the limits from the left (the specified initial values y_1^0 and y_2^0) need not agree. In fact, one finds:

$$\text{(5)} \quad \begin{aligned} y_1(0^+) &= \frac{a_{22}}{H} f_1(0^+) - \frac{a_{12}}{H} f_2(0^+) + y_1^0 \frac{C}{H} + y_2^0 \frac{G}{H} \\ y_2(0^+) &= -\frac{a_{21}}{H} f_1(0^+) + \frac{a_{11}}{H} f_2(0^+) - y_1^0 \frac{B}{H} + y_2^0 \frac{D}{H}\,. \end{aligned}$$

These values depend upon both the specified initial values and the behaviour of the excitation functions,

Theorem 22.1. *Suppose that* 1. (20.5) *designates an anomalous system, that is* $A = 0$; 2. $C + D = H \neq 0$; 3. *the excitations are locally integrable functions which have limits as* $t \to +0$. *Then we conclude that the functions* (20.17, 19) *satisfy the completed distribution-derivative equations* (4). *Arbitrary values may be specified for the initial values* y_1^0 *and* y_2^0, *provided we interpret these as limits from the left. In general, the limits from the right* (5) *do not agree with the specified initial values.*

---

[1] Obviously, the verification follows the steps of pp. 122, 123.

Consequently, a physical system described by (20.5) *which, at* $t = 0$, *is specified by* y_1^0 *and* y_2^0, *responds with jumps at the "inception" of the excitations* f_1 *and* f_2.

The solution is formally obtained by the application of the $\mathfrak{L}$-transformation employing the classical rule regarding differentiation, using the specified initial values instead of the limits from the right.

2. $H = 0$, $E \neq 0$; *that is* $\Delta(s)$ *is a constant* $E \neq 0$

For this special case one derives the solutions of the image equations of the distribution-derivative equations from (20.14) by setting $\Delta(s) = E$ and $y_1^0 = y_2^0 = 0$; thus, we find:

$$
(6) \qquad \begin{aligned}
Y_1(s) &= \frac{1}{E}\{\ \ F_1(s)\,(a_{22}\,s + b_{22}) - F_2(s)\,(a_{12}\,s + b_{12})\}\\
Y_2(s) &= \frac{1}{E}\{-F_1(s)\,(a_{21}\,s + b_{21}) + F_2(s)\,(a_{11}\,s + b_{11})\}.
\end{aligned}
$$

Corresponding to (6) one finds in the original space:

$$
(7) \qquad \begin{aligned}
y_1 &= \frac{1}{E}\{\ \ a_{22}\,D\,f_1 + b_{22}\,f_1 - a_{12}\,D\,f_2 - b_{12}\,f_2\}\\
y_2 &= \frac{1}{E}\{-a_{21}\,D\,f_1 - b_{21}\,f_1 + a_{11}\,D\,f_2 + b_{11}\,f_2\}.
\end{aligned}
$$

These solutions differ from the solutions of case 1 ($H \neq 0$) insofar that here the *convolutions* involving the excitations *are absent*, and also that *besides the excitations at least one distribution-derivative of the excitations* occurs. This shows that even for the special case that f_1 and f_2 represent functions, the solutions may include true distributions. For instance, the function $f_1 = u(t - t_0)$ has the distribution-derivative $D f_1 = \delta(t - t_0)$. For applications it is important that physical systems with the characteristics: $A = 0$, $H = 0$ may respond to *jumps at the input* with *impulses at the output.*

As under 1, we want to specify *arbitrary* initial values (limits from the left) and, for this purpose, we start with the completed distribution-derivative equations (4); for this situation we must, according to (20.14), add to the above shown $\mathfrak{L}$-transforms (6) the respective terms:

$$
(8) \qquad \frac{1}{E}(C\,y_1^0 + G\,y_2^0) \qquad \text{and} \qquad \frac{1}{E}(-B\,y_1^0 + D\,y_2^0),
$$

and, correspondingly, we add to the solutions (7) the respective terms:

$$
(9) \qquad \frac{1}{E}(C\,y_1^0 + G\,y_2^0)\,\delta \qquad \text{and} \qquad \frac{1}{E}(-B\,y_1^0 + D\,y_2^0)\,\delta.
$$

The solutions include impulses which result from the mismatch at $t = 0$ of the respective limits from the left and from the right.

It is not difficult to verify that the solutions (7) completed by the respective terms (9) satisfy the relation (20.9), thus establishing them as solutions.

As before we want to investigate how the solutions satisfy the completed distribution-derivative equations (4), although the solutions derived here have an entirely different structure than the ones derived for $H \neq 0$. We form the left hand side of the first equation of (4) using the solutions (7) plus (9). Upon collecting terms with identical distributions, one obtains, when disregarding the factor $1/E$, the coefficients of the several distributions:

$$\begin{array}{ll}
D^2 f_1: & a_{11}a_{22} - a_{12}a_{21} = A = 0, \\
D f_1: & a_{11}b_{22} + b_{11}a_{22} - a_{12}b_{21} - b_{12}a_{21} = C + D = H = 0, \\
f_1: & b_{11}b_{22} - b_{12}b_{21} = E, \\
D^2 f_2: & -a_{11}a_{12} + a_{12}a_{11} = 0, \\
D f_2: & -a_{11}b_{12} - b_{11}a_{12} + a_{12}b_{11} + b_{12}a_{11} = 0, \\
f_2: & -b_{11}b_{12} + b_{12}b_{11} = 0, \\
\delta': & (a_{11}C - a_{12}B)\, y_1^0 + (a_{11}G + a_{12}D)\, y_2^0 = -(a_{11}D + a_{12}B)\, y_1^0 + (a_{11}G - a_{12}C)\, y_2^0 \\
 & = b_{11}A\, y_1^0 - b_{12}A\, y_2^0 = 0 \quad \text{by (20.22, 25).}
\end{array}$$

Summation of these terms and subsequent division by E yields f_1. The impulse δ has the factor:

$$\frac{1}{E}\{(b_{11}C - b_{12}B)\, y_1^0 + (b_{11}G + b_{12}D)\, y_2^0\} - a_{11}y_1^0 - a_{12}y_2^0.$$

Eq. (20.20) implies that $b_{11}C - b_{12}D = a_{11}E$; explicit representation of the determinants yields:

(10) $$b_{11}G + b_{12}D = a_{12}E.$$

It follows that the factor of δ is zero. The second equation of (4) can be verified in a similar manner.

Theorem 22.2. *Suppose that* 1. *the system* (20.5) *is an anomalous system, that is* $A = 0$; 2. $C + D = H = 0$; 3. $E \neq 0$; 4. *the excitation functions are locally integrable. Then we conclude that the distributions given by* (7) *plus* (9) *satisfy the completed distribution-derivative equations* (4). *The initial values* y_1^0 *and* y_2^0 *must be interpreted as limits from the left:* $y_1(0-)$ *and* $y_2(0-)$. *When existing, in general, the limits from the right differ from those from the left.*

3. $H = 0$, $E = 0$; *that is,* $\Delta(s) \equiv 0$

For brevity we begin this investigation with the completed distribution-derivative equations (4). The corresponding image equations agree with (20.11). The determinant of this algebraic system is zero; hence we must distinguish between homogeneous case and inhomogeneous case.

I. *Homogeneous System of Image Equations*

This situation is encountered if and only if

$$F_1(s) = -(a_{11}y_1^0 + a_{12}y_2^0), \quad F_2(s) = -(a_{21}y_1^0 + a_{22}y_2^0)$$

When f_1 and f_2 designate distributions,[2] then it follows that:

$$f_1 = -(a_{11}y_1^0 + a_{12}y_2^0)\,\delta, \qquad f_2 = -(a_{21}y_1^0 + a_{22}y_2^0)\,\delta. \tag{11}$$

For a homogeneous system of two linear equations with zero determinant of the matrix of coefficients, one may arbitrarily select one of the two unknowns; the other unknown is then determined. Instead of using this principle with the image equations and then returning the result to the original space, one may more quickly use it with the relation (20.9) of the originals. Because of $f_1(0^+) = f_2(0^+) = 0$ (δ is equal to zero for $t > 0$), we find:

$$B\,y_1 + C\,y_2 = 0.$$

In connection with (20.10) we argued that B and C could not both be zero. Suppose that $B \neq 0$; then y_2 is any distribution of $\mathscr{D}'_0$ and

$$y_1 = -\frac{C}{B}\,y_2. \tag{12}$$

For the purpose of verification, we form the left hand side of the first equation of (4):

$$a_{11}\left(-\frac{C}{B}\,D y_2 - y_1^0\,\delta\right) - b_{11}\,\frac{C}{B}\,y_2 + a_{12}(D y_2 - y_2^0\delta) + b_{12}y_2$$
$$= \frac{1}{B}\,\{D y_2(a_{12}B - a_{11}C) + y_2(b_{12}B - b_{11}C)\} - (a_{11}y_1^0 + a_{12}y_2^0)\,\delta.$$

Using $C = -D$ in the first round bracket yields, by (20.22) and (11),

$$\frac{1}{B}\,\{D y_2 b_{11}A - y_2 a_{11}E\} + f_1 = f_1,$$

since $A = E = 0$.

The left hand side of the second equation of (4) yields:

$$a_{21}\left(-\frac{C}{B}\,D y_2 - y_1^0\,\delta\right) - b_{21}\,\frac{C}{B}\,y_2 + a_{22}(D y_2 - y_2^0\,\delta) + b_{22}y_2$$
$$= \frac{1}{B}\,\{D y_2(a_{22}B - a_{21}C) + y_2(b_{22}B - b_{21}C)\} - (a_{21}y_1^0 + a_{22}y_2^0)\,\delta.$$

Explicit representation of the determinants shows that

$$a_{22}B - a_{21}C = a_{22}B - a_{21}D = b_{21}A,$$
$$b_{22}B - b_{21}C = -a_{21}E;$$

hence, with (11),

$$\frac{1}{B}\,\{D y_2 b_{21}A - y_2 a_{21}E\} + f_2 = f_2.$$

Theorem 22.3. *For the special situation in which* $A = C + D = E = 0$ *and in*

[2] For the case that f_1 and f_2 represent functions, it follows that $F_1 \equiv F_2 \equiv 0$, hence $f_1 \equiv f_2 \equiv 0$, and also

$$a_{11}y_1^0 + a_{12}y_2^0 = 0, \qquad a_{21}y_1^0 + a_{22}y_2^0 = 0.$$

Since $A = 0$, y_2^0 is arbitrary and $y_1^0 = -\dfrac{a_{12}}{a_{11}}\,y_2^0$ (from the beginning, we supposed that $a_{11} \neq 0$).

which the excitations f_1 and f_2 are given by (11), *we have infinitely many solutions of the completed distribution-derivative equations* (4): y_2 *arbitrary and* $y_1 = -(C/B)y_2$, *when* $B \neq 0$; y_1 *arbitrary and* $y_2 = -(B/C)y_1$, *for the case that* $C \neq 0$.

Remark: The case presented here may actually be encountered. If, for instance, the matrix of (4) has the form

$$\begin{Vmatrix} 1 & 2 & 1 & 2 \\ 1 & 3 & 1 & 3 \end{Vmatrix},$$

then $A = 0$, $C = 1$, $D = -1$, $E = 0$.

II. *The Inhomogeneous System of Image Equations*

When the system (20.11) is inhomogeneous, and when $\Delta(s) \equiv 0$, then we have, in general, incompatible equations. In order to have solutions, the right hand sides of (20.14) must be zero. This condition implies that:

$$\begin{aligned} F_1(s)(a_{22}s + b_{22}) - F_2(s)(a_{12}s + b_{12}) &= -Cy_1^0 - Gy_2^0 \\ -F_1(s)(a_{21}s + b_{21}) + F_2(s)(a_{11}s + b_{11}) &= By_1^0 - Dy_2^0. \end{aligned} \tag{13}$$

Upon inverse transformation into the original space we find there the compatibility conditions:

$$\begin{aligned} a_{22}Df_1 + b_{22}f_1 - a_{12}Df_2 - b_{12}f_2 &= -(Cy_1^0 + Gy_2^0)\,\delta \\ -a_{21}Df_1 - b_{21}f_1 + a_{11}Df_2 + b_{11}f_2 &= (By_1^0 - Dy_2^0)\,\delta. \end{aligned} \tag{14}$$

These equations could have been derived immediately from the system (4). Because of $E = 0$, one can eliminate the quantities y_1 and y_2 from the system (4); multiplication of the first equation by b_{21} and of the second equation by b_{11}, and subsequent subtraction yields:

$$(a_{11}b_{21} - a_{21}b_{11})(Dy_1 - y_1^0\delta) + (a_{12}b_{21} - a_{22}b_{11})(Dy_2 - y_2^0\delta) = b_{21}f_1 - b_{11}f_2$$

or

$$B(Dy_1) - D(Dy_2) = b_{21}f_1 - b_{11}f_2 + (By_1^0 - Dy_2^0)\,\delta.$$

Using this equation together with equation (20.9)[3] which may be written

$$By_1 - Dy_2 = -a_{21}f_1 + a_{11}f_2,$$

since $C = -D$, one recognizes that, necessarily,

$$-a_{21}Df_1 + a_{11}Df_2 = b_{21}f_1 - b_{11}f_2 + (By_1^0 - Dy_2^0)\,\delta.$$

This agrees with the second equation of (14); the first equation of (14) can be obtained in a similar manner.

The compatibility conditions (14) indicate that the distribution-derivative equations (4) have solutions provided that the excitations satisfy a similar pair

[3] Initially (20.9) was obtained from (20.5) by the elimination of y'_1 and y'_2; it also follows from (4) by the elimination of $Dy_1 - y_1^0\delta$ and $Dy_2 - y_2^0\delta$.

of distribution-derivative equations, the right hand sides of which are determined by the initial values.

For the determinants of the coefficient matrix of (14) employing the symbols used earlier, but stroked, one finds:

$$A' = \begin{vmatrix} a_{22} & -a_{12} \\ -a_{21} & a_{11} \end{vmatrix} = A, \quad E' = \begin{vmatrix} b_{22} & -b_{12} \\ -b_{21} & b_{11} \end{vmatrix} = E,$$

$$C' = \begin{vmatrix} a_{22} & -b_{12} \\ -a_{21} & b_{11} \end{vmatrix}, \quad D' = \begin{vmatrix} b_{22} & -a_{12} \\ -b_{21} & a_{11} \end{vmatrix}$$

The relation

$$C' + D' = C + D$$

can easily be verified. We also have $A' = C' + D' = E' = 0$, since $A = C + D = E = 0$. Therefore, the system of equations for f_1 and f_2 is of the same type 3 as the system of equations for y_1 and y_2.

When the conditions (13) are satisfied, then (20.11) has infinitely many solutions; one may arbitrarily select one of the unknowns Y_1 or Y_2; the other one is then determined through one of the equations (20.11). Rather than performing these steps and then inversely transforming the result to the original space, we simply employ (20.9). One finds, for any arbitrarily selected y_2,

$$y_1 = -\frac{C}{B} y_2 - \frac{a_{21}}{B} f_1 + \frac{a_{11}}{B} f_2, \tag{15}$$

provided $B \neq 0$.[4]

Theorem 22.4. *For the special situation in which* $A = C + D = E = 0$, *and for arbitrary excitations and arbitrary initial values, the Eqs.* (4) *are, in general, incompatible and, consequently, fail to have solutions. If the compatibility conditions* (14) *are satisfied, then the equations can be solved, yielding an infinite manyfold of solutions.*

For physical problems which naturally require a unique solution, the situation 3 is not encountered.

---

[4] C cannot be zero. Otherwise, since $C = -A$, we would have $A = C = D = 0$, and the coefficients of the second line of (20.5) would be proportional to the ones of the first line; compare the remark following (20.10). Therefore, to a different y_2 there corresponds a different y_1.

23. The Behaviour of the Laplace Transform near Infinity

In Chapter 6 we demonstrated that every $\mathfrak{L}$-transform $F(s)$ represents an analytic function in a region of infinite extent, at least in a half-plane. In the sequel we shall frequently have to evaluate integrals involving $F(s)$ along curves that tend towards infinity. Thus we need to know the behaviour of $F(s)$ near infinity.

In the subsequent Chapters, up to and including Chapter 28, we shall restrict our discussion to $\mathfrak{L}$-transforms of functions, barring those of distributions.

A $\mathfrak{L}$-transform may be holomorphic at $s = \infty$, like $\mathfrak{L}\{u(t)\} = 1/s$. In general, however, $s = \infty$ is a singular point of $F(s)$ like, for instance, of $\mathfrak{L}\{u(t-a)\} = e^{-as}/s\,(a > 0)$. When approaching a singular point of a function along rays, one might observe that the behaviour of the function depends upon the chosen direction of approach: The limit may fail to exist; or, if limits do exist for different directions of approach, they may be unequal. The $\mathfrak{L}$-transform $F(s)$ is defined in a right half-plane, thus we need consider only those rays starting at some specified point s_0 of the half-plane of convergence which aim towards infinity and which form with the positive real axis an angle which is bounded by $-\pi/2$ and $+\pi/2$ respectively: $|\operatorname{arc}(s-s_0)| \leqq \pi/2$. We shall show that along all rays that are not vertical, i.e. for $|\operatorname{arc}(s-s_0)| < \pi/2$, $F(s)$ tends towards zero; indeed, $F(s)$ converges uniformly towards zero in every fixed angular region: $|\operatorname{arc}(s-s_0)| \leqq \psi < \pi/2$. This is to be interpreted as follows: For every specified $\varepsilon > 0$, however small, one can find an $R > 0$ so that $|F(s)| < \varepsilon$ for all s with $|\operatorname{arc}(s-s_0)| \leqq \psi < \pi/2$, and $|s-s_0| > R$. We shall express this compactly thus: $F(s)$ *tends towards zero, when, in the angular region* $|\operatorname{arc}(s-s_0)| \leqq \psi < \pi/2$, s *tends two-dimensionally towards* ∞.[1]

A similar property for vertical lines can be verified only in the half-plane of absolute convergence. Along vertical lines in the half-plane of convergence, we shall only produce an estimate for $F(s)$.

First we derive a theorem which is needed in the subsequent development.

Theorem 23.1. *If* $\mathfrak{L}\{f\} = F(s)$ *converges at the point* s_0, *then the integral* $\mathfrak{L}\{F\}$ *converges uniformly in every angular region* $|\operatorname{arc}(s-s_0)| \leqq \psi < \pi/2$.

Proof: We have, by (3.1),

$$\int_0^\omega e^{-st} f(t)\,dt = e^{-(s-s_0)\omega}\varphi(\omega) + (s-s_0)\int_0^\omega e^{-(s-s_0)t}\varphi(t)\,dt,$$

with

$$\varphi(t) = \int_0^t e^{-s_0\tau} f(\tau)\,d\tau.$$

[1] When the independent variable varies in a plane, one needs, besides the concept of one-dimensional convergence (which can kinematically be visualized as a "walk along some path"), also the concept of *two-dimensional convergence* (for which this kinematical visualization fails). The above presented definition is for the case that the point ∞ is the point towards which s two-dimensionally converges. The analogous definition for a finite point a of convergence is as follows: "$F(s)$ tends towards A, when s tends two-dimensionally towards a" implies: For every given $\varepsilon > 0$, one can find an r so that $|F(s) - A| < \varepsilon$ for all s with $|a - s| < r$. The corresponding definition for the case that s tends towards a in some specified angular region follows easily by appropriate amendment.

One easily verifies the equation

$$0 = F(s_0) - e^{-(s-s_0)\omega} F(s_0) - (s - s_0) \int_0^\omega e^{-(s-s_0)t} F(s_0)\, dt;$$

adding it to the prior one, yields:

$$\int_0^\omega e^{-st} f(t)\, dt = F(s_0) + e^{-(s-s_0)\omega} [\varphi(\omega) - F(s_0)] + (s - s_0) \int_0^\omega e^{-(s-s_0)t} [\varphi(t) - F(s_0)]\, dt.$$

Consider two values ω_1 and ω_2, with $0 < \omega_1 < \omega_2$, and find, forming the difference,

$$\int_{\omega_1}^{\omega_2} e^{-st} f(t)\, dt = e^{-(s-s_0)\omega_2} [\varphi(\omega_2) - F(s_0)] - e^{-(s-s_0)\omega_1} [\varphi(\omega_1) - F(s_0)]$$

$$+ (s - s_0) \int_{\omega_1}^{\omega_2} e^{-(s-s_0)t} [\varphi(t) - F(s_0)]\, dt.$$

We use the fact that $\lim_{t\to\infty} \varphi(t) = F(s_0)$; consequently, for every $\varepsilon > 0$ there exists an Ω such that

$$|\varphi(t) - F(s_0)| < \varepsilon \quad \text{for } t > \Omega.$$

Selecting $\Omega < \omega_1 < \omega_2$, and $\Re s > \Re s_0$, we find

$$\left| \int_{\omega_1}^{\omega_2} e^{-st} f(t)\, dt \right| \leqq 2\varepsilon + |s - s_0|\, \varepsilon \int_{\omega_1}^{\omega_2} e^{-\Re(s-s_0)t}\, dt \leqq 2\varepsilon + |s - s_0|\, \varepsilon \int_0^\infty e^{-\Re(s-s_0)t}\, dt$$

$$= \varepsilon \left(2 + \frac{|s - s_0|}{\Re(s - s_0)}\right),$$

since $|e^{-(s-s_0)\omega}| < 1$. In the angular region $|\operatorname{arc}(s - s_0)| \leqq \psi < \pi/2$, excepting the point s_0, we have:

$$\frac{\Re(s - s_0)}{|s - s_0|} = \cos[\operatorname{arc}(s - s_0)] \geqq \cos\psi,$$

hence

$$\left| \int_{\omega_1}^{\omega_2} e^{-st} f(t)\, dt \right| \leqq \varepsilon \left(2 + \frac{1}{\cos\psi}\right);$$

that is, it is arbitrarily small irrespective of the value of s. That means: $\mathfrak{L}\{f\}$ converges uniformly in the angular region, excepting the point s_0. Obviously, the conclusion is not affected, when s_0 is included with the angular region.

Now we have the tool needed in the verification of

Theorem 23.2. *If* $\mathfrak{L}\{f\} = F(s)$ *converges at* s_0, *then* $F(s)$ *tends towards zero, when, in the angular region* $|\operatorname{arc}(s - s_0)| \leqq \psi < \pi/2$, s *tends two-dimensionally towards* ∞.

Proof: We partition $F(s)$ in the following manner:

$$F(s) = \int_0^{T_1} + \int_{T_1}^{T_2} + \int_{T_2}^{\infty} e^{-st} f(t)\, dt,$$

and we choose, for a given $\varepsilon > 0$, firstly a sufficiently small number T_1, so that

$$\left| \int_0^{T_1} e^{-st} f(t)\, dt \right| \leqq \int_0^{T_1} |f(t)|\, dt < \frac{\varepsilon}{3} \quad \text{for } \Re s \geqq 0,$$

and secondly a sufficiently large number T_2 so that, by Theorem 23.1,

$$\left| \int_{T_2}^{\infty} e^{-st} f(t)\, dt \right| < \frac{\varepsilon}{3} \quad \text{for all } s \text{ in the angular region;}$$

having thus fixed numbers T_1 and T_2, we select a suffiently large $x_0 > 0$ so that

$$\left| \int_{T_1}^{T_2} e^{-st} f(t)\, dt \right| \leqq e^{-x_0 T_1} \int_{T_1}^{T_2} |f(t)|\, dt < \frac{\varepsilon}{3} \quad \text{for } \Re s \geqq x_0.$$

Consequently,

$$|F(s)| < \varepsilon \quad \text{for all } s \text{ in the angular region, with } \Re s \geqq x_0.$$

This is the conclusion of Theorem 23.2, since the manner of delineation of a neighbourhood of ∞ in this angular region is irrelevant: either outside some specified circle, or to the right of some vertical line; for we might consider the points with $\Re s \geqq x_0$,which fall outside a specified circle.

Supplement: *The above Theorem* 23.2 *is valid for every arbitrary point* s_0.

This conclusion is based upon the following geometric observation: Any angular region with arbitrary vertex s_0 is, beyond a certain abscissa, completely contained in an angular region which has a point of convergence as vertex and a slightly larger angle of opening.

Theorem 23.2 is frequently used to demonstrate that some given function cannot be a $\mathfrak{L}$-transform of a function. We discovered in Chapter 2 that powers s^{α} with negative exponent α represent $\mathfrak{L}$-transforms. Theorem 23.2 shows clearly that powers s^α cannot be $\mathfrak{L}$-transforms of functions if the exponent α is zero or positive.[2] Also the functions

$$\exp(-s^\alpha) \quad \text{with} \quad \alpha \geqq 1$$

cannot be $\mathfrak{L}$-transforms, although they do tend towards zero when the real-valued s tends towards ∞. For $\alpha = 1$, we have established this conclusion on p. 25. Let $\alpha > 1$; define $s = re^{i\varphi}$, and find:

$$s^\alpha = r^\alpha e^{i\alpha\varphi}, \quad \Re s^\alpha = r^\alpha \cos\alpha\varphi,$$

hence

$$|\exp(-s^\alpha)| = \exp(-r^\alpha \cos\alpha\varphi).$$

Now, if we select a fixed φ, such that

$$\frac{1}{\alpha}\frac{\pi}{2} < \varphi < \operatorname{Min}\left(\frac{\pi}{2}, \frac{\pi}{\alpha}\right),$$

then

$$\frac{\pi}{2} < \alpha\varphi < \operatorname{Min}\left(\alpha\frac{\pi}{2}, \pi\right), \text{ hence } \cos\alpha\varphi < 0.$$

Thus, along a ray from the origin which forms the angle $\varphi < \pi/2$ with the real axis, the function $\exp(-s^\alpha)$ does not tend towards 0, in fact, it tends towards ∞.

For the investigation of the behaviour of $F(s) = \mathfrak{L}\{f\}$ along a vertical line, we need the Riemann-Lebesgue Lemma which is well known in the theory of Fourier series. The Riemann-Lebesgue Lemma concludes that the Fourier coefficients

$$a_n = \frac{1}{\pi}\int_0^{2\pi} f(t) \cos n\,t\,dt, \quad b_n = \frac{1}{\pi}\int_0^{2\pi} f(t) \sin n\,t\,dt$$

tend towards zero, for increasing n, provided $f(t)$ is an *absolutely* integrable function. More compactly, we can restate the conclusion thus:

$$\int_0^{2\pi} e^{-int} f(t)\,dt \to 0 \quad \text{as } n \to \infty.$$

For our purpose we need to extend this Lemma to an arbitrary, finite interval $(0, T)$, and to a continuously growing variable y instead of the discretely growing parameter n.

[2] This is true for functions. Powers with non-negative exponents are, by (13.11), the $\mathfrak{L}$-transforms of certain distributions.

Theorem 23.3 (Riemann-Lebesgue Lemma). *If $f(t)$ is absolutely integrable in the interval* $(0, T)$, *then*

$$\int_0^T e^{-iyt} f(t)\, dt$$

tends towards zero, when $y \to \pm\infty$.

Proof: We may presume $y > 0$, and $f(t)$ to be real-valued. One can find an n such that

$$T = n\frac{\pi}{y} + \delta, \quad 0 \leqq \delta < \frac{\pi}{y},$$

that is, π/y is n times contained in T. We partition the integral in the following manner:

$$\int_0^T e^{-iyt} f(t)\, dt = \sum_{k=0}^{n-1} \int_{k\pi/y}^{(k+1)\pi/y} e^{-iyt} f(t)\, dt + \int_{n\pi/y}^{T} e^{-iyt} f(t)\, dt.$$

In the terms with odd $k = 1, 3, 5, \cdots$, we substitute $t = u + (\pi/y)$:

$$\int_{k\pi/y}^{(k+1)\pi/y} e^{-iyt} f(t)\, dt = \int_{(k-1)\pi/y}^{k\pi/y} e^{-iyu-i\pi} f\left(u + \frac{\pi}{y}\right) du = -\int_{(k-1)\pi/y}^{k\pi/y} e^{-iyt} f\left(t + \frac{\pi}{y}\right) dt,$$

and we join these with the corresponding terms with even $k = 0, 2, 4, \cdots$. Setting

$$2m = \begin{cases} n & \text{for even } n \\ n-1 & \text{for odd } n, \end{cases}$$

we find (compare Fig. 10):

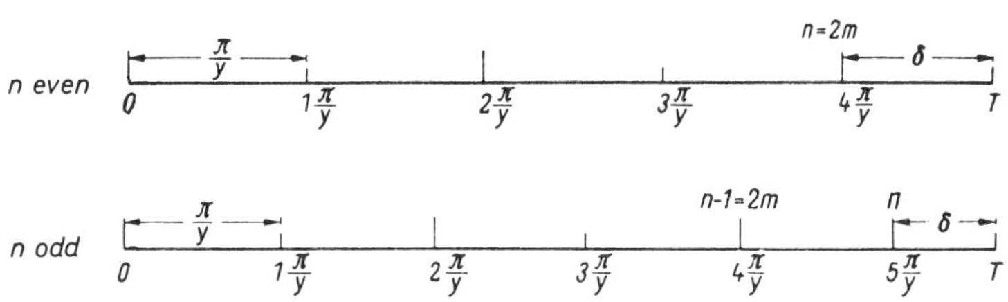

Figure 10

(1) $$\sum_{l=0}^{m-1} \int_{2l\pi/y}^{(2l+1)\pi/y} e^{-iyt}\left[f(t) - f\left(t + \frac{\pi}{y}\right)\right] dt + \int_{2m\pi/y}^{n\pi/y} e^{-iyt} f(t)\, dt + \int_{n\pi/y}^{T} e^{-iyt} f(t)\, dt.$$

This provides the estimation:

$$\left|\int_0^T e^{-iyt} f(t)\,dt\right| \leqq \int_0^{(2m-1)\pi/y} \left| f(t) - f\left(t+\frac{\pi}{y}\right)\right| dt + \int_{2m\pi/y}^{n\pi/y} |f(t)|\,dt + \int_{n\pi/y}^{T} |f(t)|\,dt.$$

Consider the right hand side of this equation; the second integral has an interval of integration the length of which is either 0 or else π/y; the third integral has one of length $T - (n\pi/y) = \delta < \pi/y$. It follows that either integral tends towards 0, when $y \to \infty$. We replace the upper limit of the first integral by T, thus increasing the value of the integral, defining $f(t) = 0$, for $t > T$, so that $f(t + \pi/y)$ is defined in the entire interval:

$$\int_0^T \left| f(t) - f\left(t+\frac{\pi}{y}\right)\right| dt.$$

This integral tends towards 0, when $y \to \infty$, that is when $\pi/y \to 0$, by the Theorem cited in the footnote on p. 48. This completes the proof.

The above presented form of the lemma still is not suitable for our purposes; we extend it by the following two Theorems.

Theorem 23.4. *If $f(t)$ is absolutely integrable in the interval $(0, T)$, then the integral with variable upper limit*

$$\int_0^t e^{-iy\tau} f(\tau)\,d\tau$$

tends, uniformly in $0 \leqq t \leqq T$, towards zero, when $y \to \pm\infty$.

Proof: In the previous proof, we replace T by t. Hence, in (1), both numbers n and m depend upon t. For the first term, we use the integral

$$\int_0^T \left| f(\tau) - f\left(\tau+\frac{\pi}{y}\right)\right| d\tau \qquad (f(\tau) = 0 \text{ for } \tau > T)$$

as a bound; this integral converges to 0, when $y \to \infty$, independently of t. The second and third term, taken together, are, in absolute value,

$$\leqq \int_{t_0}^{t} |f(\tau)|\,d\tau,$$

where $t_0 = 2m\pi/y$ is a number which differs from t by, at most, $2\pi/y$.

Now, we use the fact that, in the closed interval $0 \leqq t \leqq T$,

$$\int_0^t |f(\tau)|\,d\tau$$

represents a continuous, and therefore uniformly continuous, function of t. Thus, for every $\varepsilon > 0$ we can find an η so that for every pair of values t_0 and t, with $0 < t - t_0 < \eta$,

$$\int_{t_0}^{t} | f(\tau) | \, d\tau < \varepsilon.$$

Hence, for all values of y with $2\pi/y < \eta$, the second and the third term of (1) are together, in absolute value, smaller than ε. It follows that these terms also tend, uniformly in t, towards 0.

Theorem 23.5. *If $f(t)$ is absolutely integrable in the interval $(0, T)$, then*

$$\int_0^T e^{-iyt} \, e^{-xt} \, f(t) \, dt$$

tends towards zero, when $y \to \pm \infty$, uniformly for $x \geqq x_0$, where x_0 is an arbitrary but fixed number.

Proof: Integration by parts yields:

$$\int_0^T e^{-xt} \left[e^{-iyt} f(\tau) \right] dt = e^{-xT} \int_0^T e^{-iy\tau} f(\tau) \, d\tau + x \int_0^T e^{-xt} \int_0^t e^{-iy\tau} f(\tau) \, d\tau \, ,$$

hence

$$\left| \int_0^T e^{-xt} \, e^{-iyt} f(t) \, dt \right| \leqq e^{-x_0 T} \left| \int_0^T e^{-iy\tau} f(\tau) \, d\tau \right|$$

$$+ \, | x | \cdot \operatorname*{Max}_{0 \leqq t \leqq T} \left| \int_0^t e^{-iy\tau} f(\tau) \, d\tau \right| \cdot \int_0^T e^{-xt} \, dt.$$

The first term of the right hand side tends, by Theorem 23.3, towards zero, when $y \to \pm \infty$, independently of x. For the second term, we have

$$| x | \int_0^T e^{-xt} \, dt = | 1 - e^{-xT} | \leqq 1 + e^{-x_0 T};$$

it follows, by Theorem 23.4, that this term also tends, uniformly in x, towards zero.

Based upon these results, one easily obtains a Theorem concerning the behaviour of $F(s) = \mathfrak{L}\{f\}$ along a vertical ray, provided the ray is in the half-plane of absolute convergence.

Theorem 23.6. *If $\mathfrak{L}\{f\} = F(s)$ converges absolutely for $s = x_0$ (real) and, consequently, for $\Re s \geqq x_0$, then $F(x + iy)$ tends, uniformly in $x \geqq x_0$, towards zero, when $y \to \pm \infty$.*

Proof: Let $s = x + iy$, and $x \geqq x_0$. By hypothesis, for every given $\varepsilon > 0$, one can find a T so that

$$\left| \int_T^\infty e^{-st} f(t)\,dt \right| \leqq \int_T^\infty e^{-xt} \,|\, f(t) \,|\, dt \leqq \int_T^\infty e^{-x_0 t} \,|\, f(t) \,|\, dt < \frac{\varepsilon}{2}\,.$$

By Theorem 23.5, there exists a Y so that

$$\left| \int_0^T e^{-st} f(t)\,dt \right| < \frac{\varepsilon}{2} \quad \text{for } |y| > Y,\ x \geqq x_0.$$

Hence,

$$\left| \int_0^\infty e^{-st} f(t)\,dt \right| < \varepsilon \quad \text{for } |y| > Y,\ x \geqq x_0.$$

From Theorems 23.2 and 23.6, we derive

Theorem 23.7. *If $\mathfrak{L}\{f\} = F(s)$ converges absolutely for $\Re s \geqq x_0$, then $F(s)$ tends towards zero, when, in the half-plane $\Re s \geqq x_0$, s tends two-dimensionally towards ∞. In particular, one can select sufficiently large values X and Y so that, for all s with $\Re s \geqq X$ or respectively for all s with $|\Im s| \geqq Y$, $|F(s)|$ is arbitrarily small.*

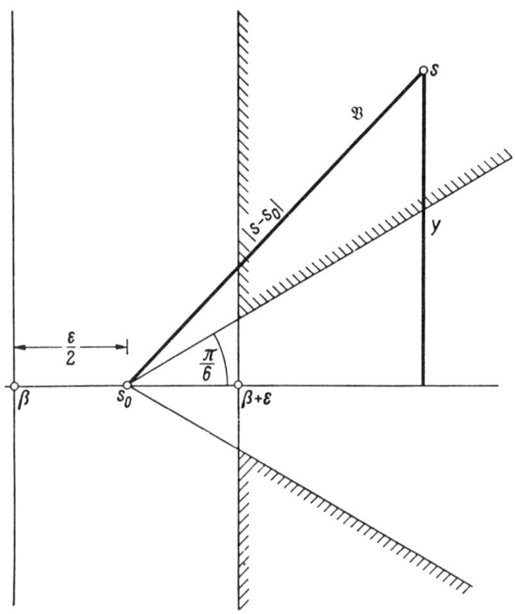

Figure 11

When $\mathfrak{L}\{f\} = F(s)$ converges only conditionally in a half-plane, or in some strip, then we can use the Fundamental Theorem 3.4 to express $F(s)$ by another, absolutely converging $\mathfrak{L}$-integral, to which we may apply Theorem 23.6.

Theorem 23.8. *Let β designate the abscissa of convergence of $\mathfrak{L}\{f\} = F(s)$, then in every half-plane $x \geqq \beta + \varepsilon$ ($\varepsilon > 0$, arbitrarily small), with $s = x + iy$:*

$$F(s) = o(y) \quad \text{as } |y| \to \infty, \text{ uniformly in } x.$$

In the case that $\beta = -\infty$, the conclusion is true in every fixed right half-plane.

Proof: For $\beta = -\infty$, let s_0 be any fixed real point; for finite β, use $s_0 = \beta + (\varepsilon/2)$. By Theorem 23.2, in the angular region $|\operatorname{arc}(s - s_0)| \leqq \pi/6$, indeed we have $F(s) = o(1)$. Thus, we must verify the conclusion only for the remainder $\mathfrak{V}$ of the half-plane $x \geqq \beta + \varepsilon$ ($x \geqq s_0 + (\varepsilon/2)$, in the case that $\beta = -\infty$) shown in Fig. 11. For $x > s_0$, that is for $x \geqq \beta + \varepsilon$ we have, by Theorem 3.4,

$$F(s) = (s - s_0) \int_0^\infty e^{-(s-s_0)t} \varphi(t)\, dt;$$

the integral converges absolutely, hence, by Theorem 23.6,

$$\int_0^\infty e^{-(s-s_0)t} \varphi(t)\, dt = o(1) \quad \text{for } |y| \to \infty, \text{ uniformly in } x \geqq \beta + \varepsilon.$$

In $\mathfrak{V}$, it is true that

$$\frac{|y|}{|s - s_0|} \geqq \sin\frac{\pi}{6} = \frac{1}{2}, \quad \text{hence} \quad |s - s_0| \leqq 2\,|y|.$$

It follows that, in $\mathfrak{V}$, $F(s) = o(y)$, uniformly in x.

The conclusion of this Theorem cannot be improved; this fact can be demonstrated by means of examples.

Supplement. *Since $|y| \leqq |s|$, we certainly have: $F(s) = o(s)$.*

The fact that $F(s)/s$ tends towards zero, when s moves vertically up or down, plays an important rôle in many applications of the theory of functions to the $\mathfrak{L}$-transformation.

24. The Complex Inversion Formula for the Absolutely Converging Laplace Transformation. The Fourier Transformation

Hitherto, without exception, we determined the image function $F(s)$ of some given original function $f(t)$. Clearly, often one is faced with the inverse problem, that is to find the corresponding original function $f(t)$ of some given function $F(s)$, which is known to be a $\mathfrak{L}$-transform. A large number of so-called "inversion formulae" is available which solve this problem, each being applicable under specific hypotheses. For practical applications by far the most important is formula (1.10) which was mentioned early in the beginning of this book; it was then derived from formulae (1.5) and (1.6) which pertain to the Fourier integral, without explicit enumeration of the necessary hypotheses. We shall now present these.

For this purpose, we shall use the symbols g and G instead of f and F for the functions, and x and y for the respective variables which are considered as *real-valued*. First we form with $g(x)$ the function

$$G(y) = \int_{-\infty}^{+\infty} e^{-iyx}\, g(x)\, dx. \tag{1}$$

The correspondence generated in this manner is the **Fourier Transformation,** for which we introduce the symbolic notation $\mathfrak{F}$:

$$G(y) = \mathfrak{F}\{g(x)\}.$$

The equation (1) has meaning, provided the integral converges at least for some values of y. Requiring that $g(x)$ be absolutely integrable in the interval $(-\infty, +\infty)$, briefly[1]

$$\int_{-\infty}^{+\infty} |g(x)|\, dx < \infty, \tag{2}$$

is the simplest hypothesis which guarantees the existence of integral (1). This hypothesis is very restrictive indeed. However, it does offer the advantage that the corresponding $\mathfrak{F}\{g\}$ actually converges absolutely for all real y. Moreover, there is no other hypothesis which is equally simple and useful. Granted this hypothesis, we derive the following properties for the $\mathfrak{F}$-transform $G(y)$.

Theorem 24.1. *If $g(x)$ is absolutely integrable in $(-\infty, +\infty)$, as shown in (2), then $G(y) = \mathfrak{F}\{g(x)\}$ is, in $-\infty < y < \infty$, bounded, and uniformly continuous for all y.*

---

[1] The integrand is non-negative, hence $\int_{-Y}^{+Y} |g(x)|\, dx$ converges either to a finite number or, in the improper sense, towards $+\infty$, when $X \to \infty$. When the integral converges in the proper sense, the limit is finite; this can be written in the form: the limit is $< \infty$.

Proof: We have

$$|G(y)| = \left|\int_{-\infty}^{+\infty} e^{-iyx} g(x)\, dx\right| \leqq \int_{-\infty}^{+\infty} |g(x)|\, dx\,;$$

hence, $G(y)$ is bounded in $-\infty < y < \infty$. Furthermore,

$$|G(y+\delta) - G(y)| = \left|\int_{-\infty}^{+\infty} e^{-iyx}(e^{-i\delta x} - 1)\, g(x)\, dx\right| \leqq \int_{-\infty}^{+\infty} |e^{-i\delta x} - 1|\, |g(x)|\, dx.$$

Since $\qquad e^{-i\delta x} - 1 = (\cos \delta x - 1) - i \sin \delta x,$

$$|e^{-i\delta x} - 1|^2 = (\cos \delta x - 1)^2 + \sin^2 \delta x = 2\,(1 - \cos \delta x) = 4 \sin^2 \frac{\delta x}{2}\,,$$

it follows that

$$|G(y+\delta) - G(y)| \leqq 2 \int_{-\infty}^{+\infty} \left|\sin \frac{\delta x}{2}\right| |g(x)|\, dx$$

$$\leqq 2\left(\int_{-\infty}^{-X} + \int_{X}^{+\infty}\right) |g(x)|\, dx + \int_{-X}^{X} \left|\delta x \frac{\sin \frac{\delta x}{2}}{\delta x/2}\right| |g(x)|\, dx$$

$$\leqq 2\left(\int_{-\infty}^{-X} + \int_{X}^{+\infty}\right) |g(x)|\, dx + |\delta|\, X \int_{-X}^{X} |g(x)|\, dx.$$

For a given $\varepsilon > 0$ one can find, by (2), a sufficiently large X so that

$$2\left(\int_{-\infty}^{-X} + \int_{X}^{+\infty}\right) |g(x)|\, dx < \frac{\varepsilon}{2}.$$

Using this fixed value X, we obtain for all sufficiently small values δ

$$|\delta|\, X \int_{-X}^{X} |g(x)|\, dx < \frac{\varepsilon}{2}\,,$$

hence

$$|G(y+\delta) - G(y)| < \varepsilon.$$

Thus, $G(y)$ is continuous for $-\infty < y < \infty$, indeed uniformly continuous, since y did not enter the evaluation.

Our aim now is to determine if and when $g(x)$ may be recovered from $G(y)$, using the formula

(3) $$g(x) = \frac{1}{2\pi} \int_{-\infty}^{+\infty} e^{ixy} G(y)\, dy$$

either for all values of x or, at least, for certain values of x.

The integral

$$G(y) = \int_{-\infty}^{+\infty} e^{-iy\xi} \cdot g(\xi)\, d\xi$$

converges, uniformly for all y, because of (2). Thus, when integrating $G(y)$ over the finite interval $-Y \leq y \leq +Y$, we may interchange the order of integration;[2] this is also permissible if we first multiply the integrand by the bounded function e^{ixy}, where x is an arbitrary, fixed number:

$$\frac{1}{2\pi} \int_{-Y}^{+Y} e^{ixy} G(y)\, dy = \frac{1}{2\pi} \int_{-Y}^{+Y} e^{ixy}\, dy \int_{-\infty}^{+\infty} e^{-iy\xi}\, g(\xi)\, d\xi = \frac{1}{2\pi} \int_{-\infty}^{+\infty} g(\xi)\, d\xi \int_{-Y}^{+Y} e^{iy(x-\xi)}\, dy$$

$$(4) \qquad = \frac{1}{2\pi} \int_{-\infty}^{+\infty} g(\xi)\, \frac{e^{iY(x-\xi)} - e^{-iY(x-\xi)}}{i(x-\xi)}\, d\xi = \frac{1}{\pi} \int_{-\infty}^{+\infty} \frac{\sin Y(x-\xi)}{x-\xi}\, g(\xi)\, d\xi .$$

Let δ be some fixed number so that $0 < \delta < 1$, and $X > |x| + 1$ be some value the ultimate determination of which we must defer. With these, we partition the integral as follows (compare Fig. 12):

$$\int_{-\infty}^{+\infty} = \int_{-\infty}^{-X} + \int_{-X}^{x-\delta} + \int_{x-\delta}^{x+\delta} + \int_{x+\delta}^{X} + \int_{X}^{\infty} = I_1 + I_2 + I_3 + I_4 + I_5 .$$

In the integrals I_1 and I_5, we have $|x - \xi| > 1$, and $|\sin Y(x - \xi)| \leq 1$; hence, for all values of Y:

$$|I_1| \leq \int_{-\infty}^{-X} |g(\xi)|\, d\xi, \quad |I_5| \leq \int_{X}^{\infty} |g(\xi)|\, d\xi .$$

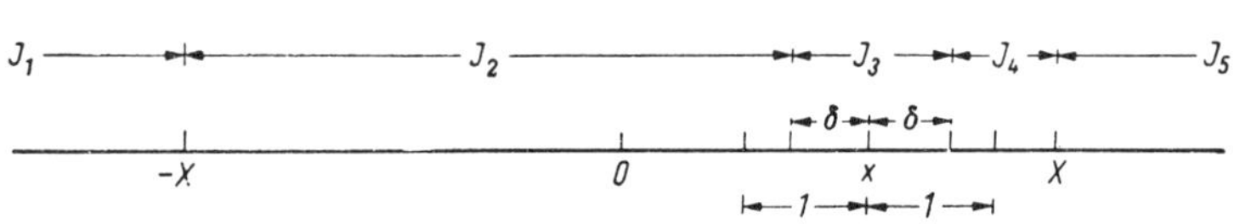

Figure 12

By (2), for a given ε, we can choose the value X so large that, for all Y:

$$(5) \qquad \frac{1}{\pi} |I_1 + I_5| < \frac{\varepsilon}{3} .$$

[2] When integrating a uniformly converging series over a finite interval, we may interchange summation and integration. A similar rule applies when the series is replaced by an integral over an infinite interval.

The interval of integration of the integral

$$I_2 = \int_{\delta}^{x+X} \sin Y u \, \frac{g(x-u)}{u} \, du$$

does not include the origin; hence $g(x - u)/u$ is absolutely integrable in that interval. Thus, by the Riemann-Lebesgue Lemma (Theorem 23.3.), we conclude that

$$I_2 \to 0 \text{ and similarly } I_4 \to 0 \quad \text{as } Y \to \infty.$$

Consequently, for all sufficiently large Y:

(6)
$$\frac{1}{\pi} | I_2 + I_4 | < \frac{\varepsilon}{3}.$$

The remaining integral

(7)
$$\frac{1}{\pi} I_3 = \frac{1}{\pi} \int_{x-\delta}^{x+\delta} \frac{\sin Y(x-\xi)}{x-\xi} g(\xi) \, d\xi = \frac{1}{\pi} \int_{-\delta}^{+\delta} \frac{\sin Y u}{u} g(x-u) \, du$$

is the well-known *Dirichlet integral* of the theory of Fourier series, where the following property is verified: When forming the Fourier series of a function g, which has period 2π, then the partial sum converges at some fixed point x towards some limit l, if and only if the integral (7) has the limit l as $Y \to \infty$. The value of δ may be an arbitrarily small, fixed number.

Thus, every hypothesis concerning the behaviour of g in the interval $(x - \delta, x + \delta)$, which guarantees the convergence of the integral (7) towards a limit l, serves simultaneously as a sufficient hypothesis for the convergence of the Fourier series at the point x, to the limit l, which may or may not coincide with the value $g(x)$. Several such hypotheses are known in the theory of Fourier series.

Here, we shall utilize these results. The criterion which Dirichlet himself has presented, and which is the simplest to formulate, is as follows: If g is monotonic in the interval $(x - \delta, x + \delta)$, then the integral (7) converges towards[3]

(8)
$$l = \frac{g(x^+) + g(x^-)}{2}.$$

The same conclusion follows from the far more general hypothesis which merely

---

[3] For a monotonic function g, we have at every point x limits from the left and from the right respectively: $g(x^-)$ and $g(x^+)$; the value of g at the point can be any value between these limits: $g(x^-) \leqq g(x) \leqq g(x^+)$.

requires bounded variation[4] of g. This results directly from the fact that every function of bounded variation may be presented as the difference of two monotonic functions.

We summarize thus: If g has the property of bounded variation in an (arbitrarily small) interval about the point x, and if the number δ, which so far has only been restricted by $0 < \delta < 1$, is chosen sufficiently small so that $(x - \delta, x + \delta)$ is entirely in that interval; then I_3/π (compare (7)) tends towards the value (8), when $Y \to \infty$. Thus, for sufficiently large values of Y:

$$\left| \frac{1}{\pi} I_3 - \frac{g(x+) + g(x-)}{2} \right| < \frac{\varepsilon}{3}. \tag{9}$$

Combining (4), (5), (6), and (9), we obtain for all sufficiently large Y:

$$\left| \frac{1}{2\pi} \int_{-Y}^{+Y} e^{ixy} G(y)\, dy - \frac{g(x+) + g(x-)}{2} \right| < \varepsilon.$$

This implies that

$$\lim_{Y \to \infty} \frac{1}{2\pi} \int_{-Y}^{+Y} e^{ixy} G(y)\, dy = \frac{g(x+) + g(x-)}{2}. \tag{10}$$

Special attention is called to the fact that the left side of (10) cannot be written directly as

$$\frac{1}{2\pi} \int_{-\infty}^{+\infty} e^{ixy} G(y)\, dy;$$

for this integral is defined by:

$$\lim_{Y_1 \to -\infty,\, Y_2 \to +\infty} \frac{1}{2\pi} \int_{Y_1}^{Y_2} e^{ixy} G(y)\, dy, \tag{11}$$

where Y_1 and Y_2 independently approach the respective limits $-\infty$ and $+\infty$. It may well happen that (11) does not exist, while (10) does converge. For instance, for the odd function $G(y) = y/(1 + y^2)$, with $x = 0$, we have, for all $Y > 0$:

---

[4] A function g is said to be of bounded variation in the interval (a, b), if for every finite sequence of intermediate points x_k: $a = x_0 < x_1 < \cdots < x_{n-1} < x_n = b$, we find:

$$\sum_{\nu=1}^{n} |g(x_\nu) - g(x_{\nu-1})| \leqq M,$$

where M represents some fixed number. The importance of this class of functions follows from the observation that the continuous functions which are of bounded variation on some finite interval define exactly those curves which have finite length on the finite interval.

$$\int_{-Y}^{+Y} e^{ixy} G(y)\,dy = \int_{-Y}^{+Y} \frac{y}{1+y^2}\,dy = 0;$$

consequently, we have for the limit, as $Y \to \infty$, the value zero.

However,

$$\int_{Y_1}^{0} \frac{y}{1+y^2}\,dy \quad \text{and} \quad \int_{0}^{Y_2} \frac{y}{1+y^2}\,dy$$

do not have limits as $Y_1 \to -\infty$, and $Y_2 \to +\infty$ respectively.

Whenever

$$\lim_{Y\to\infty} \int_{-Y}^{+Y} \cdots dy$$

exists, irrespective of the existence of

$$\lim_{Y_1\to-\infty,\, Y_2\to+\infty} \int_{Y_1}^{Y_2} \cdots dy,$$

we say that *Cauchy's Principal Value* of $\int_{-\infty}^{+\infty} \cdots dy$ exists; it is designated by

$$V.P. \int_{-\infty}^{+\infty} \cdots dy.$$

We comprehend why formula (10) does not yield the value $g(x)$ but, instead, the mean of the two limits of g at x, from the right and from the left respectively. For, if one would alter g at the point x, then $G(y)$ which is an integral would remain unchanged. Hence, it cannot be possible to calculate a determined value $g(x)$, by means of $G(y)$; whereas, the mean value is determined by the behaviour of g in an entire (arbitrarily small) neighbourhood of x; it is this behaviour which does determine the $G(y)$.

When, however, g is continuous at x, then $g(x^+) = g(x^-) = g(x)$, and formula (10) produces $g(x)$.

In general, the value of g at some point x is not important; therefore, one "*normalizes*" the function at those points x, where $g(x^+)$ and $g(x^-)$ exist, by means of the definition:

$$g(x) = \frac{g(x^+) + g(x^-)}{2},$$

thus possibly altering the value of the function at x.

We summarize as follows.

Theorem 24.2. *Let* $\int_{-\infty}^{+\infty} |g(x)|\, dx < \infty$; *hence* $G(y) = \mathfrak{F}\{g\}$ *exists for all real* y. *At every point* x, *where* g *is of bounded variation in some (arbitrarily small) neighbourhood of* x, *we have the inversion formula for the Fourier transformation:*

$$\frac{g(x+) + g(x-)}{2} = V.P.\frac{1}{2\pi}\int_{-\infty}^{+\infty} e^{ixy} G(y)\,dy. \tag{12}$$

For normalized g, *or at those points* x *where, moreover,* g *is continuous, formula (12) yields the value* $g(x)$.

Example: $g(x) \equiv e^{-|x|}$ is absolutely integrable in $(-\infty, +\infty)$, hence $\mathfrak{F}\{g\} = G(y)$ exists for all y; it is given by

$$G(y) = \int_{-\infty}^{+\infty} e^{-iyx} e^{-|x|}\,dx = \int_0^{\infty} e^{-iyx} e^{-x}\,dx + \int_{-\infty}^{0} e^{-iyx} e^{x}\,dx$$

$$= \int_0^{\infty} e^{-iyx} e^{-x}\,dx + \int_0^{\infty} e^{iyx} e^{-x}\,dx = 2\int_0^{\infty} e^{-x}\cos y\,x\,dx.$$

The last integral is the $\mathfrak{L}$-transform of $\cos xy$, evaluated at $s = 1$, that is:

$$G(y) = \frac{2}{y^2+1} = \mathfrak{F}\{e^{-|x|}\}.$$

The function $e^{-|x|}$ is of bounded variation in every finite interval; moreover, it is continuous everywhere. Thus, we may employ the inversion formula; it produces the result $e^{-|x|}$ for all x. The inversion integral converges in the ordinary sense, in fact absolutely; hence the $V.P.$ may be omitted, and we find:

$$e^{-|x|} = \frac{1}{\pi}\int_{-\infty}^{+\infty} e^{ixy}\frac{1}{y^2+1}\,dy.$$

This last expression may be written as a $\mathfrak{F}$-transform, that is, upon replacing x by $-x$,

$$\mathfrak{F}\left\{\frac{1}{x^2+1}\right\} = \pi\, e^{-|y|}.$$

---

[5] Obviously, a function of bounded variation need not be continuous. Conversely, a continuous function need not be of bounded variation, a fact that is demonstrated by the function

$$g(x) = \begin{cases} 0 & \text{for } x = 0 \\ x\sin\frac{1}{x} & \text{for } x \neq 0, \end{cases}$$

which is not of bounded variation on an interval which includes 0.

Substituting the explicit expression of $G(y)$ in formula (12) yields:

$$g(x) = V.P. \frac{1}{2\pi} \int_{-\infty}^{+\infty} e^{ixy}\, dy \int_{-\infty}^{+\infty} e^{-iy\xi}\, g(\xi)\, d\xi$$

$$= V.P. \frac{1}{2\pi} \int_{-\infty}^{+\infty} dy \int_{-\infty}^{+\infty} e^{iy(x-\xi)}\, g(\xi)\, d\xi .$$

This representation of a function by an iterated integral is known as the *Fourier Integral Theorem*. The hypotheses for its validity are: $\int_{-\infty}^{+\infty} |g(x)|\, dx < \infty$; g is of bounded variation in a neighbourhood of x; g is normalized in x.

We may now derive an inversion formula for the $\mathfrak{L}$-transformation under exactly specified hypotheses, using Theorem 24.2. In order to fully exploit the possibilities offered by the Theorem 24.2., we firstly generalize the $\mathfrak{L}$-transformation by extending the interval of integration of the $\mathfrak{L}$-integral from the interval $(0, \infty)$ employed so far, to the entire real axis $(-\infty, +\infty)$. We shall call the new transformation the **"two-sided Laplace Transformation"** and we shall represent it by the operator $\mathfrak{L}_{\mathrm{II}}$:

$$F(s) = \int_{-\infty}^{+\infty} e^{-st} f(t)\, dt \equiv \mathfrak{L}_{\mathrm{II}}\{f\}.$$

The notation $\mathfrak{L}_{\mathrm{I}}$ will be used in place of $\mathfrak{L}$ whenever we want to emphasize that the $\mathfrak{L}$-transformation in the earlier sense, the *"one-sided Laplace Transformation"* is meant.

The $\mathfrak{L}$-integral between 0 and $+\infty$ converges, if at all, in a right half-plane; the $\mathfrak{L}$-integral between $-\infty$ and 0 converges in a left half-plane, as demonstrated by:

$$\int_{-\infty}^{0} e^{-st} f(t)\, dt = \int_{0}^{\infty} e^{st} f(-t)\, dt = \int_{0}^{\infty} e^{-(-s)t} f(-t)\, dt .$$

When both half-planes do have a strip in common, then this strip is the region of convergence of the $\mathfrak{L}_{\mathrm{II}}$-integral. For instance, for $f(t) \equiv e^{-|x|}$, the first integral converges for $\Re s > -1$, the second one for $\Re s < +1$; hence the $\mathfrak{L}_{\mathrm{II}}$-integral converges in the strip $-1 < \Re s < +1$. However, for $f(t) \equiv 1$, the two integrals converge for $\Re s > 0$ and $\Re s < 0$ respectively; that is, we cannot find a strip common to both half-planes.

The convergence of the $\mathfrak{L}_{\mathrm{II}}$-integral at a single point does not enable us to deduce its behaviour at other points. However, when it is known that the $\mathfrak{L}_{\mathrm{II}}$-integral converges at two points s_1 and s_2, with $\Re s_1 < \Re s_2$, then convergence in the strip $\Re s_1 < \Re s < \Re s_2$ follows; for the integral evaluated between 0 and $+\infty$ converges for $\Re s_1 < \Re s$, and the one evaluated between $-\infty$ and 0 converges for $\Re s < \Re s_2$.

More can be concluded in case of absolute convergence. When the $\mathfrak{L}_{\mathrm{II}}$-integral converges absolutely at some point s_0, then, by Theorem 3.1, it converges absolutely on the entire line $\mathfrak{R}s = \mathfrak{R}s_0$. This observation is important for the subsequent deductions.

Let us consider the $\mathfrak{L}_{\mathrm{II}}$-integral at the points s of a vertical line:

$$s = x + i\,y, \quad x = \text{const},$$

then

$$F(x + i\,y) = \mathfrak{L}_{\mathrm{II}}\{f\} = \int_{-\infty}^{+\infty} e^{-iyt}\left[e^{-xt} f(t)\right] dt = \mathfrak{F}\{e^{-xt} f(t)\}. \tag{13}$$

On the line $\mathfrak{R}s = x$, the $\mathfrak{L}_{\mathrm{II}}\{f\}$ is the $\mathfrak{F}$-transform of $e^{-xt}f(t)$. In order to apply Theorem 24.2. to this function, we require that

$$\int_{-\infty}^{+\infty} e^{-xt}\,|\,f(t)\,|\,dt < \infty;$$

that is, that $\mathfrak{L}_{\mathrm{II}}\{f\}$ converges absolutely for $s = x$, and hence for $\mathfrak{R}s = x$, and that $e^{-xt}f(t)$ is of bounded variation in some neighbourhood of the fixed point t. For the latter request it suffices that $f(t)$ is of bounded variation, for the product of two functions of bounded variation, $f(t)$ and $g(t)$, is of bounded variation. This is shown by the following development:

$$\begin{aligned} &|f(t_\nu)\,g(t_\nu) - f(t_{\nu-1})\,g(t_{\nu-1})| \\ &\qquad = |f(t_\nu)\,g(t_\nu) - f(t_\nu)\,g(t_{\nu-1}) + f(t_\nu)\,g(t_{\nu-1}) - f(t_{\nu-1})\,g(t_{\nu-1})| \\ &\qquad \leqq |f(t_\nu)|\cdot|g(t_\nu) - g(t_{\nu-1})| + |g(t_{\nu-1})|\cdot|f(t_\nu) - f(t_{\nu-1})|. \end{aligned}$$

In the finite interval with the partition points t_ν, we have

$$|\,f(t_\nu)\,| < M_1, \quad |\,g(t_{\nu-1})\,| < M_2,$$

for a function of bounded variation is necessarily bounded. Thus, it follows that

$$\begin{aligned} &\sum_{\nu=1}^{n} |\,f(t_\nu)\,g(t_\nu) - f(t_{\nu-1})\,g(t_{\nu-1})| \\ &\qquad \leqq M_1 \sum_{\nu=1}^{n} |\,g(t_\nu) - g(t_{\nu-1})\,| + M_2 \sum_{\nu=1}^{n} |\,f(t_\nu) - f(t_{\nu-1})\,|. \end{aligned}$$

The sums of the right hand side are bounded, and so is the sum of the left hand side.

Applying the inversion formula (12) to formula (13), one finds, for normalized f,

$$e^{-xt} f(t) = V.P.\,\frac{1}{2\pi} \int_{-\infty}^{+\infty} e^{ity}\,F(x + iy)\,dy$$

or

$$f(t) = V.P. \frac{1}{2\pi} \int_{-\infty}^{+\infty} e^{t(x+iy)} F(x+iy)\, dy,$$

which can be written with $s = x + iy$ as follows:

$$f(t) = V.P. \frac{1}{2\pi i} \int_{x-i\infty}^{x+i\infty} e^{ts} F(s)\, ds.$$

This integral is to be evaluated along the line $\Re s = x =$ const, in the complex plane.

We summarize:

Theorem 24.3. *Suppose that* $\mathfrak{L}_{\mathrm{II}}\{f\} = F(s)$ *converges absolutely for* $s = x$ *(real) and, consequently, for* $\Re s = x$, *that is*

$$\int_{-\infty}^{+\infty} e^{-xt} |f(t)|\, dt < \infty.$$

We conclude that the "Complex Inversion Formula"

(14) $$\frac{f(t^+) + f(t^-)}{2} = V.P. \frac{1}{2\pi i} \int_{x-i\infty}^{x+i\infty} e^{ts} F(s)\, ds = \lim_{Y\to\infty} \frac{1}{2\pi i} \int_{x-iY}^{x+iY} e^{ts} F(s)\, ds$$

holds at every point t *where* f *is of bounded variation in some neighbourhood of* t. *When* f *is continuous at* t *or normalized in* t, *then* (14) *yields the value* $f(t)$.

The two-sided $\mathfrak{L}$-transformation comprises also the one-sided $\mathfrak{L}$-transformation, provided we define for the latter $f(t)$, which is given only for $t \geqq 0$, by zero for $t < 0$; in this case, we have $\mathfrak{L}_{\mathrm{II}}\{f\} = \mathfrak{L}_{\mathrm{I}}\{f\}$.

When $\mathfrak{L}_{\mathrm{I}}\{f\}$ converges absolutely for $s = x_0$ (real), then the same is true for all $s = x \geqq x_0$, indeed, for all s with $\Re s \geqq x_0$. Consequently, one can write the inversion formula with any $x \geqq x_0$. Furthermore, observe that if the given $f(t)$ is of bounded variation in an interval on the right of $t = 0$, then the for $t < 0$ by zero completed function is of bounded variation in an entire interval which contains $t = 0$. We have $f(0^-) = 0$; therefore, at $t = 0$ the mean of the limits from the left and from the right equals $f(0^+)/2$. For $t < 0$, $f(t) \equiv 0$ is eo ipso of bounded variation, hence the inversion formula converges always and yields the value zero. This fact is not of immediate interest for the $\mathfrak{L}_{\mathrm{I}}$-transformation, although it can, occasionally, be utilized.

We are now in a position to formulate the Inversion Theorem of the $\mathfrak{L}_{\mathrm{I}}$-transformation.

Theorem 24.4 (Inversion Theorem). *Suppose that* $\mathfrak{L}_{\mathrm{I}}\{f\} = F(s)$ *converges absolutely for* $s = x_0$ *(real) and, consequently, for* $\Re s \geqq x_0$, *that is*

$$\int_0^{\infty} e^{-x_0 t} |f(t)|\, dt < \infty.$$

We conclude that the "complex inversion formula"

$$\frac{f(t^+) + f(t^-)}{2} = V.P.\,\frac{1}{2\pi i}\int_{x-i\infty}^{x+i\infty} e^{ts}\,F(s)\,ds \quad (x \geqq x_0) \tag{15}$$

holds at all those points $t > 0$ *where* $f(t)$ *is of bounded variation in some neighbourhood of* t. *If* $f(t)$ *is of bounded variation in an interval on the right of* $t = 0$, *then*

$$\frac{f(0^+)}{2} = V.P.\,\frac{1}{2\pi i}\int_{x-i\infty}^{x+i\infty} F(s)\,ds \quad (x \geqq x_0)\,. \tag{16}$$

For all $t < 0$, *we find*

$$V.P.\,\frac{1}{2\pi i}\int_{x-i\infty}^{x+i\infty} e^{ts}\,F(s)\,ds = 0 \quad (x \geqq x_0)\,. \tag{17}$$

We notice that every vertical line with the abscissa $x \geqq x_0$ may be selected as the path of integration for the stated formulae, yielding identical values for the integrals, independently of the selected particular abscissa. An argumentation of the theory of functions leads directly to the same conclusion. This argumentation recurs frequently when working with the $\mathfrak{L}$-transformation; thus, we take this opportunity to demonstrate this argumentation. This also serves as the first example of the utility of the fact that $F(s)$ is an analytic function.

In the theory of functions, it is a common operation to replace the path of integration by another more convenient one with the same end points, invoking

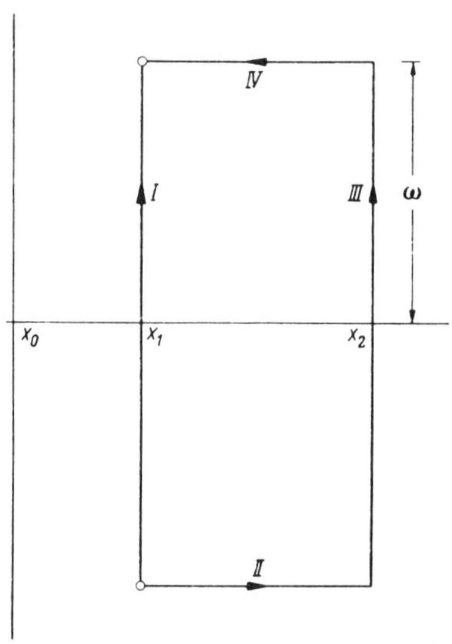

Figure 13

Cauchy's theorem. In the present case, the end points of the path of integration are removed towards infinity, a fact which necessitates a limiting process.

Let x_1 and x_2 be two abscissae with $x_0 < x_1 < x_2$, and let ω be a positive number. We construct the rectangle having the vertical sides: $\Re s = x_1$ and $\Re s = x_2$, and the horizontal sides: $\Im s = +\omega$ and $\Im s = -\omega$, calling these four sides I, II, III, and IV, to which we assign the respective orientations as shown in Fig. 13. The functions e^{ts} and $F(s)$ are analytic inside of and on the boundary of the rectangle. Hence, by Cauchy's theorem:

$$\int_{\mathrm{I}} e^{ts} F(s)\, ds = \int_{\mathrm{II}} + \int_{\mathrm{III}} + \int_{\mathrm{IV}} e^{ts} F(s)\, ds. \tag{18}$$

On the horizontal side IV, we have

$$|e^{ts}| = e^{t \Re s} \leqq \begin{cases} e^{t x_2} & \text{for } t \geqq 0 \\ e^{t x_1} & \text{for } t < 0; \end{cases}$$

that is, $|e^{ts}|$ has an upper bound which does not depend upon the value of ω. By Theorem 23.6, $F(s)$ tends, uniformly in $x_1 \leqq \Re s \leqq x_2$, towards 0, when $\omega \to \infty$. The length of the path of integration along the upper side IV is, invariantly, $(x_2 - x_1)$. Thus, the integral along the side IV converges towards zero, when $\omega \to \infty$. The same conclusion is valid for the integral along the lower horizontal line II. Hence we find from (18): .

$$\lim_{\omega \to \infty} \int_{\mathrm{I}} e^{ts} F(s)\, ds = \lim_{\omega \to \infty} \int_{\mathrm{III}} e^{ts} F(s)\, ds,$$

that is,

$$V.P. \int_{x_1 - i\infty}^{x_1 + i\infty} e^{ts} F(s)\, ds = V.P. \int_{x_2 - i\infty}^{x_2 + i\infty} e^{ts} F(s)\, ds.$$

Observe that the invariance with x of the integrals (15), (16), and (17) has been derived here only for $x > x_0$, not for $x \geqq x_0$. This is due to the fact that the classical Cauchy theorem that we used in the argumentation requires that the integrand be analytic in the interior of the boundary as well as on the boundary. This hypothesis cannot be guaranteed along $\Re s = x_0$. A more modern version of Cauchy's theorem requires on the boundary merely (two-dimensional) continuity of the function towards the interior. A theorem which we cannot verify here, enables us to conclude that the function $F(s)$ is indeed two-dimensionally continuous towards the right on the line $\Re s = x_0$, provided $\mathfrak{L}\{f\} = F(s)$ converges absolutely at $s = x_0$. With this additional information, we may extend the conclusion of Theorem 24.4 to include $x = x_0$.

We shall employ Theorem 24.4 to derive a formula which is widely used in mathematics and theoretical physics. The transform $\mathfrak{L}\{u\} = \mathfrak{L}\{1\} = 1/s$ converges absolutely for $\Re s > 0$. Consequently, we may select any $x_0 > 0$, and any abscissa $x > 0$ for the vertical line of integration. The function $u(t)$ is of bounded variation in some neighbourhood of any point t, hence

$$V.P.\frac{1}{2\pi i}\int_{x-i\infty}^{x+i\infty}\frac{e^{ts}}{s}\,ds=\begin{cases}1 & \text{for } t>0\\ \frac{1}{2} & \text{for } t=0\\ 0 & \text{for } t<0\end{cases}\quad (x>0). \tag{19}$$

With (19) we derived an analytic representation of the "*discontinuous factor*", a function which assumes the value one for positive arguments t, and the value zero for negative arguments t.

Moreover, when employing considerations similar to those developed on p. 12 for $\int_1^\infty (\sin u/u)\,du$, we can demonstrate that the $V.P.$ may be omitted for $t \neq 0$, since each of the integrals $\int_x^{x+i\infty}$ and $\int_{x-i\infty}^{x}$ converges. However, for $t = 0$, we must retain the $V.P.$, because the integral

$$\int_{x-i\infty}^{x+i\infty}\frac{ds}{s}$$

does not exist, whereas the following limit does exist:

$$\lim_{Y\to\infty}\frac{1}{2\pi i}\int_{x-iY}^{x+iY}\frac{ds}{s}=\lim_{Y\to\infty}\frac{1}{2\pi}\int_{-Y}^{+Y}\frac{dy}{x+iy}=\lim_{Y\to\infty}\frac{1}{2\pi}\int_{-Y}^{+Y}\frac{x-iy}{x^2+y^2}\,dy.$$

The real part of the integrand is an even function of y, hence $\int_{-Y}^{+Y} = 2\int_0^{+Y}$; the imaginary part of the integrand is an odd function of y, hence $\int_{-Y}^{+Y} = 0$. Thus, we conclude that

$$\lim_{Y\to\infty}\frac{1}{\pi}\int_0^{Y}\frac{x}{x^2+y^2}\,dy=\lim_{Y\to\infty}\frac{1}{\pi}\int_0^{Y/x}\frac{du}{1+u^2}=\lim_{Y\to\infty}\frac{1}{\pi}\operatorname{arctg}\frac{Y}{x}=\frac{1}{2}.$$

In (19) we must select $x > 0$. The function $1/s$ is not integrable at the origin, hence the imaginary axis $\Re s = 0$ cannot be used as the line of integration. However, if we replace the imaginary axis near the origin by a semicircle to the right of arbitrary radius δ, then we may, following the argumentation of p. 159, replace the path of integration, $\Re s = x > 0$, by the indented contour of Fig. 14, which includes as straight line sections parts of the imaginary axis $\Re s = 0$.

When considering, in the sense of Chapter 1, the $\mathfrak{L}$-transform $F(s) = F(x+iy)$ as the spectral density of the function $e^{-xt}f(t)$, and when regarding the inversion formula in the form

$$e^{-xt} f(t) = \frac{1}{2\pi} \int_{-\infty}^{+\infty} e^{iyt} F(x+iy)\, dy \tag{20}$$

as a representation of $e^{-xt}f(t)$ as a superposition of oscillations, then, of course, this is possible only when the path of integration of the integral is a straight line, that is, when we can set $s = x + iy$ with $x =$ const. Whenever an indented contour is being used, the inversion formula cannot be brought into the form of formula (20). This means that, by (19), a spectral representation of the function $e^{-xt}u(t)$ can be developed, for $x > 0$; one cannot use (19) with the indented path of integration to construct a spectral representation of $u(t)$ itself. We call attention to this limitation which is often overlooked in technical investigations.[6]

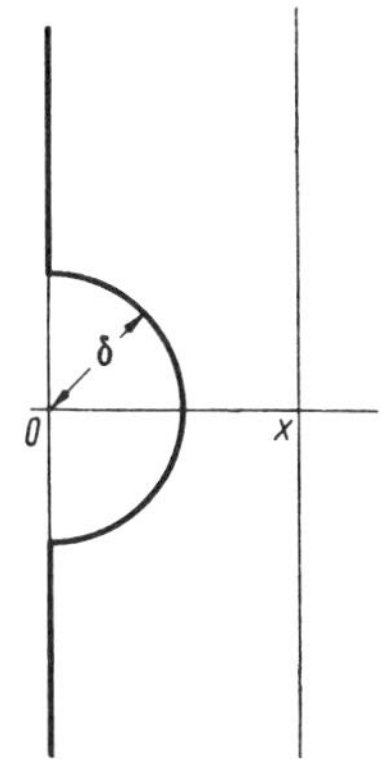

Figure 14

25. Deformation of the Path of Integration of the Complex Inversion Integral

The $\mathfrak{L}$-transform represents an analytic function. Thus we may, by Cauchy's theorem, alter the straight line path of integration of the complex inversion integral in a certain manner. The last example of Chapter 24 served to demonstrate such an alteration. For practical applications, one frequently employs the following modification: The vertical line is shifted to the left until the first singular point s_0 of $F(s)$ is met. Then, the vertical line is replaced near s_0 by an arc of a circle to the right, at the same time inclining the remaining straight sections towards the left as shown in Fig. 15. The newly created path of integration offers favourable conditions of convergence for the integral, for the factor e^{ts} converges, for $t > 0$, rapidly towards zero along the straight line sections biased towards the left; whereas this factor oscillates between finite limits along the vertical lines.

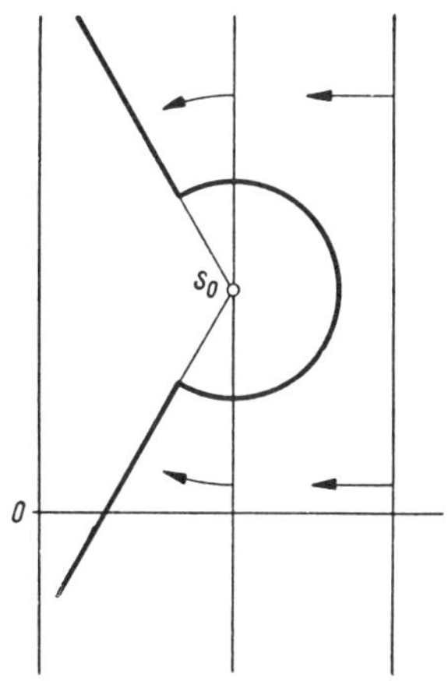

Figure 15

[6] The function $u(t)$ has a spectral density in the theory of distributions; however, the spectral density of $u(t)$ ist not $(1/s)_{x=0} = 1/iy$, instead it is PF $1/iy + \pi\delta(y)$. To demonstrate this, we would have to generalize the $\mathfrak{F}$-transformation to include distributions; this would be beyond the scope of this text.

Clearly, not every $\mathfrak{L}$-transform $F(s)$ permits this type of deformation of the path of integration (compare p. 245). However, by means of the following Theorem 25.1, we can show that such alterations are permissible in many cases. This Theorem has a wide range of application, for it does not require $F(s)$ to be a $\mathfrak{L}$-transform, indeed, it does not even require $F(s)$ to be analytic.

Theorem 25.1. *Consider some point s_0, a vertical line through s_0, and a sequence of semicircles $\mathfrak{H}_n$ $(n = 0, 1, 2, \ldots)$ to the* ***left*** *of the vertical line, with centre s_0 and radii ϱ_n which satisfy the conditions: $\varrho_0 < \varrho_1 < \ldots \to \infty$. Suppose that, on these semicircles the function $F(s)$ is integrable and bounded by bounds which converge towards zero, when $n \to \infty$:*

$$|F(s)| \leqq \delta_n \text{ on } \mathfrak{H}_n, \quad \delta_n \to 0 \quad as \quad n \to \infty.$$

Then we conclude that

$$\int_{\mathfrak{H}_n} e^{ts} F(s) ds \to 0 \quad with\ \mathbf{t} > \mathbf{0} \quad as \quad n \to \infty.$$

When, in particular, on every left semicircle $\mathfrak{H}_\varrho$ having radius ϱ and centre s_0

$$|F(s)| \leqq \delta_\varrho \, where\ \delta_\varrho \to 0 \quad as \quad \varrho \to \infty,$$[1]

then it follows that

$$\int_{\mathfrak{H}_\varrho} e^{ts} F(s)\, ds \to 0 \quad with\ \mathbf{t} > \mathbf{0} \quad as \quad \varrho \to \infty.$$

This Theorem remains valid when, instead of the semicircles any portions of these[2] *are used as paths of integration, and when $|F(s)| \leqq \delta_\varrho \to 0$ is guaranteed on these portions.*

In the case that the semicircles are to the ***right*** *of the vertical line through s_0, the above stated conclusions are true, under the same hypotheses, for $\mathbf{t} < \mathbf{0}$.*

Proof: Let $s = s_0 + \sigma$, then the integral on the right hand side of

$$\int_{\mathfrak{H}_n} e^{ts} F(s)\, ds = e^{ts_0} \int e^{t\sigma} F(s_0 + \sigma)\, d\sigma$$

is to be evaluated in the σ-plane along the semicircle to the left ou the imaginary axis having radius ϱ_n, and the centre at the origin. The sequence of integrals of the left hand side tends towards zero, provided the sequence of integrals of the right hand side does converge to zero. Therefore, it suffices to verify the Theo-

---

[1] This means that in the left half-plane $F(s)$ tends, uniformly with respect to $\operatorname{arc}(s - s_0)$, towards 0.

[2] In the case that the arcs are in an angular region $\pi/2 < \psi \leqq \operatorname{arc}(s - s_0) \leqq 2\pi - \psi < 3\pi/2$, the Theorem is trivially true, since e^{ts} tends, uniformly with respect to arc $(s - s_0)$, more strongly towards zero than the length of the arc grows towards ∞. The strength of the Theorem lies in the fact that the semicircles join the vertical line $\Re s = \Re s_0$, where e^{ts} does not tend towards zero.

rem for $s_0 = 0$. For this latter case we may write for $\mathfrak{H}_n : s = \varrho_n e^{i\vartheta}$, with $\pi/2 \leqq \vartheta \leqq 3\pi/2$, and we find, with $ds = \varrho_n i e^{i\vartheta} d\vartheta$:

$$\text{(1)} \qquad \left| \int_{\mathfrak{H}_n} e^{ts} F(s)\, ds \right| \leqq \delta_n \int_{\pi/2}^{3\pi/2} e^{t\varrho_n \cos\vartheta} \varrho_n\, d\vartheta \qquad \left(\vartheta = \frac{\pi}{2} + \varphi\right)$$

$$= \delta_n \varrho_n \int_0^{\pi} e^{-t\varrho_n \sin\varphi}\, d\varphi = 2\, \delta_n \varrho_n \int_0^{\pi/2} e^{-t\varrho_n \sin\varphi}\, d\varphi .$$

In the entire interval $0 \leqq \varphi \leqq \pi/2$, the curve $y = \sin\varphi$ is above the secant $y = (2/\pi)\varphi$; consequently one has:

$$\sin\varphi \geqq \frac{2}{\pi}\varphi \quad \text{for } 0 \leqq \varphi \leqq \frac{\pi}{2} ;$$

hence,

$$\left| \int_{\mathfrak{H}_n} e^{ts} F(s)\, ds \right| \leqq 2\, \delta_n \varrho_n \int_0^{\pi/2} e^{-t\varrho_n (2/\pi)\varphi}\, d\varphi$$

$$= 2\, \delta_n \frac{1 - e^{-t\varrho_n}}{t(2/\pi)} \to 0 \quad \text{with } t > 0 \quad \text{as } n \to \infty.$$

The above verification is certainly true for a sequence of portions of the semicircles. Following the outline of the above proof, one would find in the corresponding estimation of (1), for the sequence of fixed portions of the semicircles: $\pi/2 \leqq \vartheta_1 \leqq \vartheta \leqq \vartheta_2 \leqq 3\pi/2$, the dominating sequence:

$$\delta_n \int_{\vartheta_1}^{\vartheta_2} e^{t\varrho_n \cos\vartheta} \varrho_n\, d\vartheta ;$$

the integrand being positive, this sequence itself is dominated by the sequence derived for the entire semicircles.

When using a sequence of semicircles $\mathfrak{H}_n$ to the right of the imaginary axis, with $t < 0$, we set $s = -\sigma$ and we find

$$\int_{\mathfrak{H}_n} e^{ts} F(s)\, ds = -\int e^{-t\sigma} F(-\sigma)\, d\sigma ,$$

the integrals on the right hand side being evaluated over the corresponding sequence of semicircles to the left of the imaginary axis, with $-t > 0$. The earlier part of this proof establishes convergence toward zero, when $n \to \infty$.

1. As a first application of Theorem 25.1, we shall present *another verification* of formula (24.19), for a path of integration that is indented towards the right at $s = 0$.

For $t > 0$, we complete the contour by the addition of a semicircle $\mathfrak{H}$ to the left of the imaginary axis, having radius ϱ, and the centre at the origin (see Fig. 16).

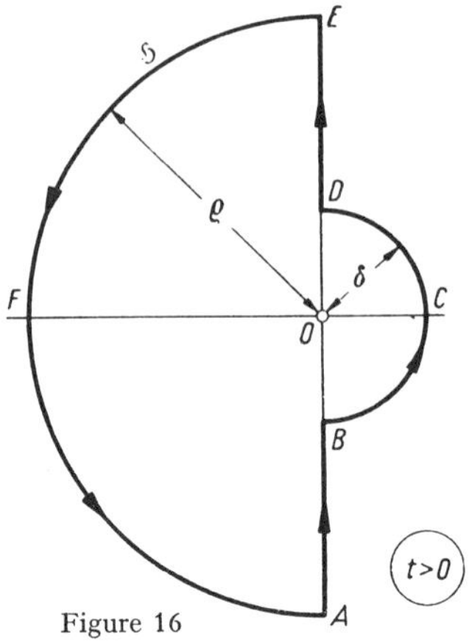

Figure 16

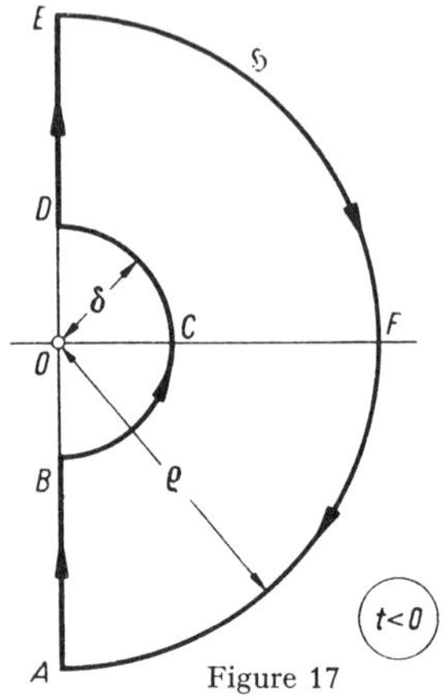

Figure 17

We investigate the integral

$$\frac{1}{2\pi i}\int e^{ts}\frac{1}{s}\,ds\,,$$

along the contour A B C D E F A. By Cauchy's formula, the value of this integral will be the value of the analytic function e^{ts} at $s = 0$, that is, 1. On the semicircle $\mathfrak{H}$ we find: $|1/s| = 1/\varrho \to 0$, when $\varrho \to \infty$; thus, by Theorem 25.1, the integral along $\mathfrak{H}$ tends towards zero, when $\varrho \to \infty$. Consequently, the remainder is

$$\lim_{\varrho\to\infty}\frac{1}{2\pi i}\int\limits_{\text{ABCDE}} e^{ts}\frac{1}{s}\,ds = 1 \quad \text{for} \quad t > 0.$$

For $t < 0$, we complete the contour by the addition of a semicircle to the right (see Fig. 17). In this case, the function $e^{ts}(1/s)$ is analytic inside the closed contour, and the value of the integral is zero. By Theorem 25.1, the integral along $\mathfrak{H}$ vanishes in the limit, when $\varrho \to \infty$, and we are left with:

$$\lim_{\varrho\to\infty}\frac{1}{2\pi i}\int\limits_{\text{ABCDE}} e^{ts}\frac{1}{s}\,ds = 0 \quad \text{for} \quad t < 0.$$

For $t = 0$, we can immediately evaluate the integral. We set $s = iy$ along the straight line portions, and $s = \delta e^{i\varphi}$ on the semicircle; thus we find:

$$\int\limits_{\text{ABCDE}}\frac{1}{s}\,ds = \int\limits_{-\varrho}^{-\delta}\frac{1}{iy}\,i\,dy + \int\limits_{-\pi/2}^{+\pi/2}\delta^{-1}e^{-i\varphi}\,\delta\,i\,e^{i\varphi}\,d\varphi + \int\limits_{\delta}^{\varrho}\frac{1}{iy}\,i\,dy$$

$$= -\int\limits_{\delta}^{\varrho}\frac{1}{y}\,dy + i\int\limits_{-\pi/2}^{+\pi/2}d\varphi + \int\limits_{\delta}^{\varrho}\frac{1}{y}\,dy = i\pi\,,$$

hence

$$\frac{1}{2\pi i}\int\limits_{\text{ABCDE}}\frac{1}{s}\,ds = \frac{1}{2}\,.$$

It follows that the limit too is 1/2, when $\varrho \to \infty$. The above evaluation clearly shows how the value 1/2 originates.

2. The following example demonstrates the advantage of the earlier suggested replacement of the vertical line by an indented one with the rays biased towards the left. We begin with the correspondence

$$\mathfrak{L}\left\{\frac{t^{a-1}}{\Gamma(a)}\right\} = s^{-a} \quad (a > 0)\,.$$

For $\Re s > 0$, the $\mathfrak{L}$-integral converges absolutely; the original function is of bounded variation and continuous, for $t > 0$. Hence, we may use Theorem 24.3, and we find:

$$\text{(2)} \qquad V.P.\,\frac{1}{2\pi i}\int_{x-i\infty}^{x+i\infty} e^{ts}\, s^{-a}\, ds = \begin{cases} \dfrac{t^{a-1}}{\Gamma(a)} & \text{for} \quad t > 0 \\ 0 & \text{for} \quad t < 0. \end{cases} \qquad (a > 0\,, \quad x > 0)\,.$$

Similarly, as was done for the case $\alpha = 1$ (see p. 160), we could replace the vertical line, with abscissa x, by an indented line, with abscissa 0. However, we advance beyond that, and we consider the contour shown in Fig. 18. The only singular point of $e^{ts}s^{-a}$, that is $s = 0$, is outside the contour; consequently, the value of the integral along the entire contour is zero. Next, we investigate the behaviour of the integral, for $t > 0$, along the several portions of the path of integration when the radius ω (see Fig. 18) tends towards ∞. The function $F(s) = s^{-a}$ tends, uniformly with regard to arc s, towards zero, when $|s| \to \infty$. By Theorem

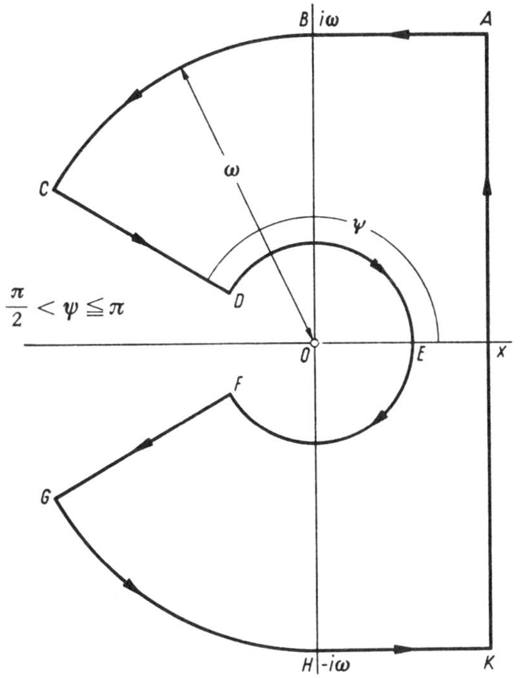

Figure 18

25.1, for $t > 0$, the integrals along the circular arcs BC and GH respectively also vanish, when $\omega \to \infty$. The integrals along the line elements AB and HK also vanish, when $\omega \to \infty$, since s^{-a} converges uniformly towards zero, while the function e^{ts} is bounded, and the length of the line elements, invariably, equals x:

$$\lim_{\omega \to \infty} \int_{x-i\omega}^{x+i\omega} e^{ts} s^{-a} ds + \lim_{\omega \to \infty} \int_{\mathrm{CDEFG}} e^{ts} s^{-a} ds = 0 .$$

We invert the orientation of the curve C D E F G, and we let the points C and G each tend towards infinity, designating the limiting curve by $\mathfrak{W}$. The angle ψ (see Fig. 18) is in the interval $\pi/2 < \psi \leq \pi$, hence $\mathfrak{W}$ has the shape shown in Fig. 19: Fig. 19a) for $\psi < \pi$, and Fig. 19b) for $\psi = \pi$.[3] The last equation indicates

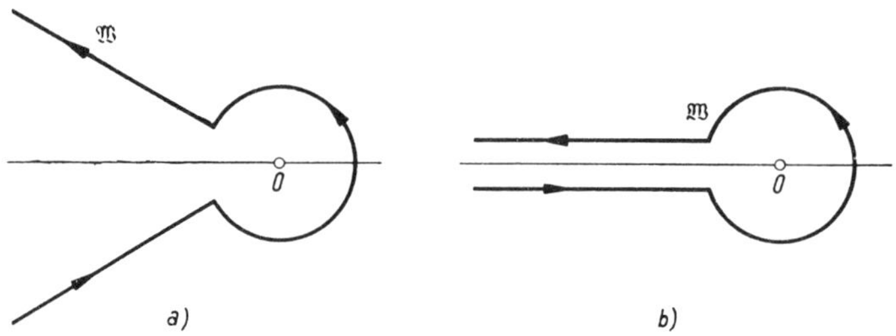

Figure 19

that the integral along the vertical line at x has the same value as the integral which is obtained along the curve $\mathfrak{W}$. Thus, we obtain, because of (2),

(3) $$\frac{1}{2\pi i} \int_{\mathfrak{W}} e^{ts} s^{-a} ds = \frac{t^{a-1}}{\Gamma(a)} \quad \text{for} \quad t > 0,\ a > 0.$$

The presented argumentation is not valid for $t < 0$.

When creating the curve $\mathfrak{W}$ from the curve C D E F G by the limiting process: $\omega \to \infty$, both points, C and G, must simultaneously approach infinity; however, the integral would also converge if each of the two points would independently approach infinity.

Thus, we may replace, in formula (2), the vertical line by the angular, indented path $\mathfrak{W}$, provided $t > 0$. This is very important indeed, for the integral (3) obviously converges for every complex value of α. This, of course, does not in itself demonstrate the value of the integral as $t^{a-1}/\Gamma(\alpha)$. However, this may be shown in the following manner: For $t = 1$, the integral (3) is a function of α only:

(4) $$\frac{1}{2\pi i} \int_{\mathfrak{W}} e^{s} s^{-a} ds = G(a) .$$

[3] For $\psi = \pi$, the horizontal rays are located on the upper and on the lower border of the plane which has been cut along the negative real axis.

Using integration by parts, one finds for $\alpha \neq 1$:

$$G(\alpha) = \frac{1}{2\pi i}\,\frac{1}{\alpha-1}\int\limits_{\mathfrak{W}} e^s\, s^{-\alpha+1}\, ds = \frac{1}{\alpha-1}\, G(\alpha-1)\,.$$

The function $G(\alpha)$ satisfies the same functional equation as does $1/\Gamma(\alpha)$, which is:

$$\frac{1}{\Gamma(\alpha)} = \frac{1}{\alpha-1}\,\frac{1}{\Gamma(\alpha-1)}\,.$$

Moreover, for $\alpha > 0$, $G(\alpha)$ agrees with $1/\Gamma(\alpha)$; hence the two functions are identical for all values of α.

For $\alpha = 0, -1, -2, \ldots$, we have $\Gamma(\alpha) = \infty$, and $1/\Gamma(\alpha) = G(\alpha) = 0$; this observation also follows directly from (4). For, if we construct a closed curve composed of a finite portion of $\mathfrak{W}$, and a circular arc as shown in Fig. 20, then we

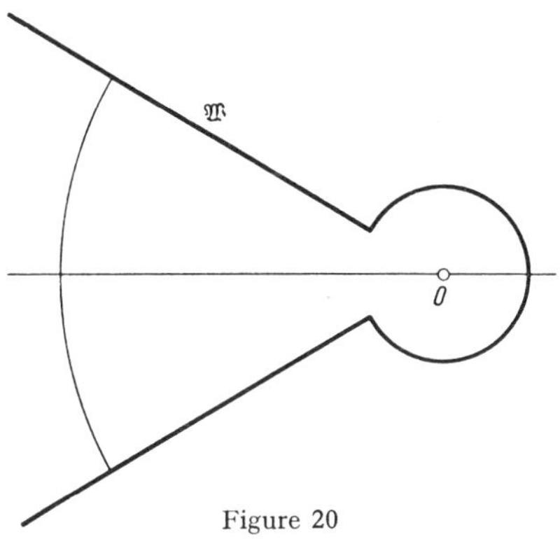

Figure 20

find $e^s s^{-\alpha}$ to be analytic in the interior of the region bounded by this curve, for $\alpha = 0, -1, -2, \ldots$. It follows that the respective integral has the value zero. Letting the radius of the circular arc tend towards infinity then, obviously, the value of the integral along this arc vanishes, and we are left with the integral along $\mathfrak{W}$, which is 0.

Substituting in the formula derived above

(5) $$\frac{1}{2\pi i}\int\limits_{\mathfrak{W}} e^\sigma\, \sigma^{-\alpha}\, d\sigma = \frac{1}{\Gamma(\alpha)} \quad (\alpha \text{ arbitrary})$$

$\sigma = ts$, with $t > 0$, we find:

$$\frac{1}{2\pi i}\int e^{ts}\, t^{-\alpha+1}\, s^{-\alpha}\, ds = \frac{1}{\Gamma(\alpha)}\,;$$

this integral is evaluated along a curve which is generated from the curve $\mathfrak{W}$ by a similarity transformation with the ratio $1/t$. The radius of the circular part of $\mathfrak{W}$ is arbitrary; consequently, the new curve too is a curve of type $\mathfrak{W}$.

Thus, we finally obtain:

(6) $$\frac{1}{2\pi i}\int\limits_{\mathfrak{W}} e^{ts}\, s^{-\alpha}\, ds = \frac{t^{\alpha-1}}{\Gamma(\alpha)} \quad \text{for } t > 0, \text{ and } \alpha \text{ complex.}$$

The right hand side is zero for $\alpha = 0, -1, -2, \ldots$.

Specifying $t = 1$ in formula (6) yields formula (5) which is called *Hankel's Formula*; it offers a particularly elegant method of defining the Gamma-function for all values of α. We shall often use formula (6) in the sequel.

3. We shall demonstrate the possibility of a further reduction of the path of integration with the following example. Starting with the correspondence for the Bessel function:

$$\mathfrak{L}\{J_0(t)\} = \frac{1}{\sqrt{s^2+1}},$$

we conclude, by Theorem 24.4, that

$$\text{(7)} \qquad V.P.\ \frac{1}{2\pi i} \int_{x-i\infty}^{x+i\infty} \frac{e^{ts}}{\sqrt{s^2+1}}\, ds = J_0(t) \quad \text{for } t > 0 \quad (x > 0).$$

The function $(s^2+1)^{-1/2}$ tends, uniformly for all directions, towards zero, when $s \to \infty$; hence, by Theorem 25.1, we may replace the path of integration by the contour shown as heavy line in Fig. 21a), since the contribution to the value of the integral along the dashed part of the contour vanishes in the limit. The contour bypasses the two singular points $\pm i$ by circles. The function $(s^2+1)^{-1/2}$

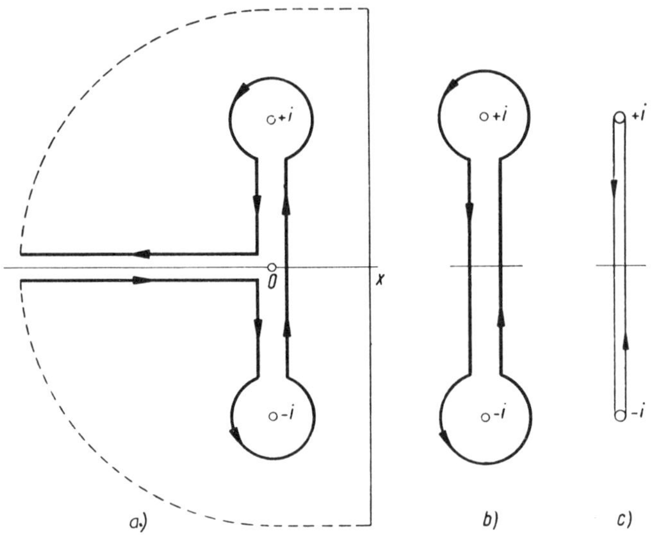

Figure 21

alters its sign whilst circling the point $+i$ and the point $-i$ respectively; hence is resumes its former sign after circling both points. Consequently, it has the same value on both rays from 0 to $-\infty$, when the in Fig. 21a) separately shown rays merge with the negative real axis. However, the two rays are passed in opposite directions, and the contributions cancel one another. Thus, we are left with the path of integration shown in Fig. 21b). Upon introducing polar coordinates, one can easily show that the contributions to the value of the integral along the circles about the points $+i$ and $-i$ respectively tend towards 0 for vanishing radii. Thus, we may employ the contour shown in Fig. 21c). As mentioned above, the function $(s^2+1)^{-1/2}$ does alter its sign when circling about $+i$, and about $-i$; thus, it assumes on the down-going line the negative of the value it attains on the up-going line, when the two separately drawn lines merge with the section of the imaginary axis. These lines are passed, during the integration, in opposite orientation; hence, instead of (7), we find:

$$J_0(t) = \frac{1}{\pi i} \int_{-i}^{+i} \frac{e^{ts}}{\sqrt{s^2+1}}\, ds \quad \text{for} \quad t > 0,$$

or, with $s = iy$,

$$J_0(t) = \frac{1}{\pi} \int_{-1}^{+1} \frac{e^{ity}}{\sqrt{1-y^2}}\, dy .$$

The left hand side, as well as the right hand side, are even functions of t; consequently, the presented formula is also valid for $t < 0$. The same formula was derived on p. 57 in an entirely different manner.

26. The Evaluation of the Complex Inversion Integral by Means of the Calculus of Residues

The complex inversion formula, either with the original, straight line path of integration or with some other, deformed path, is itself neither suited for the immediate numerical evaluation of the sought original function nor qualified to provide the desired insight into the behaviour of this original function $f(t)$. Nevertheless, this formula has practical value, for it serves as the starting point for the development of other representations which are better qualified for the above mentioned purposes.

When applying the $\mathfrak{L}$-transformation, in particular when solving partial differential equations, one often encounters the situation that the analytic continuation of the image function $F(s)$ is a meromorphic function; that is, it has only poles on the finite part of the complex plane, these poles being necessarily placed in a left half-plane. Whenever such conditions are encountered, one can use the method of residues to evaluate the inversion integral

$$f(t) = \lim_{\omega \to \infty} \frac{1}{2\pi i} \int_{a-i\omega}^{a+i\omega} e^{ts} F(s)\, ds. \tag{1}$$

A limit point of poles is a singular point, however not a pole. Hence, the poles of $F(s)$ cannot have a finite limit point. Thus, we have, at most, countably many isolated poles: $s_0, s_1, s_2, \ldots$. We may order these by increasing absolute values[1]

$$|s_0| \leqq |s_1| \leqq \cdots.$$

In the half-plane $\Re s \leq \alpha$ we draw curves $\mathfrak{C}_n$ which connect the points $\alpha + i\omega_n$ and $\alpha - i\omega_n$, so that $\mathfrak{C}_n$, together with the part of the line $\Re s = \alpha$ between these terminal points, encloses exactly the first n poles: $s_0, s_1, s_2, \ldots, s_n$, as shown in

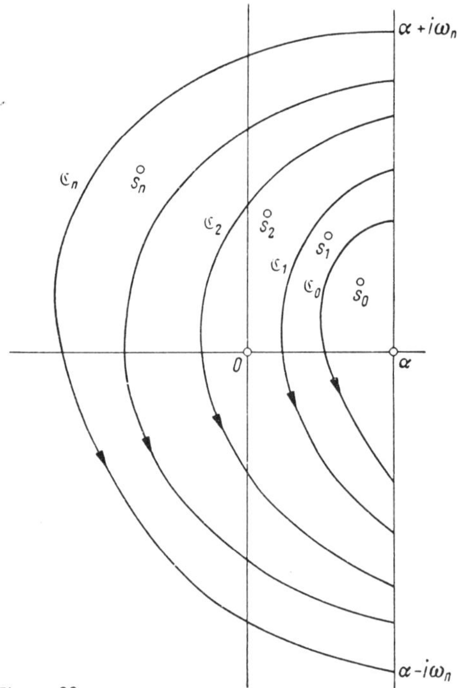

Figure 22

Fig. 22. Let $r_\nu(t)$ designate the residue of $e^{ts}F(s)$ at s_ν; then we find by means of Cauchy's residue theorem

$$(2) \qquad \frac{1}{2\pi i}\int_{a-i\omega_n}^{a+i\omega_n} e^{ts}F(s)\,ds + \frac{1}{2\pi i}\int_{\mathfrak{C}_n} e^{ts}F(s)\,ds = \sum_{\nu=0}^{n} r_\nu(t).$$

In the sequel, we shall need two hypotheses:

H1. *The curves $\mathfrak{C}_n$ are chosen so that $\omega_n \to \infty$, when $n \to \infty$.*

Consequently, we have, by (1),

$$\lim_{n\to\infty} \frac{1}{2\pi i}\int_{a-i\omega_n}^{a+i\omega_n} e^{ts}\,F(s)\,ds = f(t).$$

[1] This principle of ordering is suggested here merely for the sake of definiteness. Other principles may prove useful in practical applications.

H 2. *$F(s)$ has the property that*

$$\lim_{n\to\infty} \int_{\mathfrak{C}_n} e^{ts} F(s)\, ds = 0.$$

Then, we have, by (2),

$$f(t) = \sum_{\nu=0}^{\infty} r_\nu(t)\,; \tag{3}$$

that is, $f(t)$ is represented by an infinite series, a representation which is particularly well suited for numerical evaluation. For the attempted evaluation by (3), we need to express the residues $r_\nu(t)$ explicitly. The pole at the point s_ν is characterized by the principal part of the Laurent expansion about s_ν; for a pole s_ν of order m_ν, the principal part has the form

$$\frac{a_1^{(\nu)}}{s-s_\nu} + \cdots + \frac{a_{m_\nu}^{(\nu)}}{(s-s_\nu)^{m_\nu}}. \tag{4}$$

In order to determine $r_\nu(t)$, we multiply (4) by $e^{ts}/(2\pi i)$ and then integrate the expression along a small circle which is centred at s_ν. In this manner we obtain, when disregarding the coefficients $a_\lambda^{(\nu)}$, a sum of terms of the form

$$\frac{1}{2\pi i}\int \frac{e^{ts}}{(s-s_\nu)^k}\, ds \quad (k = 1, \cdots, m_\nu).$$

By Cauchy's formula for the derivative of an analytic function, this integral is equal to

$$\frac{1}{(k-1)!}\left(\frac{d^{k-1}}{ds^{k-1}}\, e^{ts}\right)_{s=s_\nu} = \frac{t^{k-1}}{(k-1)!}\, e^{ts_\nu}.$$

Combining these results, one finds:

$$r_\nu(t) = \left(a_1^{(\nu)} + a_2^{(\nu)}\frac{t}{1!} + \cdots + a_{m_\nu}^{(\nu)}\frac{t^{m_\nu-1}}{(m_\nu-1)!}\right) e^{s_\nu t}. \tag{5}$$

The expression (5) is, in fact, the original function of the sum (4). Accordingly, one may visualize the representation (3) of $f(t)$ as being generated by the following process: One writes formally the "partial fraction expansion" of $F(s)$:

$$F(s) \sim \sum_{\nu=0}^{\infty}\left(\frac{a_1^{(\nu)}}{s-s_\nu} + \cdots + \frac{a_{m_\nu}^{(\nu)}}{(s-s_\nu)^{m_\nu}}\right), \tag{6}$$

which is composed of its principal parts exhibiting its poles with their multiplicities; this expression is then returned to the original space, term by term. Observe that this formal process is permissible only when the specified hypotheses H 1 and H 2 are satisfied, and that the series (6) need not represent $F(s)$; indeed, it may fail to converge.

Clearly, there are functions $F(s)$ for which the hypotheses H1 and H2 are satisfied; for instance, when $F(s)$ is a rational function for which the series (6) has but a finite number of terms; in this case, (6) is the conventional partial fraction expansion of $F(s)$ (compare p. 76). Moreover, (6) may also converge for non-rational, meromorphic functions $F(s)$ with infinitely many poles, and represent this function. For instance, we know that

$$\frac{\operatorname{ctgh}\sqrt{s}}{\sqrt{s}} = \frac{1}{s} + 2\sum_{\nu=1}^{\infty} \frac{1}{s+\nu^2\pi^2}. \tag{7}$$

However, this is by no means decisive for the validity of the representation (3) of the original function $f(t)$; the essential requirement is the verification of the hypotheses H1 and H2.

This fact is emphasized here, for one frequently encounters, particularly in the technical literature, the application of the following, illegitimate procedure: The poles s_ν of the meromorphic function $F(s)$, and the corresponding coefficients $a_\lambda^{(\nu)}$ are determined. Incidentally, this determination is particularly easy when $F(s)$ is given in the form:

$$F(s) = \frac{p(s)}{q(s)},$$

where both $p(s)$ and $q(s)$ are entire functions, which have no common zeros; and the poles of $F(s)$, that is the zeros of $q(s)$, are all simple poles. This is, for example, true for the function (7) with $p(s) = \cosh\sqrt{s}$, and $q(s) = \sqrt{s}\sinh\sqrt{s}$. In this case only coefficients $a_1^{(\nu)}$ are encountered, which may be determined by:

$$a_1^{(\nu)} = \lim_{s\to s_\nu} \frac{p(s)}{q(s)}(s-s_\nu) = \lim_{s\to s_\nu} \frac{p(s)}{\dfrac{q(s)-q(s_\nu)}{s-s_\nu}} = \frac{p(s_\nu)}{q'(s_\nu)}.$$

With these coefficients, $F(s)$ is set equal to the expression (6); that is, in the case that only simple poles are encountered:

$$F(s) = \sum_{\nu=0}^{\infty} \frac{a_1^{(\nu)}}{s-s_\nu} = \sum_{\nu=0}^{\infty} \frac{p(s_\nu)}{q'(s_\nu)} \frac{1}{s-s_\nu},$$

just as if $F(s)$ were a rational function, which has merely a finite number of poles (compare (15.8)); finally, this representation is returned to the original space, term by term. That is, for the special case involving only simple poles:

$$f(t) = \sum_{\nu=0}^{\infty} \frac{p(s_\nu)}{q'(s_\nu)} e^{s_\nu t}. \tag{8}$$

The thus executed, termwise inverse transformation involves the interchange of two limiting processes: infinite summation and integration; this is not always permissible. Moreover, the representation (6) for $F(s)$ is not correct in all cases; a fact that can easily be demonstrated by simple examples. For instance, an entirely possible, non-rational meromorphic function with a finite number of poles would, by (6), be identified with a rational function. The true facts are:

the difference between $F(s)$ and the series (6) does not have finite singular points; that is, it is an entire function which ought to be added to (6). This is true in general: although the infinite series (6) may converge, a behaviour which depends upon both the poles s_ν and the coefficients $a_\lambda^{(\nu)}$, it may differ from $F(s)$ by an entire function. The determination of the latter is particularly difficult, for it reflects the character of the singularity of $F(s)$ at infinity. Consequently, any attempt to legalize the above described, naive process is a difficult task indeed. By contrast, the initially explained proper method avoids the question as to whether or not $F(s)$ can be expanded into a series (6); instead, it constructs $f(t)$ directly, using the residues.

Naturally, the suggested method does not provide an infallable recipe. For, to begin with, the expansion (3) of $f(t)$ need not exist; a fact that follows from the above discussion regarding the entire function that may be contained in $F(s)$. Moreover, although the expansion may exist, it may be impossible to find, for $F(s)$, the necessary family of curves $\mathfrak{C}_n$ which satisfy the hypothesis H2; for the curves $\mathfrak{C}_n$ must be located with proper consideration of the character of the function $F(s)$, so that the behaviour of the latter can be predicted along these curves. Usually one starts with circular arcs to the left, centred at the origin, which, for $\alpha > 0$, are to be extended in an appropriate manner to the line $\Re s = \alpha$. For, if we can show that the maximum of $|F(s)|$ on the circular arcs tends towards 0, when the radius grows, then the part of the integral evaluated along these semicircles tends towards zero, by Theorem 25.1. Usually, it is not difficult to handle the integrals along the above mentioned extensions towards the line $\Re s = \alpha$. Occasionally, the use of rectangles, parabolas, and the like in place of the above suggested circular arcs proves fruitful.

As an example, for which portions of circles suffice, consider the function[2]

$$F(s) = \frac{1}{s^2}\,\frac{\sinh x s}{\cosh a s} \qquad (0 \leqq x \leqq a),$$

a function which may be encountered while solving hyperbolic boundary value problems by means of the $\mathfrak{L}$-transformation. Here, we have to select $\alpha > 0$. The numerator $\sinh x s$ has a zero at $s = 0$; consequently, $F(s)$ has a simple pole at $s = 0$. Further simple poles of $F(s)$ are the zeros of $\cosh as$; these are: $\pm(2\nu - 1)(\pi/2a)\,i$ for $\nu = 1, 2, 3, \cdots$. We select circles centred at the origin having the radii $\varrho_\gamma = \nu(\pi/a)$, so that exactly paired poles are located between successive circles; consequently, in the sum (3) we must combine in *one* term the respective residues of each pair of poles. To generate suitable curves $\mathfrak{C}_n$, we use the portions of these circles to the left of $\Re s = \alpha$, as shown in Fig. 23; the finite number of circles which do not intersect the line $\Re s = \alpha$ may be disregarded.

---

[2] The function

$$\frac{\sinh x s}{\cosh a s} = \frac{e^{xs} - e^{-xs}}{e^{as} + e^{-as}} = \frac{e^{-(a-x)s} - e^{-(a+x)s}}{1 + e^{-2as}}$$

is bounded in every half-plane $\Re s \geqq x > 0$. It follows, by Theorem 28.3, that $F(s)$ is a $\mathfrak{L}$-transform which may be returned to the original space by means of the complex inversion formula.

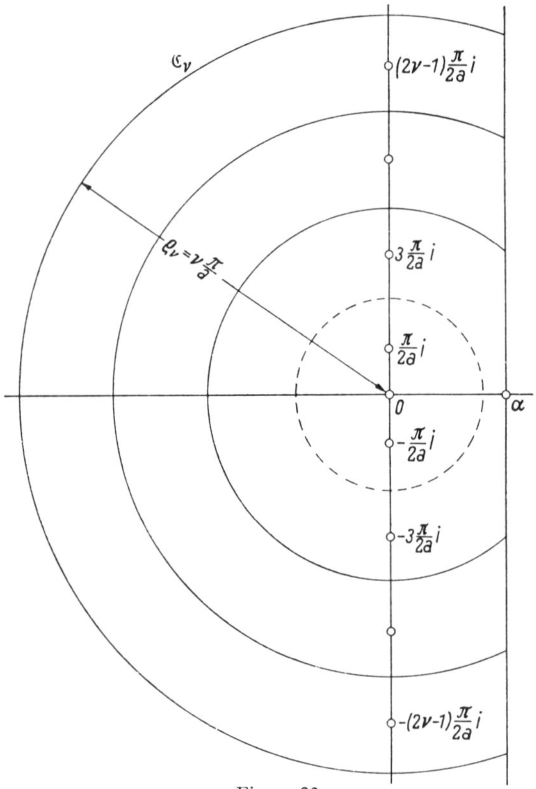

Figure 23

For this particular example, we can demonstrate that the function

$$\varphi(s) = \frac{\sinh x s}{\cosh a s} \quad (0 \leqq x \leqq a)$$

is bounded on the entire circles about the origin having radii ϱ_ν, and that the maximum of $|F(s)|$ on these circles tends towards 0, when $\varrho_\nu \to \infty$. When using $s = \varrho_\nu e^{i\vartheta}$, we find for the numerator of $F(s)$:

$$\sinh(x\,\varrho_\nu\,e^{i\vartheta}) = \sinh(x\,\varrho_\nu\cos\vartheta + i\,x\,\varrho_\nu\sin\vartheta)$$

$$= \sinh(x\,\varrho_\nu\cos\vartheta)\cosh(i\,x\,\varrho_\nu\sin\vartheta) + \cosh(x\,\varrho_\nu\cos\vartheta)\sinh(i\,x\,\varrho_\nu\sin\vartheta)$$

$$= \sinh(x\,\varrho_\nu\cos\vartheta)\cos(x\,\varrho_\nu\sin\vartheta) + \cosh(x\,\varrho_\nu\cos\vartheta)\,i\sin(x\,\varrho_\nu\sin\vartheta)\,,$$

and, therefore,

$$2\,|\sinh(x\,\varrho_\nu\,e^{i\vartheta})|^2$$

$$= 2\sinh^2(x\,\varrho_\nu\cos\vartheta)\cos^2(x\,\varrho_\nu\sin\vartheta) + 2\cosh^2(x\,\varrho_\nu\cos\vartheta)\sin^2(x\,\varrho_\nu\sin\vartheta)$$

$$= [\cosh(2\,x\,\varrho_\nu\cos\vartheta) - 1]\cos^2(x\,\varrho_\nu\sin\vartheta)$$
$$+ [\cosh(2\,x\,\varrho_\nu\cos\vartheta) + 1]\sin^2(x\,\varrho_\nu\sin\vartheta)$$

$$= \cosh(2\,x\,\varrho_\nu\cos\vartheta) - \cos(2\,x\,\varrho_\nu\sin\vartheta)\,. \tag{9}$$

Similarly, for the denominator of $F(s)$ we find:

$$(10)\qquad 2\,|\cosh(a\,\varrho_\nu\,e^{i\vartheta})|^2 = \cosh(2\,a\,\varrho_\nu\cos\vartheta) + \cos(2\,a\,\varrho_\nu\sin\vartheta) = \cosh(2\,\nu\,\pi\cos\vartheta) + \cos(2\,\nu\,\pi\sin\vartheta) = d_\nu\,.$$

A lower bound for the last expression is needed. Without loss of generality we may restrict the independent variable to the first quadrant, $0 \leqq \vartheta \leqq \pi/2$, since, obviously, identical values of d_ν are encountered in the other quadrants. We have $\cosh \geqq 1$, and $-1 \leqq \cos \leqq 1$. The expression (10) may approach the dangerous value zero only when both the cosh is approximately 1 and the cos is near -1; for the former, we require ϑ to be near $\pi/2$; however, near $\pi/2$, $\cos(2\pi\nu\sin\vartheta)$ is positive, hence $d_\nu \geqq \cosh(2\nu\pi\cos\vartheta)$. We can describe the situation more precisely: In the interval $0 \leqq \vartheta \leqq \pi/2$, where $0 \leqq 2\nu\pi\sin\vartheta \leqq 2\nu\pi$, we have $\cos(2\nu\pi\sin\vartheta) = 0$ for

$$2\,\nu\,\pi\sin\vartheta = \frac{\pi}{2},\ \frac{3}{2}\pi,\ \ldots,\ \left(2\,\nu - \frac{1}{2}\right)\pi,$$

that is, for

$$\sin\vartheta = \frac{1}{4\nu},\ \frac{3}{4\nu},\ \ldots,\ \frac{4\nu-1}{4\nu}.$$

Let ϑ_0 designate the ϑ of the last zero; it is given by

$$\sin\vartheta_0 = \frac{4\nu-1}{4\nu} \quad\text{and}\quad \cos\vartheta_0 = \frac{\sqrt{8\nu-1}}{4\nu}.$$

Then, we have

$$\cos(2\,\nu\,\pi\sin\vartheta) \geqq 0 \qquad \text{for } \vartheta_0 \leqq \vartheta \leqq \frac{\pi}{2},$$

and, in the same interval,

$$d_\nu \geqq \cosh(2\,\nu\,\pi\cos\vartheta)\,.$$

On the remainder of the first quadrant: $0 \leqq \vartheta < \vartheta_0$, we find

$$\begin{aligned} d_\nu &= \cosh(2\,\nu\,\pi\cos\vartheta)\left[1 + \frac{\cos(2\nu\pi\sin\vartheta)}{\cosh(2\nu\pi\cos\vartheta)}\right] \\ &\geqq \cosh(2\,\nu\,\pi\cos\vartheta)\left[1 - \frac{1}{\cosh(2\nu\pi\cos\vartheta_0)}\right] \\ &= \cosh(2\,\nu\,\pi\cos\vartheta)\left[1 - \frac{1}{\cosh\left(\frac{\pi}{2}\sqrt{8\nu-1}\right)}\right] \\ &\geqq \cosh(2\,\nu\,\pi\cos\vartheta)\left[1 - \frac{1}{\cosh\left(\frac{\pi}{2}\sqrt{7}\right)}\right] \\ &= C\cosh(2\,\nu\,\pi\cos\vartheta) \text{ for } \nu \geqq 1\,. \end{aligned}$$

The last lower bound is also a lower bound on the interval $\vartheta_0 \leq \vartheta \leq \pi/2$, because of $C < 1$; hence, it may serve as a lower bound on the entire quadrant: $0 \leqq \vartheta \leqq \pi/2$.

From (9) we conclude that:

$$2 \left| \sinh \left(x\, \varrho_\nu\, e^{i\vartheta}\right) \right|^2 \leqq \cosh \left(2\, \nu\, \pi \frac{x}{a} \cos \vartheta\right),$$

and, consequently,

$$\left| \varphi \left(\varrho_\nu\, e^{i\vartheta}\right) \right|^2 \leqq \frac{\cosh\left(2\nu\pi \frac{x}{a} \cos\vartheta\right)}{C \cosh\left(2\nu\pi\cos\vartheta\right)} \leqq \frac{1}{C} \quad \text{for } 0 \leqq \frac{x}{a} \leqq 1\,.$$

We have thus discovered that

$$|F(s)| \leqq \frac{1}{C\varrho_\nu^2} \to 0, \text{ when } n \to \infty,$$

on the left semicircles of radii ϱ_ν. It follows, by Theorem 25.1, that the contribution to the integral on these semicircles tends towards zero, when $n \to \infty$.

$F(s)$ tends, uniformly on the extensions of the semicircles to the line $\Re s = \alpha$, towards 0; both the function e^{ts} and the lengths of these extensions are bounded. Thus, we conclude that the contribution to the integrals along these extensions tends towards zero, when $n \to \infty$.

Thus, we have shown that the expansion (3) does indeed represent $f(t)$; for the presented problem we have only simple poles, hence (8) may be used to evaluate (3). We find, for $\nu \leqq 1$,

$$p(s_\nu) = \sinh x\left(\pm\, (2\,\nu - 1)\, \frac{\pi}{2\,a}\, i\right) = \pm\, i \sin\,(2\,\nu - 1)\, \frac{x}{a}\, \frac{\pi}{2}\,;$$

$$q'(s) = 2\, s \cosh a\, s + s^2\, a \sinh a\, s\,,$$

$$\begin{aligned} q'(s_\nu) &= \pm\, 2(2\,\nu - 1)\, \frac{\pi}{2a}\, i \cosh a\left(\pm\, (2\,\nu - 1)\, \frac{\pi}{2a}\, i\right) \\ &\quad - (2\,\nu - 1)^2\, \frac{\pi^2}{4\,a^2}\, a \sinh a\left(\pm\, (2\,\nu - 1)\, \frac{\pi}{2a}\, i\right) \\ &= \mp\, 2(2\,\nu - 1)\, \frac{\pi}{2a} \cos\,(2\,\nu - 1)\, \frac{\pi}{2} \mp i(2\,\nu - 1)^2\, \frac{\pi^2}{4\,a} \sin\,(2\,\nu - 1)\, \frac{\pi}{2} \\ &= \pm\, i(-1)^\nu\, (2\,\nu - 1)^2\, \frac{\pi^2}{4\,a}\,. \end{aligned}$$

Combining both residues corresponding to the same ν one finds:

$$\begin{aligned} &\frac{\sin(2\nu - 1)\, \dfrac{x}{a}\, \dfrac{\pi}{2}}{(-1)^\nu\, (2\nu - 1)^2\, \dfrac{\pi^2}{4\,a}} \left(e^{(2\nu - 1)(\pi/2a)\, it} + e^{-(2\nu - 1)(\pi/2a)\, it}\right) \\ &= \frac{8\,a}{\pi^2}\, \frac{(-1)^\nu}{(2\nu - 1)^2} \sin \frac{(2\nu - 1)\,\pi x}{2\,a} \cos \frac{(2\nu - 1)\,\pi t}{2\,a}\,. \end{aligned}$$

The residue at $s = 0$ is given by

$$\lim_{s \to 0} \frac{\frac{\sinh x s}{s}}{\cosh a s} e^{ts} = x .$$

Hence, we find:

$$f(t) = x + \frac{8a}{\pi^2} \sum_{\nu=1}^{\infty} \frac{(-1)^\nu}{(2\nu-1)^2} \sin \frac{(2\nu-1)\pi x}{2a} \cos \frac{(2\nu-1)\pi t}{2a} \quad (0 \leqq x \leqq a) .$$

This function has the period $4a$ in t. We should expect this, since $F(s)$, when extended by $(e^{as} - e^{-as})e^{-2as}$, yields:

$$F(s) = \frac{1}{s^2} \frac{e^{xs} - e^{-xs}}{e^{as} + e^{-as}} = \frac{1}{s^2} \frac{e^{-(a-x)s} - e^{-(a+x)s} - e^{-(3a-x)s} + e^{-(3a+x)s}}{1 - e^{-4as}} .$$

Comparing this last expression with formula (7.2), we conclude that $f(t)$ has the period $4a$, and that the 'finite' $\mathfrak{L}$-transform evaluated over the interval $(0, 4a)$ has the form:

$$\int_0^{4a} e^{-st} f(t)\, dt = \frac{1}{s^2} \left\{ \left(1 - e^{-(a+x)s}\right) - \left(1 - e^{-(a-x)s}\right) - \left(1 - e^{-(3a+x)s}\right) + \left(1 - e^{-(3a-x)s}\right) \right\} .$$

This expression enables us to determine $f(t)$ explicitly. To simplify the problem, we first consider a modified image function, replacing the factor $1/s^2$ by $1/s$. We use

$$\frac{1}{s} \left(1 - e^{-(a+x)s}\right) = \int_0^{a+x} e^{-st}\, dt \circ\!\!-\!\!\bullet \begin{cases} 1 & \text{for } 0 < t < a + x \\ 0 & \text{for } a + x < t < 4a . \end{cases}$$

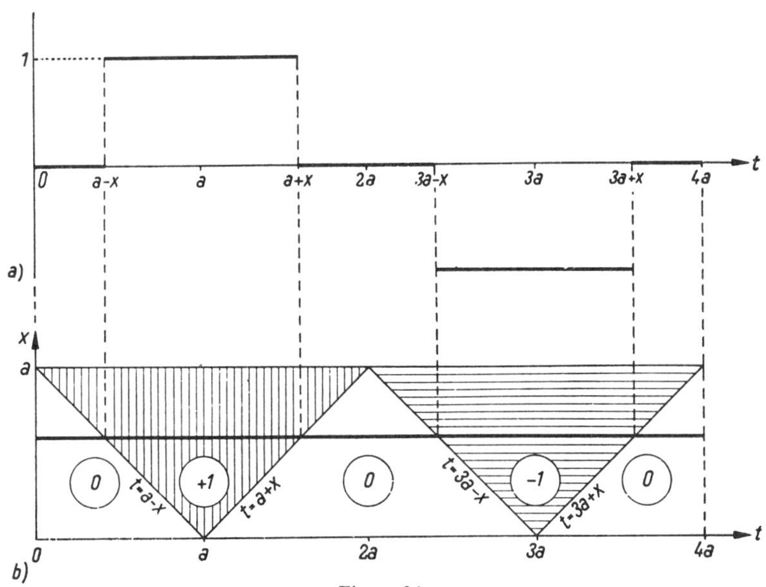

Figure 24

Proceeding similarly with the other terms, and superimposing the thus produced functions using the appropriate signs, we generate the original function of the modified image function for a fixed x ($0 \leqq x \leqq a$):

t-Interval	$(0, a-x)$	$(a-x, a+x)$	$(a+x, 3a-x)$	$(3a-x, 3a+x)$	$(3a+x, 4a)$
Value of the Function	0	+1	0	−1	0

The original function of the modified image function is shown in Fig. 24; in Fig. 24a) it is shown for some fixed x, in Fig. 24b) it is shown as a function of two independent variables, x and t. The image function actually investigated is obtained from the modified image function through multiplication by $(1/s)$; accordingly, we need to integrate the original function of Fig. 24 between 0 and t in order to generate the actually sought original function.

27. The Complex Inversion Formula for the Simply Converging Laplace Transformation

The inversion formula of Theorem 24.4 is contingent on two restricting hypotheses: $\mathfrak{L}\{f\}$ must have a half-plane of absolute convergence, and f must be of bounded variation near the point t where $f(t)$ is to be evaluated. Using the Integration Theorem 8.1, we can develop an inversion formula which is not contingent on these two hypotheses.

Whenever $\mathfrak{L}\{f\} = F(s)$ converges for a real $s = x_0 > 0$, then by Theorem 8.1, $\mathfrak{L}\{\varphi\}$, with

$$\varphi(t) = \int_0^t f(\tau)\, d\tau,$$

converges absolutely for $\Re s > x_0$, and $\mathfrak{L}\{\varphi\} = F(s)/s$. This conclusion is also true for $x_0 = 0$, since for every s with $\Re s > 0$ we can insert an x'_0, with $0 < x'_0 < \Re s$, where $\mathfrak{L}\{f\}$ converges. Moreover, $\varphi(t)$ is of bounded variation on every finite interval $0 \leqq t \leqq T$; this fact is demonstrated by the following argumentation. Selecting any arbitrary partitioning $0 = t_0 < t_1 < \cdots < t_{n-1} < t_n = T$, we find[1]:

$$\sum_{\nu=1}^{n} |\varphi(t_\nu) - \varphi(t_{\nu-1})| = \sum_{\nu=1}^{n} \left| \int_{t_{\nu-1}}^{t_\nu} f(\tau)\, d\tau \right| \leqq \sum_{\nu=1}^{n} \int_{t_{\nu-1}}^{t_\nu} |f(\tau)|\, d\tau = \int_0^T |f(\tau)|\, d\tau.$$

[1] Here we recognize two facts: Firstly, for theoretical considerations, the condition of bounded variation for Theorem 24.4 is far more practical than the original Dirichlet condition of monotonic behaviour. Secondly, the condition that $f(t)$ must be *absolutely* integrable in every finite interval (see p. 12) is, in fact, needed.

Also, $\varphi(t)$ is continuous, and $\varphi(0^+) = 0$. Applying Theorem 24.4 to $\varphi(t)$, we obtain

Theorem 27.1. *When* $\mathfrak{L}\{f\} = F(s)$ *converges (simply) for some real* $s = x_0 \geqq 0$, *then we find, with* $x > x_0 \geqq 0$:

$$\text{(1)} \qquad V.P.\ \frac{1}{2\pi i}\int_{x-i\infty}^{x+i\infty} e^{ts}\frac{F(s)}{s}\,ds = \begin{cases} \int_0^t f(\tau)\,d\tau & \textit{for } t \geqq 0 \\ 0 & \textit{for } t < 0. \end{cases}$$

The expression (1) is formally obtained from the previous inversion formula (24.15) by means of the following process: First we normalize $f(t)$; that is, at every point where the graph of $f(t)$ exhibits a jump, we specify:

$$f(t) = \frac{f(t^+) + f(t^-)}{2},$$

thereby not altering the value of the integral of f. Next, we define $f(t) = 0$, for $t < 0$; with this latter modification we repeat the step that was taken when (24.14) was altered into (24.15). Finally, we integrate formula (24.15) not between the limits 0 and t but between $-\infty$ and t, on the right hand side under the integral symbol, where now $x = \Re s$ must be positive as specified, in fact, by Theorem 27.1, for only then is

$$\int_{-\infty}^{t} e^{\tau s}\,d\tau = \frac{e^{ts}}{s};$$

observe that this integral would diverge for $\Re s \leqq 0$.

It is self-evident (compare p. 153) that the value of $f(t)$ cannot be calculated from $F(s)$ without further hypotheses. By contrast, the integral of $f(t)$ can be evaluated without further restrictions. At every point of continuity, the function $f(t)$ can be obtained by differentiation of $\varphi(t)$.[2] In the Lebesgue theory, one concludes that $\int_0^t f(\tau)\,d\tau$ is differentiable almost everywhere, and that the derivative is indeed equal to $f(t)$ almost everywhere.

Observe that the integral (1) must be evaluated along a vertical line with *positive* abscissa; this requirement must be satisfied although the half-plane of convergence of $\mathfrak{L}\{f\}$ may extend into the left half of the complex plane. Using argumentation similar to that employed on p. 153 which is borrowed from the theory of functions, we can find what value the integral assumes, in the case that the integral is evaluated along a line with *negative* abscissa.

---

[2] Differentiation of an integral with respect to the upper limit may also be permissible at points of discontinuity of the integrand, and it may correctly produce the value of the integrand. As an example consider the function:

$$f(t) = 1 \text{ for } t = \frac{1}{n}\ (n = 1, 2, \cdots), \quad f(t) = 0 \quad \text{elsewhere.}$$

We have $\int_0^t f(\tau)\,d\tau \equiv 0$ for all t, and consequently $d/dt \int_0^t f(\tau)\,d\tau \equiv 0$; at the point $t = 0$, where $f(t)$ is discontinuous, differentiation yields the correct value $f(0) = 0$.

Suppose that $\mathfrak{L}\{f\}$ converges for real $s = x_0 < 0$ and, consequently, for all $\Re s > x_0$. Let us select two abscissae x_1 and x_2 so that $x_0 < x_1 < 0 < x_2$, and construct with these a rectangle as shown in Fig. 25. The function $F(s)/s$ is analytic inside of and on the rectangle with the exception of the interior point $s = 0$, where, in general, $F(s)/s$ has a simple pole; only when $F(0) = 0$ is $F(s)/s$ holomorphic at $s = 0$. The integral

$$\frac{1}{2\pi i}\int e^{ts}\frac{F(s)}{s}\,ds,$$

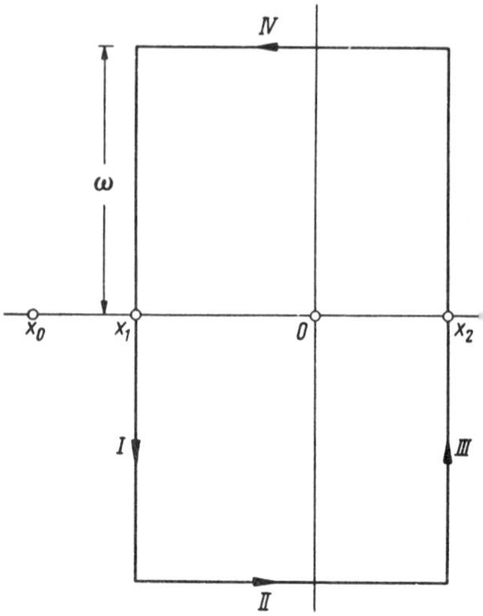

Figure 25

evaluated along the boundary of the rectangle in the positive sense yields the residue of $e^{ts}F(s)/s$ at 0; that is, $F(0)$, when $s = 0$ is a simple pole.[3] The value of the integral is zero when the integrand is holomorphic at $s = 0$, that is when $F(0) = 0$. In either case, the value of the integral is $F(0)$, that is:

$$\frac{1}{2\pi i}\left(\int_{\mathrm{I}} + \int_{\mathrm{II}} + \int_{\mathrm{III}} + \int_{\mathrm{IV}} e^{ts}\frac{F(s)}{s}\,ds\right) = F(0).$$

In Theorem 23.8 we have shown that the function $F(s)/s = F(x+iy)/(x+iy)$ tends, uniformly in $x_1 \leqq x \leqq x_2$, towards zero, when $y \to \pm\infty$. Thus, we may recall the arguments of p. 159, and we conclude that the contribution to the value of the integral along the sides II and IV of the rectangle vanishes in the limit, when $\omega \to \infty$. The integral along the right vertical line III of the rectangle converges, by Theorem 27.1, towards $\int_0^t f(\tau)\,d\tau$ for $t \geqq 0$, and towards 0 for $t < 0$, for we have $x_2 > 0$. It also follows that the integral along the left vertical line I of the rectangle has a limit. We have

$$\lim_{\omega\to\infty}\frac{1}{2\pi i}\int_{x_1+i\omega}^{x_1-i\omega} e^{ts}\frac{F(s)}{s}\,ds + \begin{Bmatrix} \int_0^t f(\tau)\,d\tau & \text{for } t \geqq 0 \\ 0 & \text{for } t < 0 \end{Bmatrix} = F(0).$$

Because of

$$F(0) = \int_0^\infty f(\tau)\,d\tau$$

[3] Suppose that $\varphi(s)$ is analytic at $s = 0$; then

$$\frac{\varphi(s)}{s} = \frac{\varphi(0)}{s} + \frac{\varphi'(0)}{1!} + \frac{\varphi''(0)}{2!}s + \cdots.$$

At $s = 0$, the residue of $\varphi(s)/s$ is $\varphi(0)$, the coefficient of $1/s$ of the expansion.

we may rewrite the last expression thus:

$$\lim_{\omega\to\infty} \frac{1}{2\pi i} \int_{x_1-i\omega}^{x_1+i\omega} e^{ts} \frac{F(s)}{s}\, ds = \begin{cases} -\int_t^\infty f(\tau)\, d\tau & \text{for } t \geqq 0 \\ -\int_0^\infty f(\tau)\, d\tau & \text{for } t < 0. \end{cases}$$

We summarize the above conclusions in

Theorem 27.2. *When $\mathfrak{L}\{f\} = F(s)$ converges (simply) for some real $s = x_0 < 0$, then we have, for $x_0 < x < 0$,*

(2) $$V.P. \frac{1}{2\pi i} \int_{x-i\infty}^{x+i\infty} e^{ts} \frac{F(s)}{s}\, ds = \begin{cases} -\int_t^\infty f(\tau)\, d\tau & \textit{for } t \geqq 0 \\ -\int_0^\infty f(\tau)\, d\tau & \textit{for } t < 0. \end{cases}$$

When defining once and for all $f(t) = 0$ for $t < 0$, then we may write in formula (1): $\int_0^t f(\tau)\, d\tau$, and in formula (2): $-\int_t^\infty f(\tau)\, d\tau$ for all $t \lesseqgtr 0$.

Theorem 27.1 provides a *new verification of the Uniqueness Theorem* 5.1, since $F(s) \equiv 0$ yields $\int_0^t f(\tau)\, d\tau \equiv 0$ for $t > 0$.

As an example which demonstrates the applicability of the above Theorems 27.1 and 27.2, we develop conditions which will enable us to determine whether or not some given image function $F(s)$ is, in fact, a "finite" $\mathfrak{L}$-transform.

Theorem 27.3. *The conditions*

a) *$F(s)$ is an entire function*

$$\left.\begin{array}{l} \text{b) } |F(x+iy)| \leqq C \\ \text{c) } |F(-x+iy)| \leqq C\, e^{hx} \end{array}\right\} \textit{for } x \geqq 0$$

are necessary and sufficient for the fact that the original function $f(t)$ of the $\mathfrak{L}$-transform $F(s) = F(x+iy)$ is a nullfunction for $t > h \geqq 0$, that is, that

$$\int_h^t f(\tau)\, d\tau = 0 \quad \textit{for } t > h.$$

Proof: 1. *Necessity.* When $f(t)$ is a nullfunction for $t > h$, that is, when $f(t)$ vanishes almost everywhere on $t > h$, then $F(s)$ has the form:

$$F(s) = \int_0^h e^{-st} f(t)\, dt.$$

By Theorem 6.2, $F(s)$ is an entire function. Moreover,

$$|F(s)| \leqq \int_0^h e^{-xt} |f(t)| \, dt \leqq \begin{cases} \int_0^h |f(t)| \, dt = C & \text{for } x \geqq 0 \\ e^{-hx} \int_0^h |f(t)| \, dt = C e^{-hx} & \text{for } x \leqq 0. \end{cases}$$

2. *Sufficiency*. Suppose that $F(s) = \mathfrak{L}\{f\}$ converges (simply) for $s = x_0 \geqq 0$; then, by Theorem 27.1, we find, for $t \geqq 0$,

(3) $$\int_0^t f(\tau) \, d\tau = \lim_{\omega \to \infty} \frac{1}{2\pi i} \int_{\xi - i\omega}^{\xi + i\omega} e^{ts} \frac{F(s)}{s} \, ds \quad (\xi > x_0).$$

We consider the contour of integration $\mathfrak{C}$ shown in Fig. 26. It is composed of the portion of the vertical line between $\xi - i\omega$ and $\xi + i\omega$, the left semicircle $\mathfrak{h}$ centred

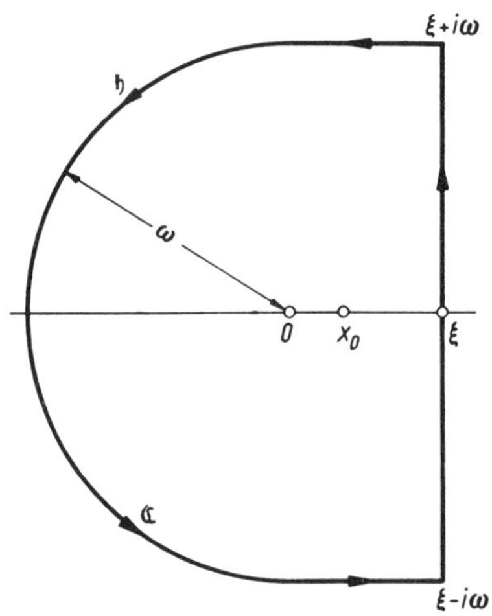

Figure 26

at the origin having radius ω, and the horizontal line sections that close the contour. The function $e^{ts} F(s)$ is analytic inside of and on $\mathfrak{C}$ for every $\omega > 0$; hence, by Cauchy's formula,

(4) $$\frac{1}{2\pi i} \int_{\mathfrak{C}} e^{ts} \frac{F(s)}{s} \, ds = F(0).$$

For the upper horizontal line segment, hypothesis b) guarantees that

(5) $$\left| \int_{\xi + i\omega}^{i\omega} e^{ts} \frac{F(s)}{s} \, ds \right| \leqq \int_0^{\xi} e^{tx} \frac{C}{\omega} \, dx = \frac{C}{\omega} \, \frac{e^{t\xi} - 1}{t} \to 0$$

when $\omega \to \infty$, for every fixed value $t > 0$. The same conclusion follows for the lower line segment. On the left semicircle $\mathfrak{h}$, we may set

$$s = \omega e^{i\vartheta}, \quad \pi/2 \leqq \vartheta \leqq 3\pi/2, \quad ds/s = i\, d\vartheta,$$

hence, by hypothesis c), with $\vartheta = \pi/2 + \varphi$,

$$\left| \int_{\mathfrak{h}} e^{ts} \frac{F(s)}{s} ds \right| \leqq \int_{\pi/2}^{3\pi/2} e^{t\omega\cos\vartheta} C e^{-h\omega\cos\vartheta} d\vartheta = C \int_0^{\pi} e^{-(t-h)\omega\sin\varphi} d\varphi$$

$$= 2C \int_0^{\pi/2} e^{-(t-h)\omega\sin\varphi} d\varphi.$$

The sine-curve is, in the interval $0 \leq \varphi \leq \pi/2$, above the straight line $2\varphi/\pi$; that is, in this interval, $\sin\varphi \geqq 2\varphi/\pi$.
Thus, we have:

$$\text{(6)} \qquad \left| \int_{\mathfrak{h}} e^{ts} \frac{F(s)}{s} ds \right| \leqq 2C \int_0^{\pi/2} e^{-(t-h)\omega\, 2\varphi/\pi} d\varphi = 2C \frac{1 - e^{-(t-h)\omega}}{(2/\pi)(t-h)\omega} \to 0$$

for $t - h > 0$, when $\omega \to \infty$. From (5) and (6) we conclude, for $t > h$, that in the limit $\omega \to \infty$ one is left in (4) with the integral along the vertical line, that is (3); consequently, we find that

$$\int_0^t f(\tau)\, d\tau = F(0) \quad \text{for } t > h,$$

and therefore,

$$\lim_{t \to h+0} \int_0^t f(\tau)\, d\tau = \int_0^h f(\tau)\, d\tau = F(0);$$

hence,

$$\int_h^t f(\tau)\, d\tau = 0 \quad \text{for } t > h.$$

Theorem 27.3 can be employed in physical applications to determine, by inspection of the image function, whether the corresponding original "signal" $f(t)$ is "active" only during a finite time interval of duration h.

28. Sufficient Conditions for the Representability as a Laplace Transform of a Function

When asking what class of functions can be represented by a power series, $\sum_{\nu=0}^{\infty} a_\nu z^\nu$, we obtain a simple answer: all those functions which are analytic on a circular disk centred at the origin. As an analogy to the above answer, one might expect that the class of functions which can be represented as $\mathfrak{L}$-transforms is composed of those functions which are analytic in right half-planes. However, merely a subset of all functions which are analytic in right half-planes constitutes, in fact, the class of functions which can be represented as $\mathfrak{L}$-transforms. Our aim is to delineate this subset. While investigating this problem, we shall have to distinguish between functions that can be represented as $\mathfrak{L}$-transforms of functions and those which can be represented as $\mathfrak{L}$-transforms of distributions. For the subset of those functions which can be represented as $\mathfrak{L}$-transforms of *functions* there exists no simple criterion in terms of the theory of analytic functions. Only *sufficient conditions* are known which describe merely a portion of the sought subset of representable functions. By contrast, the subset of functions which may be represented as $\mathfrak{L}$-transforms of *distributions* is characterized by a *necessary and sufficient condition* which is taken from the theory of functions. In this Chapter 28, we deal with the representability as $\mathfrak{L}$-transforms of functions.

For this purpose, we discuss a preliminary consideration. Any solution of the representation problem is complete only when it provides not only confirmation of the possibility of the representation $F(s) = \mathfrak{L}\{f\}$ but also instructions which enable us to find the original function $f(t)$; that is, some inversion formula $\mathfrak{L}^{-1}\{F\} = f(t)$ must be provided. Previously, when dealing with the *inversion problem*, we aimed to determine those original functions $f(t)$ whose image functions $F = \mathfrak{L}\{f\}$ can, by means of the inversion formula $\mathfrak{L}^{-1}\{F\}$, be returned to the initially considered function f; that is, what conditions regarding f do guarantee the relation

$$\mathfrak{L}^{-1}\{\mathfrak{L}\{f\}\} = f.$$

By contrast, when investigating the *representation problem*, we seek to find those image functions $F(s)$ for which the by means of the inversion formula $\mathfrak{L}^{-1}\{F\}$ produced function f permits a representation of F as $\mathfrak{L}\{f\}$; that is, we are searching for the conditions regarding F which must be satisfied such that

$$\mathfrak{L}\{\mathfrak{L}^{-1}\{F\}\} = F. \tag{1}$$

This problem may be discussed in connection with any of the inversion formulae of the $\mathfrak{L}$-transformation; thus, we shall here discuss the problem in connection with the complex inversion formula of Chapter 24, which is the only inversion formula presented in this book. It actually applies to the $\mathfrak{L}_{II}$-transformation, therefore, we shall firstly consider this. A $\mathfrak{L}_{II}$-transform is, necessarily, analytic in a strip $x_1 < \Re s < x_2$. Thus, we can restate our problem:

What (sufficient) conditions must be satisfied by a function $F(s)$, which is analytic in $x_1 < \Re s < x_2$, such that the formula

$$\int_{-\infty}^{+\infty} e^{-st}\, dt \text{ V.P. } \frac{1}{2\pi i} \int_{x-i\infty}^{x+i\infty} e^{t\sigma} F(\sigma)\, d\sigma = F(s) \quad (x_1 < \Re s < x_2,\ x_1 < x < x_2) \tag{2}$$

is valid? As a first, obvious condition we require that the inner integral of (2) must be independent of x, for this integral ought to represent $f(t)$. (Moreover, the right hand side of (2) is independent of x; hence, also the left hand side). Using Cauchy's theorem and the considerations of p. 159, we realize that the above requirement is satisfied provided $F(s)$ tends, uniformly in every narrower strip, towards zero, when $|\mathfrak{I}s| \to \infty$.

Thus, since we are free to select x, we may choose $x = \mathfrak{R}s$, hence: $s = x + iy$, $\sigma = x + i\eta$. With these new variables we find, for (2), the equation:

$$F(x+iy) = \int_{-\infty}^{+\infty} e^{-(x+iy)t}\,dt\ \text{V.P.}\ \frac{1}{2\pi}\int_{-\infty}^{+\infty} e^{t(x+i\eta)} F(x+i\eta)\,d\eta$$

(3) $$= \frac{1}{2\pi}\int_{-\infty}^{+\infty} e^{-iyt}\,dt\ \text{V.P.}\int_{-\infty}^{+\infty} e^{it\eta} F(x+i\eta)\,d\eta.$$

Formula (3) is, in fact, the Fourier theorem of p. 155 for $F(x+iy)$ as a function of the variable y, for a fixed x; the only modification being that the symbol $V.P.$ appears there before the outer integral symbol, while it is missing in front of the inner integral symbol. On p. 155 we stated three sufficient conditions for its validity: Firstly, it must be true that

$$\int_{-\infty}^{+\infty} |F(x+iy)|\,dy < \infty$$

(in the case that this condition is satisfied, the symbol $V.P.$ before the inner integral is actually no longer required); secondly, for every fixed x, $F(x+iy)$ must be of bounded variation in a neighbourhood of y; and thirdly, $F(x+iy)$ must be normalized in y. The last two requirements are satisfied eo ipso by the analytic function F for every y and for every x in the interval $x_1 < x < x_2$.[1] We conclude:

Theorem 28.1. *Let $F(s) = F(x+iy)$ be analytic in the strip $x_1 < x < x_2$, and let $F(s)$ tend, uniformly in every narrower strip $x_1 + \delta \leq x \leq x_2 - \delta$, towards zero, when $|y| \to \infty$. Moreover, suppose that, for every x in $x_1 < x < x_2$,*

$$\int_{-\infty}^{+\infty} |F(x+iy)|\,dy < \infty.$$

Then it follows that $F(s)$ can be represented as the $\mathfrak{L}_{\text{II}}$-transform

$$F(s) = V.P.\int_{-\infty}^{+\infty} e^{-st} f(t)\,dt$$

[1] $F(x+iy)$ is the definite integral of its (absolutely integrable) derivative with respect to y. Hence, by the argument of p. 178, $F(x+iy)$ is of bounded variation.

of the original function

$$f(t) = \frac{1}{2\pi i} \int_{x-i\infty}^{x+i\infty} e^{ts} F(s)\, ds \quad (x_1 < x < x_2),$$

which is independent of the choice of x in $x_1 < x < x_2$.

When especially requesting that $F(s)$ be a $\mathfrak{L}_{\mathrm{I}}$-transform or, equivalently, that $f(t) = 0$ for $t < 0$, then it is *necessary* that $x_2 = +\infty$, that is, $F(x + iy)$ must be analytic in a right half-plane $x > x_1$, and $F(s)$ must tend, in every angular region $|\operatorname{arc}(s - x_1)| \leqq \psi < \pi/2$, towards zero, when s tends two-dimensionally towards ∞ (definition on p. 139). Combination of this condition with the condition that $F(x + iy)$ must converge, uniformly in $x \geqq x_1 + \delta$, towards zero, when $|y| \to \infty$, yields the condition: $F(s)$ tends in every half-plane $x \geqq x + \delta$ towards zero, when s tends two-dimensionally towards ∞. We now claim: This condition is *sufficient* to guarantee that $f(t) = 0$ for $t < 0$. We select a positive δ so large, such that $x_0 = x_1 + \delta > 0$, and we draw a circle centred at the origin having radius $\varrho > x_0$. This circle intersects the vertical line through x_0 at $x_0 + i\omega$ and at $x_0 - i\omega$ respectively, as shown in Fig. 27. We designate the portion of the circle to the right of the vertical line through x_0, orientated in the positive sense, by $\mathfrak{B}$. Invoking Cauchy's theorem, we find that

$$\int_{x_0-i\omega}^{x_0+i\omega} e^{ts} F(s)\, ds = \int_{\mathfrak{B}} e^{ts} F(s)\, ds.$$

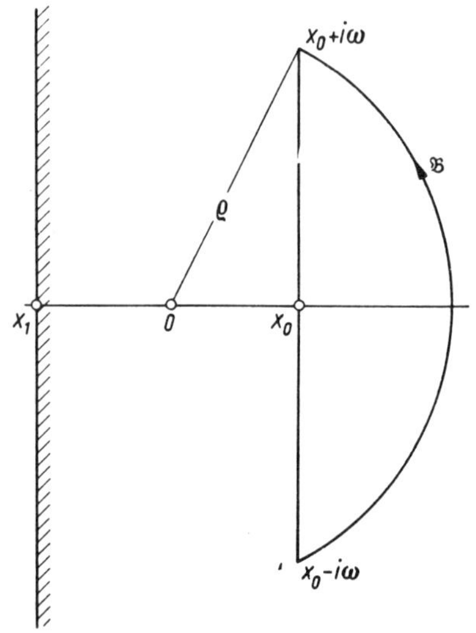

Figure 27

By Theorem 25.1, the integral on the right hand side tends, for $t < 0$, towards 0, when $\varrho \to \infty$. Hence, the same is true for the integral on the left hand side, for $\omega \to \infty$. It follows that $f(t) = 0$ for $t < 0$. We have thus arrived at the conclusion of

Theorem 28.2. *When $F(s) = F(x + iy)$ is analytic in the half-plane $x > x_1$; when $F(s)$ converges, in every half plane $x \geqq x_1 + \delta > x_1$, towards 0, when s tends two-dimensionally towards ∞; and when*

$$\int_{-\infty}^{+\infty} |F(x + iy)| \, dy < \infty \tag{4}$$

for every $x > x_1$; then it follows that $F(s)$ may be represented as the $\mathfrak{L}_{\mathrm{I}}$-transform of the original function

$$f(t) = \frac{1}{2\pi i} \int_{x-i\infty}^{x+i\infty} e^{ts} F(s) \, ds \quad (x > x_1)\,, \tag{5}$$

the integral being independent of the choice of x in $x > x_1$. For $t < 0$, the integral (5) yields the value zero.

The function $f(t)$, represented by (5) for all real t, is continuous in $-\infty < t < +\infty$.

The last statement is verified by the following consideration: We have, by (5),

$$f(t) = \frac{1}{2\pi i} \int_{x-i\infty}^{x+i\infty} e^{ts} F(s) \, ds = \frac{1}{2\pi} \int_{-\infty}^{+\infty} e^{t(x+iy)} F(x + iy) \, dy\,,$$

$$e^{-xt} f(t) = \frac{1}{2\pi} \int_{-\infty}^{+\infty} e^{ity} F(x + iy) \, dy\,.$$

Except for the unimportant sign of the exponent, the integral on the right hand side is a Fourier integral. By Theorem 24.1, $e^{-xt} f(t)$ is continuous for all t, because of (4); hence, $f(t)$ is continuous for all real t.

Remark: Observe that $f(t)$ equals zero for $t < 0$ and is continuous at $t = 0$. Hence, we must conclude that

$$f(0) = \frac{1}{2\pi i} \int_{x-i\infty}^{x+i\infty} F(s) \, ds = 0 \qquad \text{for } x > x_1\,.$$

On p. 142 we demonstrated that the function $\exp(-s^\alpha)$ cannot be a $\mathfrak{L}$-transform for $\alpha \geqq 1$. Here, we shall show, by means of Theorem 28.2, that all functions

$$F(s) = \exp(-s^\alpha) \text{ with } 0 < \alpha < 1 \tag{6}$$

are $\mathfrak{L}_{\mathrm{I}}$-transforms. With $s = \varrho e^{i\vartheta}$, we find that

$$|\exp(-s^\alpha)| = \exp(-\Re s^\alpha) = \exp(-\varrho^\alpha \cos\alpha\,\vartheta).$$

In the half-plane $\Re s \geqq 0$, we have $|\vartheta| \leqq \pi/2$, hence $\cos\alpha\vartheta \geqq \cos\alpha\pi/2 = \varepsilon > 0$, and

$$|\exp(-s^\alpha)| \leqq \exp(-\varepsilon\,\varrho^\alpha).$$

It follows that the function (6) converges towards zero, when s tends in the right half-plane $\Re s \geqq 0$ two-dimensionally towards ∞. Moreover, for $x \geqq 0$ it is true that

$$\int_{-\infty}^{+\infty} |\exp(-(x+iy)^\alpha)|\,dy \leqq \int_{-\infty}^{+\infty} \exp(-\varepsilon(x^2+y^2)^{\alpha/2})\,dy \leqq \int_{-\infty}^{+\infty} \exp(-\varepsilon|y|^\alpha)\,dy < \infty.$$

We conclude that the function (6) is the $\mathfrak{L}_\mathrm{I}$-transform of the original function

$$f(t) = \frac{1}{2\pi i}\int_{x-i\infty}^{x+i\infty} \exp(ts - s^\alpha)\,ds \quad (x \geqq 0).$$

The last integral can be evaluated by elementary means only for $\alpha = 1/2$; indeed, for $\alpha = 1/2$ the integral yields:

$$f(t) = \frac{1}{2\sqrt{\pi}\,t^{3/2}}\,e^{-1/4t} = \psi(1,t).$$

Using Theorem 28.2, we can derive Theorem 28.3 which is particularly useful in practical applications because of its simple hypotheses.

Theorem 28.3. *When $F(s)$ is analytic in the half-plane $\Re s > x_1 \geqq 0$, and when it can be represented in the form*

$$F(s) = \frac{c_1}{s^{\alpha_1}} + \cdots + \frac{c_n}{s^{\alpha_n}} + \frac{G(s)}{s^{1+\varepsilon}} \quad (0 < \alpha_\nu \leqq 1,\ \varepsilon > 0),$$

where $G(s)$ is bounded in every half-plane $\Re s \geqq x_1 + \delta > x_1$; then $F(s)$ is the $\mathfrak{L}_\mathrm{I}$-transform of the original function

$$f(t) = V.P.\,\frac{1}{2\pi i}\int_{x-i\infty}^{x+i\infty} e^{ts}F(s)\,ds \quad (x > x_1).$$

Proof: The function

$$F_1(s) = F(s) - \frac{c_1}{s^{\alpha_1}} - \cdots - \frac{c_n}{s^{\alpha_n}}$$

satisfies the conditions of Theorem 28.2; it follows that $F_1(s) = \mathfrak{L}\{f_1\}$, where

$$f_1(t) = \frac{1}{2\pi i}\int_{x-i\infty}^{x+i\infty} e^{ts}F_1(s)\,ds \quad (x > x_1)$$

$$= \frac{1}{2\pi i}\int_{x-i\infty}^{x+i\infty} e^{ts}\left(F(s) - \frac{c_1}{s^{\alpha_1}} - \cdots - \frac{c_n}{s^{\alpha_n}}\right)ds.$$

Since $x > 0$, we have, by Formula (25.2),

$$V.P. \frac{1}{2\pi i} \int_{x-i\infty}^{x+i\infty} e^{ts} \frac{c_\nu}{s^{a_\nu}}\, ds = c_\nu \frac{t^{a_\nu - 1}}{\Gamma(a_\nu)},$$

hence,

$$f_1(t) = V.P. \frac{1}{2\pi i} \int_{x-i\infty}^{x+i\infty} e^{ts} F(s)\, ds - c_1 \frac{t^{a_1-1}}{\Gamma(a_1)} - \cdots - c_n \frac{t^{a_n-1}}{\Gamma(a_n)}.$$

Setting

$$V.P. \frac{1}{2\pi i} \int_{x-i\infty}^{x+i\infty} e^{ts} F(s)\, ds = f(t),$$

we have

$$f_1(t) = f(t) - c_1 \frac{t^{a_1-1}}{\Gamma(a_1)} - \cdots - c_n \frac{t^{a_n-1}}{\Gamma(a_n)}$$

and

$$\mathfrak{L}\{f_1\} = \mathfrak{L}\{f\} - \frac{c_1}{s^{a_1}} - \cdots - \frac{c_n}{s^{a_n}}.$$

Also

$$\mathfrak{L}\{f_1\} = F_1(s) = F(s) - \frac{c_1}{s^{a_1}} - \cdots - \frac{c_n}{s^{a_n}},$$

hence,

$$F(s) = \mathfrak{L}\{f\}.$$

For an application of Theorem 28.3 see the function on p. 173, where $c_1 = \cdots = c_n = 0$, $\varepsilon = 1$.

29. A Condition, Necessary and Sufficient, for the Representability as a Laplace Transform of a Distribution

It is a characteristic of the $\mathfrak{L}$-transform of a *function* that it tends towards zero when the variable s, two-dimensionally in an angular region $|\arc s| \leqq \psi < \pi/2$, tends towards ∞ (compare Theorem 23.2, Addendum). This property is true in a whole half-plane $\Re s \geqq x_0$, provided $\mathfrak{L}\{f\}$ converges absolutely in $\Re s \geqq x_0$ (compare Theorem 23.7). The examples $\mathfrak{L}\{\delta^{(n)}\} = s^n$ $(n = 0, 1, 2, \ldots)$, $\Re s > 0$, demonstrate that the $\mathfrak{L}$-transforms of *distributions* need not possess this property. Actually, these transforms tend towards ∞ when $s \to \infty$ in $\Re s > 0$; however, not more strongly than a power of s. We shall show that the $\mathfrak{L}$-transforms of distributions are completely characterized by the property of being majorized by powers of s. The following Theorems 29.1 and 29.2 will substantiate this claim. Concepts and terminology involved in this process are explained in Chapter 12.

Theorem 29.1. *Suppose that the distribution T of order k belongs to the space $\mathcal{D}'_0$ and, consequently, $\mathfrak{L}\{T\} = F(s)$ defines a function which is analytic in a half-plane $\Re s > \sigma$. We conclude that:*

$$\mathfrak{L}\{T\} = F(s) = o(|s|^k),$$

when s, two-dimensionally in a half-plane $\Re s \geqq \sigma + \varepsilon$ $(\varepsilon > 0)$, tends towards ∞. Consequently, in every such half-plane, with the possible exception of a circular disc centred at the origin in the case that the origin belongs to the half-plane, $F(s)$ is governed by the estimation:

$$F(s) = O(|s|^k).$$

Proof: A distribution of order k in $\mathcal{D}'_0$ can be written in the form

$$T = D^k h(t) \quad (k \text{ an integer, } \geqq 0),$$

where the function $h(t)$ satisfies the conditions (12.3, 4). The transform $\mathfrak{L}\{T\}$ is defined by:

$$\mathfrak{L}\{T\} = s^k\, \mathfrak{L}\{h(t)\} \quad \text{for } \Re s > \sigma.$$

We invoke Theorem 23.7 which guarantees that

$$\mathfrak{L}\{h(t)\} = o(1),$$

when s, two-dimensionally in $\Re s \geqq \sigma + \varepsilon$, tends towards ∞; it follows that

$$\mathfrak{L}\{T\} = o(|s|^k).$$

The function $F(s)$ is bounded in every finite region; whence it is governed by the estimation:

$$F(s) = O(|s|^k),$$

in every half-plane $\Re s \geqq \sigma + \varepsilon$, provided the half-plane does not include the origin; in the case that the origin belongs to the half-plane, then the indicated estimation is valid outside a circle centred at the origin.

Theorem 29.2. *Suppose that $F(s)$ is analytic in a half-plane $\Re s > \sigma$ where it satisfies the estimation $F(s) = O(|s|^k)$, k an integer $\geqq 0$ (outside a circle centred at the origin $s = 0$ in the case that the origin belongs to the half-plane). Then we conclude that $F(s)$ is the $\mathfrak{L}$-transform of a distribution T of $\mathcal{D}'_0$.*

Proof: We specify a positive abscissa σ_1: we select $\sigma_1 > 0$, in the case that $\sigma \leqq 0$; we choose $\sigma_1 = \sigma + \varepsilon$ $(\varepsilon > 0)$, when $\sigma > 0$. In the half-plane $\Re s > \sigma_1$, we define the analytic function $H(s)$ by the expression:

(1) $$H(s) = s^{-k-2} F(s).$$

The function is governed in the specified half-plane by the estimation:

$$H(s) = O(|s|^{-2});$$

consequently, it converges towards zero when s tends, two-dimensionally, towards ∞; it satisfies the condition

$$\int_{-\infty}^{+\infty} |H(x+iy)|\, dy < \infty \qquad \text{for every} \qquad x > \sigma_1 .$$

According to Theorem 28.2, we have $H(s) = \mathfrak{L}\{h(t)\}$, where the function $h(t)$, which is defined by

$$h(t) = \frac{1}{2\pi i} \int_{x-i\infty}^{x+i\infty} e^{ts} H(s)\, ds \qquad (x > \sigma_1)\,, \tag{2}$$

is zero for $t < 0$ and is continuous for every t in $-\infty < t < +\infty$. Moreover,

$$|h(t)| = \frac{1}{2\pi} \left| \int_{-\infty}^{+\infty} e^{t(x+iy)} H(x+iy)\, dy \right| \leqq \frac{1}{2\pi} e^{tx} \int_{-\infty}^{+\infty} |H(x+iy)|\, dy$$
$$= O(e^{xt}),$$

hence, $\mathfrak{L}\{h(t)\}$ converges absolutely for $\Re s > x$ and, consequently, for $\Re s > \sigma_1$. If we define a distribution T by means of the formula:

$$T = D^{k+2} h(t), \tag{3}$$

then T belongs to $\mathscr{D}'_0$, by the definition of Chapter 12. The $\mathfrak{L}$-transform of T satisfies, by Theorem 14.3, the equation

$$\mathfrak{L}\{T\} = s^{k+2}\, \mathfrak{L}\{h(t)\} = s^{k+2} H(s).$$

Because of

$$s^{k+2} H(s) = F(s)\,,$$

we have: $\mathfrak{L}\{T\} = F(s)$. Consequently, $F(s)$ is the $\mathfrak{L}$-transform of a distribution of $\mathscr{D}'_0$.

The conclusions of the Theorems 29.1 and 29.2 may be combined and presented as a theorem.

Theorem 29.3. *Let $F(s)$ be an analytic function in some half-plane. The condition $F(s) = O(|s|^k)$, k an integer $\geqq 0$, in this half-plane (with the possible exception of a circular disc centred at the origin) is necessary and sufficient to assert that $F(s)$ can be represented as $\mathfrak{L}$-transform of some distribution T of $\mathscr{D}'_0$.*

The distribution T can be expressed explicitly by $F(s)$ as indicated by (3), using (1) and (2).

30. Determination of the Original Function by Means of Series Expansion of the Image Function

Having established a function $F(s)$ as a $\mathfrak{L}$-transform, the following technique suggests itself as a means of determining the corresponding original function: Expand $F(s)$ into a *series of image functions* having known original functions and then return this series to the original space *term by term*. Certain hypotheses must necessarily be satisfied, for the indicated process involves the interchange of $\mathfrak{L}$-transformation and infinite summation. For the special case that $F(s)$ may be represented by a partial fraction expansion, conditions taken from the theory of functions which guarantee the legality of the termwise inverse transformation have been presented in Chapter 26. In this Chapter we derive a theorem of strictly analytical character which proves extremely practical in applications. Its verification requires Lebesgue integration; thus, we introduce two lemmas from the Lebesgue theory.

Lemma 1. *Suppose that the functions $\varphi_n(t)$ are non-negative and Lebesgue integrable on the (finite or infinite) interval (a, b), and that the sequence $\varphi_n(t)$ increases monotonically, that is $\varphi_n(t) \leqq \varphi_{n+1}(t)$, and, consequently, converges towards a limit function $\varphi(t)$ which may assume the value ∞. Then we can conclude that $\varphi(t)$ too is Lebesgue integrable on the interval (a, b) if and only if the sequence of numbers $\int_a^b \varphi_n(t)\,dt$ is bounded. In fact,*

$$\int_a^b \varphi(t)\,dt = \lim_{n\to\infty} \int_a^b \varphi_n(t)\,dt.$$

This Lemma cannot be verified for Riemann integrals as demonstrated by the counter example: Arrange the rational numbers between 0 and 1 as a countable sequence: $r_0, r_1, r_2, \ldots,$ and define the sequence of functions on (0, 1) by

$$\varphi_n(t) = \begin{cases} 1 & \text{for } t = r_0, r_1, r_2, \ldots, r_n \\ 0 & \text{elsewhere.} \end{cases}$$

The limit of this sequence of functions on (0, 1) is clearly

$$\varphi(t) = \begin{cases} 1 & \text{for rational } t \\ 0 & \text{for irrational } t. \end{cases}$$

However, this limit function is not Riemann integrable, although all hypotheses of the Lemma 1 are satisfied: $\varphi_n(t) \geqq 0$, $\varphi_n(t) \leqq \varphi_{n+1}(t)$, and the sequence of numbers $\int_0^1 \varphi_n(t)\,dt = 0$ is bounded.

Lemma 2. *When* a) *the functions $\psi_n(t)$ $(n = 0, 1, 2, \ldots)$ are Lebesgue integrable on the (finite or infinite) interval (a, b);* b) $|\psi_n(t)| \leqq \varphi(t)$ *for all n, $\varphi(t)$ being Lebesgue* *integrable on* (a, b)*;* c) *the sequence* $\psi_n(t)$ *converges for almost all values of* t *to a limit function* $\psi(t)$, *when* $n \to \infty$*; then we can conclude that* $\psi(t)$ *too is Lebesgue integrable on the interval* (a, b) *and, in fact,*

$$\int_a^b \psi(t)\, dt = \lim_{n \to \infty} \int_a^b \psi_n(t)\, dt.$$

The above quoted counter example may be used once again to show that the second Lemma too cannot be verified for Riemann integration.

We can now state and verify the Theorem which describes the possible termwise inverse transformation of a series of image functions into the original space.

Theorem 30.1. *Let the function* $F(s)$ *be represented by a series of* $\mathfrak{L}$*-transforms*[1]

$$F(s) = \sum_{\nu=0}^{\infty} F_\nu(s), \quad F_\nu(s) = \mathfrak{L}\{f_\nu(t)\},$$

whereby all integrals

$$\mathfrak{L}\{f_\nu\} = \int_0^\infty e^{-st} f_\nu(t)\, dt = F_\nu(s) \quad (\nu = 0, 1, \cdots)$$

converge in a common half-plane $\Re s \geqq x_0$. *Moreover, we require that*

a) *the integrals*

$$\mathfrak{L}\{|f_\nu|\} = \int_0^\infty e^{-st} |f_\nu(t)|\, dt = G_\nu(s) \qquad (\nu = 0, 1, \cdots)$$

converge in this half-plane or, equivalently, the integrals $G_\nu(x_0)$ *exist;*

b) *the series*

$$\sum_{\nu=0}^{\infty} G_\nu(x_0)$$

converges. The last condition implies that $\sum_{\nu=0}^{\infty} F_\nu(s)$ *converges absolutely and uniformly in* $\Re s \geqq x_0$.

Then we can conclude that $\sum_{\nu=0}^{\infty} f_\nu(t)$ *converges, indeed absolutely, towards a function* $f(t)$ *for almost all* $t \geqq 0$*; this* $f(t)$ *is the original function of* $F(s)$*:*

$$\sum_{\nu=0}^{\infty} F_\nu(s) \bullet\!\!-\!\!\circ \sum_{\nu=0}^{\infty} f_\nu(t).$$

$\mathfrak{L}\{f\}$ *converges absolutely for* $\Re s \geqq x_0$.

---

[1] The functions $f_\nu(t)$ need be integrable only in the Lebesgue sense.

Proof: We set

$$e^{-x_0 t} \sum_{\nu=0}^{n} |f_\nu(t)| = \varphi_n(t).$$

Then we have

$$\varphi_n(t) \geqq 0, \quad \varphi_n(t) \leqq \varphi_{n+1}(t),$$

hence

$$\lim_{n\to\infty} \varphi_n(t) = \varphi(t)$$

exists for all t, possibly assuming the value ∞. By hypothesis b), there exists the limit

$$\lim_{n\to\infty} \sum_{\nu=0}^{n} G_\nu(x_0) = \lim_{n\to\infty} \sum_{\nu=0}^{n} \int_0^\infty e^{-x_0 t} \, | f_\nu(t)| \, dt = \lim_{n\to\infty} \int_0^\infty \varphi_n(t) \, dt,$$

which implies that the numerical sequence $\int_0^\infty \varphi_n(t)\,dt$ is bounded. We conclude, by Lemma 1, that $\varphi(t)$ is Lebesgue integrable. It follows that $\varphi(t)$ must be finite almost everywhere (otherwise the integral could not exist). This implies that

$$e^{-x_0 t} \sum_{\nu=0}^{\infty} | f_\nu(t) |, \text{thus also} \sum_{\nu=0}^{\infty} | f_\nu(t) | \text{ and all the more} \sum_{\nu=0}^{\infty} f_\nu(t)$$

converges almost everywhere. We set

$$\sum_{\nu=0}^{\infty} f_\nu(t) = f(t) \text{ almost everywhere.}$$

Also, by hypothesis a)

$$e^{-x_0 t} \sum_{\nu=0}^{n} f_\nu(t) = \psi_n(t)$$

is Lebesgue integrable on the interval $(0, \infty)$ (the integral exists not merely as a limit when the upper limit tends towards ∞, but immediately). We have

$$|\psi_n(t)| \leqq e^{-x_0 t} \sum_{\nu=0}^{n} |f_\nu(t)| = \varphi_n(t) \leqq \varphi(t),$$

where $\varphi(t)$ is Lebesgue integrable as demonstrated earlier. Moreover, $\psi_n(t)$ converges almost everywhere towards $\psi(t) = e^{-x_0 t} f(t)$. Hence, $\psi(t)$ is Lebesgue integrable in the interval $(0, \infty)$, by Lemma 2, and

$$\int_0^\infty \psi(t)\,dt = \lim_{n\to\infty} \int_0^\infty \psi_n(t)\,dt,$$

that is,

$$\int_0^\infty e^{-x_0 t} f(t)\,dt = \sum_{\nu=0}^{\infty} \int_0^\infty e^{-x_0 t} f_\nu(t)\,dt = \sum_{\nu=0}^{\infty} F_\nu(x_0) = F(x_0).$$

A function which is Lebesgue integrable in an interval is, necessarily, absolutely integrable; thus $\mathfrak{L}\{f\}$ converges absolutely for $s = x_0$, and the conclusion of Theorem 30.1 is verified for $s = x_0$. For $\Re s \geqq x_0$ we find

$$\left| e^{-st} \sum_{\nu=0}^{n} f_\nu(t) \right| \leqq \varphi(t),$$

hence, the conclusion is true also for $\Re s \geqq x_0$.

We use Theorem 30.1 to derive several theorems which are often useful in applications of the $\mathfrak{L}$-transformation.

Theorem 30.2. *When the series*

$$F(s) = \sum_{\nu=0}^{\infty} \frac{a_\nu}{s^{\lambda_\nu}} \quad (0 < \lambda_0 < \lambda_1 < \cdots \to \infty)$$

converges absolutely for $|s| > \varrho \geqq 0$, then it is the $\mathfrak{L}$-transform of the series

$$f(t) = \sum_{\nu=0}^{\infty} a_\nu \frac{t^{\lambda_\nu - 1}}{\Gamma(\lambda_\nu)},$$

which is obtained by termwise transformation of the former; the latter converges absolutely for all $t \neq 0$.

Remark: The series of $F(s)$ is transformed into a Dirichlet series

$$\sum_{\nu=0}^{\infty} a_\nu e^{-\lambda_\nu z}$$

by means of the substitution $s = e^z$. A Dirichlet series which converges at some point and, consequently, in a right half-plane need not converge absolutely anywhere.

Proof: We set

$$F_\nu(s) = \frac{a_\nu}{s^{\lambda_\nu}}, \quad f_\nu(t) = a_\nu \frac{t^{\lambda_\nu - 1}}{\Gamma(\lambda_\nu)};$$

hence, the integrals

$$\int_0^\infty e^{-st} \, |f_\nu(t)| \, dt = \frac{|a_\nu|}{\Gamma(\lambda_\nu)} \int_0^\infty e^{-st} t^{\lambda_\nu - 1} \, dt$$

converge for $\Re s > 0$. By hypothesis, the series

$$\sum_{\nu=0}^{\infty} \int_0^\infty e^{-x_0 t} \, |f_\nu(t)| \, dt = \sum_{\nu=0}^{\infty} \frac{|a_\nu|}{\Gamma(\lambda_\nu)} \int_0^\infty e^{-x_0 t} t^{\lambda_\nu - 1} \, dt = \sum_{\nu=0}^{\infty} \frac{|a_\nu|}{x_0^{\lambda_\nu}}$$

converges for every $x_0 > \varrho \geqq 0$. Thus, by Theorem 30.1,

$$\sum_{\nu=0}^{\infty} a_\nu \frac{t^{\lambda_\nu - 1}}{\Gamma(\lambda_\nu)}$$

converges absolutely for almost all $t \geqq 0$, and it represents the original function of $F(s)$. However, when such a series converges absolutely for some $t_0 > 0$, then it converges absolutely also for all complex t with $0 < |t| < t_0$ (the point $t = 0$ must be excluded in the case that some $\lambda_\nu < 1$). Consequently, the series converges for all $t \neq 0$.

An example which demonstrates the application of Theorem 30.2 is shown on p. 264.

A frequently encountered special case of Theorem 30.2 is given in the case that the exponents λ_ν are natural numbers; that is, $\lambda_\nu = \nu + 1$ (the exponents must be positive; hence, the smallest exponent is equal to 1):

(1) $$F(s) = \sum_{\nu=0}^{\infty} \frac{a_\nu}{s^{\nu+1}}.$$

Let ϱ designate the exact radius of convergence of the series, then the series converges also absolutely for $|s| > \varrho$. Thus, the corresponding original function is represented by the series

(2) $$f(t) = \sum_{\nu=0}^{\infty} \frac{a_\nu}{\nu!} t^\nu.$$

More precise statements regarding the relationship between these two series can be made in this special case. The radius of convergence of the series (1) is given by the Cauchy-Hadamard formula; it is

$$\varrho = \overline{\lim_{\nu \to \infty}} \sqrt[\nu]{|a_\nu|}.$$

Consequently, to every $\varepsilon > 0$, there exists an N, such that for $\nu \geqq N$

$$|a_\nu| < (\varrho + \varepsilon)^\nu;$$

thus, there exists an $A > 0$ such that

$$|a_\nu| < A\,(\varrho + \varepsilon)^\nu$$

for all $\nu \geqq 0$; hence, series (2) is majorized by the series

$$A \sum_{\nu=0}^{\infty} \frac{(\varrho + \varepsilon)^\nu}{\nu!} |t|^\nu.$$

We conclude that series (2) converges for all complex t, and represents an entire function $f(t)$ which is governed by the condition

(3) $$|f(t)| < A\, e^{(\varrho+\varepsilon)|t|}$$

(to every given $\varepsilon > 0$, an appropriate value A must be selected). We say $f(t)$ is "*of exponential type*".

Conversely, if an entire function (2) of exponential type is given which is subjected to the condition

$$(4) \qquad |f(t)| < A\, e^{a|t|} \quad (A > 0,\ a > 0),$$

and if $M(r)$ designates the maximum of $|f(t)|$ on the circular disc $|t| \leqq r$; then we have by the Cauchy estimation of the coefficients

$$\frac{|a_\nu|}{\nu!} \leqq \frac{M(r)}{r^\nu} < A\, \frac{e^{ar}}{r^\nu}.$$

Here r may be any positive number. For every ν we choose a fixed value of r, namely $r = \nu/a$, and we obtain

$$|a_\nu| < A\, \frac{e^\nu a^\nu \nu!}{\nu^\nu}, \quad \sqrt[\nu]{|a_\nu|} < \frac{e\,a}{\nu} \sqrt[\nu]{A\,\nu!}.$$

The formula

$$\log \nu! = \nu \log \nu - \nu + O(\log \nu) \quad \text{for} \quad \nu \to \infty$$

can be verified by elementary means. It follows that

$$\log\left(\frac{e}{\nu} \sqrt[\nu]{A\,\nu!}\right) = 1 - \log \nu + \frac{1}{\nu} \log A + \log \nu - 1 + O\left(\frac{\log \nu}{\nu}\right) = o(1),$$

hence,

$$\lim_{\nu \to \infty} \frac{e}{\nu} \sqrt[\nu]{A\,\nu!} = 1,$$

and we find for the radius of convergence ϱ of the series (1):

$$\varrho = \overline{\lim_{\nu \to \infty}} \sqrt[\nu]{|a_\nu|} \leqq a.$$

Thus, ϱ is finite and the series (1) has a region of convergence. Such a series represents a function which is holomorphic at $s = \infty$, where it vanishes. We summarize these conclusions in:

Theorem 30.3. *A function $F(s)$ which is holomorphic at $s = \infty$, where it vanishes, and which, consequently, is represented by a series of the form* (1) *with finite radius of convergence $\varrho \geqq 0$ is a $\mathfrak{L}$-transform. Its original function $f(t)$ is an entire function $f(t)$ of exponential type which is represented by the series* (2) *and which satisfies the estimation* (3). – *Conversely, every entire function $f(t)$ of exponential type which is represented by the series* (2) *and which satisfies the estimation* (4) *has a $\mathfrak{L}$-transform $F(s)$ which is holomorphic at $s = \infty$, where it vanishes and which is represented by the series* (1), *the latter having a finite radius of convergence which satisfies $\varrho \leqq a$.*

The conclusion of Theorem 30.3 may be presented in a different manner: Let
$\mathfrak{a}$ represent the class of entire functions $f(t)$ of exponential type,

$\mathfrak{A}$ represent the class of functions $F(s)$ which are analytic outside of some circle and which are holomorphic at $s = \infty$, where they vanish.

Then, every function of one class is related to a function of the other class by means of the $\mathfrak{L}$-transformation.

We have here the ideal situation of two classes of functions, each being characterized by properties taken from the theory of functions, which are related in a one-to-one manner by means of the $\mathfrak{L}$-transformation.

Another, very general theorem, which follows from Theorem 30.1, is the following:

Theorem 30.4. *Suppose that the functions $f_1(t)$ and $f_2(t)$ belong to the class $\mathfrak{J}_0$, and that the corresponding $\mathfrak{L}$-transforms $F_1(s)$ and $F_2(s)$ converge absolutely in some half-plane. Let $\varphi(z_1, z_2)$ designate a function of two variables which is holomorphic at $z_1 = z_2 = 0$, where it vanishes. Then it follows that $\varphi(F_1(s), F_2(s))$ is a $\mathfrak{L}$-transform which converges absolutely in some half-plane.*

Supplement: *An analogous theorem can be formulated for functions φ of one or arbitrarily (finitely) many independent variables.*

Proof: The hypothesis concerning φ implies that φ is a power series

$$\varphi(z_1, z_2) = \sum_{\nu_1, \nu_2 = 0}^{\infty} a_{\nu_1 \nu_2} z_1^{\nu_1} z_2^{\nu_2} \text{ with } a_{00} = 0,$$

which converges absolutely on a pair of circular discs $|z_1| < \varrho$ and $|z_2| < \varrho$; therefore it may be presented in an arbitrary countable sequential order. When $\Re s > \alpha$ is a half-plane of absolute convergence of both $\mathfrak{L}\{f_1\}$ and $\mathfrak{L}\{f_2\}$, then we conclude, by the Convolution Theorem 10.1, that all functions

$$F_1^{\nu_1}(s)\, F_2^{\nu_2}(s) \quad (\nu_1 + \nu_2 > 0)$$

are $\mathfrak{L}$-transforms which converge absolutely in $\Re s > \alpha$, and which have the original functions

$$f_1^{*\nu_1} * f_2^{*\nu_2} (\nu_1 \geqq 1,\ \nu_2 \geqq 1), \text{ and } f_1^{*\nu_1} (\nu_2 = 0), \quad f_2^{*\nu_2} (\nu_1 = 0).$$

We define

$$\int_0^{\infty} e^{-st} \, |f_1(t)| \, dt = H_1(s), \quad \int_0^{\infty} e^{-st} \, |f_2(t)| \, dt = H_2(s).$$

For a sufficiently large value $x_0 > \alpha$ we have, by Theorem 23.7,

$$|H_1(s)| < \varrho, \quad |H_2(s)| < \varrho \quad \text{for} \quad \Re s \geqq x_0,$$

hence

$$(5) \qquad \sum_{\nu_1, \nu_2 = 0}^{\infty} |a_{\nu_1 \nu_2}| \, H_1^{\nu_1}(x_0) \, H_2^{\nu_2}(x_0)$$

converges.

Corresponding to F_ν and f_ν of Theorem 30.1, here, for $\nu_1 \geqq 1$ and $\nu_2 \geqq 1$, we have the functions

$$a_{\nu_1 \nu_2} F_1^{\nu_1} F_2^{\nu_2} \quad \text{and} \quad a_{\nu_1 \nu_2} f_1^{*\nu_1} * f_2^{*\nu_2}$$

respectively; thus, the functions G_ν have the following structure:

$$G_\nu(s) = |a_{\nu_1 \nu_2}| \int_0^\infty e^{-st} \, |f_1^{*\nu_1} * f_2^{*\nu_2}| \, dt.$$

We have

$$|f_1 * f_2| = \left| \int_0^t f_1(\tau) f_2(t-\tau) \, dt \right| \leqq \int_0^t |f_1(\tau)| \cdot |f_2(t-\tau)| \, d\tau = |f_1| * |f_2|,$$

and also

$$|f_1^{*\nu_1} * f_2^{*\nu_2}| \leqq |f_1|^{*\nu_1} * |f_2|^{*\nu_2}.$$

Consequently,

$$G_\nu(x_0) \leqq |a_{\nu_1 \nu_2}| \int_0^\infty e^{-x_0 t} \, |f_1|^{*\nu_1} * |f_2|^{*\nu_2} \, dt.$$

Employing the Convolution Theorem, we find that

$$G_\nu(x_0) \leqq |a_{\nu_1 \nu_2}| \left(\int_0^\infty e^{-x_0 t} \, |f_1| \, dt \right)^{\nu_1} \left(\int_0^\infty e^{-x_0 t} \, |f_2| \, dt \right)^{\nu_2} = |a_{\nu_1 \nu_2}| \, H_1^{\nu_1}(x_0) \, H_2^{\nu_2}(x_0).$$

The derived conclusions hold also when either $\nu_1 = 0$ or $\nu_2 = 0$. The convergence of series (5) guarantees convergence of $\Sigma G_\nu(x_0)$ as well. By Theorem 30.1, $\varphi(F_1, F_2)$ is the $\mathfrak{L}$-transform of

(6) $$f(t) = \sum_{\nu_1, \nu_2 = 0}^{\infty} a_{\nu_1 \nu_2} f_1^{*\nu_1} * f_2^{*\nu_2} \quad (a_{00} = 0),$$

and $\mathfrak{L}\{f\}$ converges absolutely for $\Re s > x_0$.

Many functions can readily be recognized as $\mathfrak{L}$-transforms by the use of Theorem 30.4. Consider, for instance, the function

$$F(s) = \frac{\sinh x \sqrt{s}}{\sinh l \sqrt{s}} \quad (-l < x < +l),$$

which appears during the solution of parabolic boundary value problems by means of the $\mathfrak{L}$-transformation (compare p. 288). Writing this function in the form

$$F(s) = \frac{e^{x\sqrt{s}} - e^{-x\sqrt{s}}}{e^{l\sqrt{s}} - e^{-l\sqrt{s}}} = \frac{e^{(x-l)\sqrt{s}} - e^{-(x+l)\sqrt{s}}}{1 - e^{-2l\sqrt{s}}}$$

and defining

$$e^{(x-l)\sqrt{s}} = F_1(s), \quad e^{-(x+l)\sqrt{s}} = F_2(s),$$

we have: $F(s) = \varphi(F_1, F_2)$, where

(7) $$\varphi(z_1, z_2) = \frac{z_1 - z_2}{1 - z_1 z_2} = (z_1 - z_2) \sum_{\nu=0}^{\infty} z_1^{\nu} z_2^{\nu} \quad \text{for } |z_1 z_2| < 1.$$

The function $\varphi(z_1, z_2)$ is holomorphic at $z_1 = z_2 = 0$, and $\varphi(0, 0) = 0$. Moreover (see p. 56),

$$e^{-u\sqrt{s}} = \mathfrak{L}\{\psi(u, t)\} \quad \text{for } u > 0,$$

where $\psi(u, t)$ is a $\mathfrak{J}_0$-function, its $\mathfrak{L}$-transform being absolutely convergent for $\Re s > 0$. We see that $F_1(s)$ for $x < l$, $F_2(s)$ for $x > -l$, hence $F(s)$ for $-l < x < l$ are $\mathfrak{L}$-transforms with half-planes of absolute convergence.

For this case, we oould evaluate the convolution in the representation of $f(t)$, employing the transcendental additivity theorem (11.3) for the function $\psi(u, t)$. It is simpler, though, to write for $F(s)$ the series development (7):

$$F(s) = \left(e^{(x-l)\sqrt{s}} - e^{-(x+l)\sqrt{s}}\right) \sum_{\nu=0}^{\infty} e^{-2\nu l\sqrt{s}} = \sum_{\nu=0}^{\infty} \left(e^{-(2\nu l - x + l)\sqrt{s}} - e^{-(2\nu l + x + l)\sqrt{s}}\right)$$

and to transform, term by term:

$$f(t) = \sum_{\nu=0}^{\infty} \left(\psi(2\nu l - x + l, t) - \psi(2\nu l + x + l, t)\right)$$

$$= \frac{1}{2\sqrt{\pi}\, t^{3/2}} \left\{ \sum_{\nu=0}^{\infty} (2\nu l - x + l)\, e^{-(2\nu l - x + l)^2/4t} - \sum_{\nu=0}^{\infty} (2\nu l + x + l)\, e^{-(2\nu l + x + l)^2/4t} \right\}.$$

The second sum in the brackets can be modified, using $\nu = -\mu - 1$, thus:

$$\sum_{\nu=0}^{\infty} (-2\nu l - x - l) e^{-(-2\nu l - x - l)^2/4t} = \sum_{\mu=-1}^{-\infty} (2\mu l + l - x)\, e^{-(2\mu l + l - x)^2/4t}.$$

Thus, we find that

$$f(t) = \frac{1}{2\sqrt{\pi}\, t^{3/2}} \sum_{\nu=-\infty}^{+\infty} (2\nu l + l - x)\, e^{-(2\nu l + l - x)^2/4t}$$

(8) $$= \frac{l}{\sqrt{\pi}\, t^{3/2}} \sum_{\nu=-\infty}^{+\infty} \left(\nu + \frac{l-x}{2l}\right) \exp\left(-\frac{\left(\nu + \frac{l-x}{2l}\right)^2}{t/l^2}\right) \quad (-l < x < +l).$$

This function is closely related to the Theta function

(9) $$\vartheta_3(v, t) = \frac{1}{\sqrt{\pi t}} \sum_{\nu=-\infty}^{+\infty} e^{-(\nu+v)^2/t};$$

indeed, one finds that

(10) $$f(t) = -\frac{1}{2l^2} \left[\frac{\partial \vartheta_3(v, t/l^2)}{\partial v}\right]_{v=(l-x)/2l} \quad (-l < x < +l).$$

31. The Parseval Formula of the Fourier Transformation and of the Laplace Transformation. The Image of the Product

Suppose that the power series

$$\varphi(z) = \sum_{n=0}^{\infty} a_n z^n$$

has the radius of convergence $r > 0$; then we find, for $0 \leqq \varrho < r$:

$$\int_{-\pi}^{+\pi} \varphi(\varrho e^{i\vartheta}) \overline{\varphi(\varrho e^{i\vartheta})}\, d\vartheta = \int_{-\pi}^{+\pi} \sum_{n=0}^{\infty} a_n \varrho^n e^{in\vartheta} \sum_{m=0}^{\infty} \overline{a_m} \varrho^m e^{-im\vartheta}\, d\vartheta .$$

The series converge absolutely and uniformly; hence, they may be multiplied and integrated, term by term:

$$\sum_{n,m=0}^{\infty} a_n \overline{a_m} \varrho^{n+m} \int_{-\pi}^{+\pi} e^{i(n-m)\vartheta}\, d\vartheta .$$

Using

$$\int_{-\pi}^{+\pi} e^{i(n-m)\vartheta}\, d\vartheta = \begin{cases} 2\pi & \text{for } n = m \\ 0 & \text{for } n \neq m, \end{cases}$$

we find that

$$\frac{1}{2\pi} \int_{-\pi}^{+\pi} |\varphi(\varrho e^{i\vartheta})|^2\, d\vartheta = \sum_{n=0}^{\infty} |a_n|^2 \varrho^{2n} .$$

This expression is known as *Parseval's Formula for Power Series*. Our aim is to derive the analogous formula for the $\mathfrak{L}$-transformation.

In the above formula we consider the power series on a specified circle: $|z| = \varrho$, where, in fact, it represents a Fourier series involving the complex oscillations $e^{in\vartheta}$. Similarly, we shall consider the $\mathfrak{L}$-integral on some specified vertical line $\Re s = x$, where actually it is a Fourier integral. Therefore, we shall firstly derive the Parseval formula for Fourier integrals. For this we shall need several theorems; the first of these is an Inversion Theorem, using other conditions than the Inversion Theorem 24.2.

Theorem 31.1. *Suppose that* $\int_{-\infty}^{+\infty} |g(x)|\, dx < \infty$, *so that* $G(y) = \mathfrak{F}\{g\}$ *exists for all real* y. *When the Fourier integral*

$$h(x) = V.P.\ \frac{1}{2\pi} \int_{-\infty}^{+\infty} e^{ixy} G(y)\, dy$$

exists for some specified value x, and g is continuous at x, then we have:

$$g(x) = V.P.\frac{1}{2\pi}\int_{-\infty}^{+\infty} e^{ixy}\,G(y)\,dy.$$

Proof: Formula (24.4) implies that

$$I(x, Y) = \frac{1}{2\pi}\int_{-Y}^{+Y} e^{ixy}G(y)\,dy = \frac{1}{\pi}\int_{-\infty}^{+\infty}\frac{\sin Y(x-\xi)}{x-\xi}\,g(\xi)\,d\xi$$

$$= \frac{1}{\pi}\int_{-\infty}^{x-\delta} + \frac{1}{\pi}\int_{x-\delta}^{x+\delta} + \frac{1}{\pi}\int_{x+\delta}^{\infty} = I_1(x, Y) + I_2(x, Y) + I_3(x, Y).$$

Should we redefine $g(\xi)$ to be zero in the interval $(x-\delta, x+\delta)$, we would observe contributions in the above equation only by the first and the third integral: $I_1 + I_3$. Unquestionably, the modified function is continuous at x, and it is of bounded variation in some neighbourhood of x. As shown in the proof of Theorem 24.2, $I_1 + I_3$ tends, as $Y \to \infty$, towards the value of the modified function at x, that is towards zero. $I(x, Y)$ tends towards $h(x)$ when $Y \to \infty$, hence, the same is true for $I_2(x, Y)$:

$$I_2(x, Y) = \frac{1}{\pi}\int_{x-\delta}^{x+\delta}\frac{\sin Y(x-\xi)}{x-\xi}\,g(\xi)\,d\xi \to h(x) \text{ when } Y \to \infty.$$

Next, we form the arithmetic mean of I_2 with regard to Y:[1]

$$m(x, Y) = \frac{1}{2Y}\int_0^{2Y} I_2(x, y)\,dy = \frac{1}{2\pi Y}\int_0^{2Y} dy \int_{x-\delta}^{x+\delta}\frac{\sin y(x-\xi)}{x-\xi}\,g(\xi)\,d\xi$$

$$= \frac{1}{2\pi Y}\int_{x-\delta}^{x+\delta}\frac{g(\xi)}{x-\xi}\,d\xi\int_0^{2Y}\sin y(x-\xi)\,dy \quad {}^{2)}$$

$$= \frac{1}{2\pi Y}\int_{x-\delta}^{x+\delta} g(\xi)\,\frac{1-\cos 2Y(x-\xi)}{(x-\xi)^2}\,d\xi = \frac{1}{\pi Y}\int_{x-\delta}^{x+\delta} g(\xi)\left(\frac{\sin Y(x-\xi)}{x-\xi}\right)^2 d\xi.$$

The last integral is known as *Fejér Integral* in the theory of Fourier series. There it is known that, for arbitrary δ, this integral, divided by πY, tends towards $g(x)$, when $Y \to \infty$, provided g is continuous at x. This implies that

$$m(x, Y) \to g(x) \text{ when } Y \to \infty.$$

---

[1] The upper limit of integration $2Y$ has been chosen to avoid fractions in the subsequent expressions.

[2] The function

$$\frac{\sin y(x-\xi)}{x-\xi}\,g(\xi)$$

is integrable in the rectangle $0 \leq y \leq 2Y$, $x-\delta \leq \xi \leq x+\delta$. The double integral is representable by each of the two iterated integrals, since both exist. This implies equality for the iterated integrals; in other words: we may interchange the order of integration.

When the function $I_2(x, Y)$ has the limit $h(x)$, then, certainly, its arithmetic mean $m(x, Y)$ has the same limit; whence $g(x) = h(x)$.

Remark: Theorem 31.1 is analogous to the well known theorem: "When the Fourier series of a function $f(x)$ converges at a point of continuity of f, then it converges towards the value of the function."

The following theorem provides sufficient conditions for the convergence of $\int_{-\infty}^{+\infty} |G(y)| \, dy$.

Theorem 31.2. *Suppose that*

$$\int_{-\infty}^{+\infty} |h(x)| \, dx < \infty, \quad |h(x)| \leqq C \text{ for all } x, \text{ and that } \mathfrak{F}\{h\} = H(y) \geqq 0.$$

Then we conclude that

$$\int_{-\infty}^{+\infty} |H(y)| \, dy = \int_{-\infty}^{+\infty} H(y) \, dy \leqq 2\pi C.$$

Proof: We start with the Fejér Integral for $h(x)$, but we use $x = 0$, and the limits of integration $-\infty$ and $+\infty$, and we re-trace the above proof in the reverse direction. In this manner we find:

$$m(Y) = \frac{1}{\pi Y} \int_{-\infty}^{+\infty} h(\xi) \left(\frac{\sin Y\xi}{\xi}\right)^2 d\xi = \frac{1}{2\pi Y} \int_{-\infty}^{+\infty} h(\xi) \, \frac{1 - \cos 2Y\xi}{\xi^2} \, d\xi$$

$$= \frac{1}{2\pi Y} \int_{-\infty}^{+\infty} \frac{h(\xi)}{\xi} \, d\xi \int_0^{2Y} \sin \xi\eta \, d\eta = \frac{1}{2\pi Y} \int_0^{2Y} d\eta \int_{-\infty}^{+\infty} h(\xi) \, \frac{\sin \eta\xi}{\xi} \, d\xi.$$

The interchange of the order of integration in the fourth integral is permissible, since

$$\int_{-\infty}^{+\infty} h(\xi) \, \frac{\sin \eta\xi}{\xi} \, d\xi$$

converges uniformly for $0 \leqq \eta \leqq 2Y$, because $\int_{-\infty}^{+\infty} |h(\xi)| \, d\xi$ exists, and $|(\sin \eta\xi)/\xi| \leqq$ $\leqq \eta \leqq 2Y$. It follows that the integration with respect to η may be performed under the integral symbol; in this manner we produce the third integral.

The subsequent modifications correspond precisely to the several steps which led to equation (24.4) when taken in reverse order; they can be supported by the same argumentation:

$$\int_{-\infty}^{+\infty} h(\xi) \, \frac{\sin \eta\xi}{\xi} \, d\xi = \frac{1}{2} \int_{-\infty}^{+\infty} h(\xi) \, \frac{e^{i\eta\xi} - e^{-i\eta\xi}}{i\xi} \, d\xi$$

$$= \frac{1}{2} \int_{-\infty}^{+\infty} h(\xi) \, d\xi \int_{-\eta}^{+\eta} e^{iy\xi} \, dy = \frac{1}{2} \int_{-\eta}^{+\eta} dy \int_{-\infty}^{+\infty} e^{-iy\xi} \, h(\xi) \, d\xi$$

$$= \frac{1}{2} \int_{-\eta}^{+\eta} H(y) \, dy.$$

Substituting this in the expression for $m(Y)$, one finds:

$$m(Y) = \frac{1}{4\pi Y} \int_0^{2Y} d\eta \int_{-\eta}^{+\eta} H(y)\,dy,$$

or, upon interchanging the order of integration (compare Fig. 28),

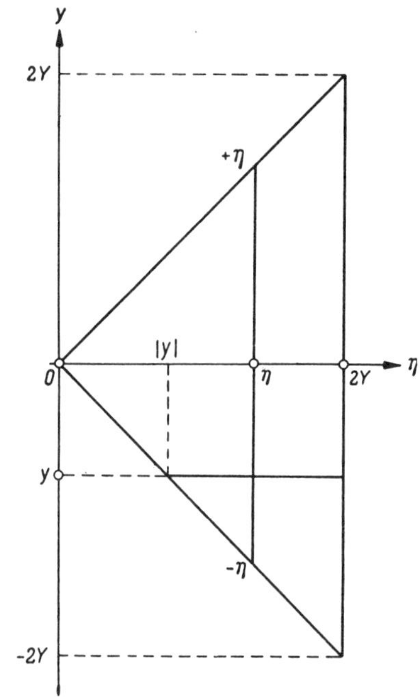

Figure 28

(1) $$m(Y) = \frac{1}{4\pi Y} \int_{-2Y}^{+2Y} H(y)\,dy \int_{|y|}^{2Y} d\eta = \frac{1}{4\pi Y} \int_{-2Y}^{+2Y} H(y)\,(2Y - |y|)\,dy.$$

The original definition of $m(Y)$ implies, since $|h(\xi)| \leqq C$, that

$$|m(Y)| = \left| \frac{1}{\pi Y} \int_{-\infty}^{+\infty} h(\xi) \left(\frac{\sin Y\xi}{\xi}\right)^2 d\xi \right| \leqq \frac{C}{\pi Y} \int_{-\infty}^{+\infty} \left(\frac{\sin Y\xi}{\xi}\right)^2 d\xi$$

$$= \frac{C}{\pi} \int_{-\infty}^{+\infty} \left(\frac{\sin u}{u}\right)^2 du = C;$$

hence, for the modified expression (1) of $m(Y)$ we produce:

$$\frac{1}{4\pi Y} \left| \int_{-2Y}^{+2Y} H(y)\,(2Y - |y|)\,dy \right| \leqq C$$

or

$$\left| \int_{-2Y}^{+2Y} H(y) \left(1 - \frac{|y|}{2Y}\right) dy \right| \leqq 2\pi C.$$

The hypothesis $H(y) \geqq 0$ enables us to rewrite the last expression as follows:

$$\int_{-2Y}^{+2Y} H(y) \left(1 - \frac{|y|}{2Y}\right) dy \leqq 2\pi C.$$

A fortiori, for $0 < {y_1 \atop y_2} < 2Y$, we have:

$$\int_{-y_1}^{y_2} H(y) \left(1 - \frac{|y|}{2Y}\right) dy \leqq 2\pi C.$$

When, for fixed values y_1 and y_2, the number Y tends towards ∞, then $1 - (|y|/2Y)$ tends, uniformly in the interval of integration, towards 1. Consequently, one obtains:

$$\int_{-y_1}^{y_2} H(y)\, dy \leqq 2\pi C,$$

where y_1 and y_2 may represent any arbitrary pair of positive numbers. Thus it follows necessarily that

$$\lim_{y_1, y_2 \to \infty} \int_{-y_1}^{y_2} H(y)\, dy = \int_{-\infty}^{+\infty} H(y)\, dy$$

exists, and is $\leqq 2\pi C$, since $\int_{-y_1}^{+y_2} H(y)\, dy$ is non-decreasing for growing y_1 and y_2, and bounded above.

Theorem 31.3 (Convolution Theorem of the Fourier Transformation). *Suppose that*

$$\int_{-\infty}^{+\infty} |g_1(x)|\, dx < \infty, \quad \int_{-\infty}^{+\infty} |g_2(x)|\, dx < \infty;$$

hence $G_1(y) = \mathfrak{F}\{g_1\}$ *and* $G_2(y) = \mathfrak{F}\{g_2\}$ *exist for all* y*; moreover, let it be assumed that*

$$\int_{-\infty}^{+\infty} |g_1(x)|^2\, dx < \infty, \quad \int_{-\infty}^{+\infty} |g_2(x)|^2\, dx < \infty.$$

Then the convolution with infinite limits of integration

$$g_1 * g_2 = \int_{-\infty}^{+\infty} g_1(\xi)\, g_2(x - \xi)\, d\xi$$

exists for all x, and we have

$$\mathfrak{F}\{g_1 * g_2\} = \mathfrak{F}\{g_1\} \cdot \mathfrak{F}\{g_2\},$$

whereby $\mathfrak{F}\{g_1 * g_2\}$ *converges absolutely.*

Proof: Here, and in the sequel, we need the

Cauchy-Schwarz Inequality:

Suppose that the functions $f(x)$ *and* $g(x)$ *are integrable in every finite subinterval of the (finite or infinite) interval* (a, b), *and suppose that:*

$$\int_a^b |f(x)|^2\,dx < \infty, \quad \int_a^b |g(x)|^2\,dx < \infty$$

then $\int_a^b f(x)g(x)\,dx$ *exists, and it is true that*

$$\left|\int_a^b f(x)\,g(x)\,dx\right|^2 \leq \left(\int_a^b |f(x)\,g(x)|\,dx\right)^2 \leq \int_a^b |f(x)|^2\,dx \cdot \int_a^b |g(x)|^2\,dx.$$

We firstly argue, for the specified hypotheses, the existence of $g_1 * g_2$, since

$$\int_{-\infty}^{+\infty} |g_1(\xi)|^2\,d\xi < \infty, \quad \int_{-\infty}^{+\infty} |g_2(x-\xi)|^2\,d\xi = \int_{-\infty}^{+\infty} |g_2(u)|^2\,du < \infty.$$

Next, we employ the substitution $u = x - \xi$ to show that

$$G_1(y)\,G_2(y) = \int_{-\infty}^{+\infty} e^{-iy\xi}\,g_1(\xi)\,d\xi \cdot \int_{-\infty}^{+\infty} e^{-iyu}\,g_2(u)\,du$$

$$= \int_{-\infty}^{+\infty} e^{-iy\xi}\,g_1(\xi)\left[\int_{-\infty}^{+\infty} e^{-iy(x-\xi)}\,g_2(x-\xi)\,dx\right]d\xi.$$

The fact that the integrals converge absolutely enables us to interchange the order of integration, hence

$$G_1(y)\,G_2(y) = \int_{-\infty}^{+\infty} e^{-iyx}\,dx \int_{-\infty}^{+\infty} g_1(\xi)\,g_2(x-\xi)\,d\xi = \mathfrak{F}\{g_1 * g_2\}$$

The resulting integral too converges absolutely.

Theorem 31.4 *Starting with the hypotheses of Theorem* 31.3, *we can conclude that* $g_1 * g_2$ *is continuous; in fact,* $g_1 * g_2$ *is uniformly continuous for all values of* x.

Proof: We form the difference of the convolution integral evaluated at two neighbouring points x and $x + \delta$; then we apply the Cauchy-Schwarz inequality to this difference. In this manner we find

$$\left| \int_{-\infty}^{+\infty} g_1(\xi)\, g_2(x + \delta - \xi)\, d\xi - \int_{-\infty}^{+\infty} g_1(\xi)\, g_2(x - \xi)\, d\xi \right|^2$$

$$= \left| \int_{-\infty}^{+\infty} g_1(\xi) \left[g_2(x + \delta - \xi) - g_2(x - \xi) \right] d\xi \right|^2$$

$$\leqq \int_{-\infty}^{+\infty} |g_1(\xi)|^2\, d\xi \cdot \int_{-\infty}^{+\infty} |g_2(x + \delta - \xi) - g_2(x - \xi)|^2\, d\xi$$

$$= \int_{-\infty}^{+\infty} |g_1(\xi)^2|\, d\xi \cdot \int_{-\infty}^{+\infty} |g_2(u + \delta) - g_2(u)|^2\, du .$$

The second integral tends towards 0,[3] when $\delta \to 0$. This necessarily implies that the difference on the left hand side also tends towards zero when $\delta \to 0$, independently of the chosen value of x.

We now have the means to derive the Parseval formula for the Fourier transformation.

---

[3] In the Lebesgue theory, it is well known that the existence of $\int_{-\infty}^{+\infty} |g(x)|^2 dx$ implies:
when $\delta \to \infty$.

$$\int_{-\infty}^{+\infty} |g(x + \delta) - g(x)|^2\, dx \to 0 \quad \text{as} \quad \delta \to 0 .$$

For Riemann integrals we arrive at the same conclusion using the theorem mentioned on p. 48. Firstly, we use

$$(|a| + |b|)^2 \leqq 2\,|a|^2 + 2\,|b|^2$$

to show that

$$\int_X^\infty |g(x + \delta) - g(x)|^2\, dx \leqq \int_X^\infty (|g(x + \delta)| + |g(x)|)^2\, dx \leqq 2 \int_X^\infty |g(x + \delta)|^2\, dx + 2 \int_X^\infty |g(x)|^2\, dx ;$$

hence, when a priori $|\delta| < 1$, then one can select a sufficiently large X so that

$$\int_X^\infty |g(x + \delta) - g(x)|^2\, dx \quad \text{and similarly} \quad \int_{-\infty}^{-X} |g(x + \delta) - g(x)|^2\, dx$$

become arbitrarily small. The points (in finite number) in $-X \leqq x \leqq +X$ where $|g(x - \delta) - g(x)|^2$ is merely improperly integrable can each be surrounded by sufficiently small intervals so that the contribution on these intervals to the integral becomes arbitrarily small. $g(x - \delta) - g(x)$ is properly integrable on the remainder, and consequently bounded; hence,

$$\int |g(x + \delta) - g(x)|^2\, dx \leqq M \int |g(x + \delta) - g(x)|\, dx .$$

Invoking the previous theorem, we conclude that this is arbitrarily small for sufficiently small values of δ.

Theorem 31.5. *Suppose that*

$$\int_{-\infty}^{+\infty} |g(x)|\,dx < \infty \quad \textit{and} \quad \int_{-\infty}^{+\infty} |g(x)|^2\,dx < \infty;$$

hence $G(y) = \mathfrak{F}\{g\}$ *exists for all real* y; *then we have, for all real* x,

$$\int_{-\infty}^{+\infty} g(\xi)\,\overline{g(\xi - x)}\,d\xi = \frac{1}{2\pi}\int_{-\infty}^{+\infty} e^{ixy}\,G(y)\,\overline{G(y)}\,dy.$$

Especially for $x = 0$, *this yields the Parseval Formula:*

(2) $$\int_{-\infty}^{+\infty} g(\xi)\,\overline{g(\xi)}\,d\xi = \frac{1}{2\pi}\int_{-\infty}^{+\infty} G(y)\,\overline{G(y)}\,dy$$

or

(3) $$\int_{-\infty}^{+\infty} |g(\xi)|^2\,d\xi = \frac{1}{2\pi}\int_{-\infty}^{+\infty} |G(y)|^2\,dy.$$

Proof: First, we show that

$$\mathfrak{F}\{\overline{g(-x)}\} = \int_{-\infty}^{+\infty} e^{-iyx}\,\overline{g(-x)}\,dx = \overline{\int_{-\infty}^{+\infty} e^{iyx}\,g(-x)\,dx} = \overline{\int_{-\infty}^{+\infty} e^{-iyx}\,g(x)\,dx} = \overline{G(y)}$$

Next, we form the convolution

$$h(x) = g(x) * \overline{g(-x)} = \int_{-\infty}^{+\infty} g(\xi)\,\overline{g(\xi - x)}\,d\xi;$$

by Theorem 31.3, it follows that $\mathfrak{F}\{h\} = H(y)$ converges absolutely; that is

$$\int_{-\infty}^{+\infty} |h(x)|\,dx < \infty,$$

and

$$H(y) = \mathfrak{F}\{g\} \cdot \mathfrak{F}\{\overline{g(-x)}\} = G(y)\,\overline{G(y)} = |G(y)|^2 \geqq 0.$$

Moreover, $h(x)$ is bounded, for the Cauchy-Schwarz inequality implies that

$$|h(x)|^2 \leqq \int_{-\infty}^{+\infty} |g(\xi)|^2\,d\xi \cdot \int_{-\infty}^{+\infty} |\overline{g(\xi - x)}|^2\,d\xi = \left(\int_{-\infty}^{+\infty} |g(\xi)|^2\,d\xi\right)^2.$$

Observe that all hypotheses of Theorem 31.2 are satisfied for $h(x)$, consequently

$$\int_{-\infty}^{+\infty} |H(y)|\, dy \quad \text{and a fortiori} \quad \frac{1}{2\pi}\int_{-\infty}^{+\infty} e^{ixy}\, H(y)\, dy$$

converge for all x. The function $h(x)$ is continuous for all x, by Theorem 31.4; hence, by Theorem 31.1, we have

$$h(x) = \frac{1}{2\pi}\int_{-\infty}^{+\infty} e^{ixy} H(y) dy = \frac{1}{2\pi}\int_{-\infty}^{+\infty} e^{ixy}\, G(y)\, \overline{G(y)}\, dy,$$

which is the conclusion of the Theorem.

Remark: Since the integral $\int_{-\infty}^{+\infty} |g(x)|^2 dx$ occurs in the Parseval formula, it is clear that the convergence of this integral must be presumed. The additional hypothesis regarding the convergence of $\int_{-\infty}^{+\infty} |g(x)|\, dx$ is introduced to guarantee the existence of $G(y)$ in the conventional sense (pointwise convergence). This hypothesis can be relaxed when convergence in the mean replaces pointwise convergence, in a manner analogous to the one employed with the Fourier series. However, this more general approach requires Lebesgue integration and other more powerful mathematical tools.

The Parseval formula can easily be extended to two functions.

Theorem 31.6. *Suppose that*

$$\int_{-\infty}^{+\infty} |g_1(x)|\, dx < \infty, \quad \int_{-\infty}^{+\infty} |g_2(x)|\, dx < \infty,$$

$$\int_{-\infty}^{+\infty} |g_1(x)|^2\, dx < \infty, \quad \int_{-\infty}^{+\infty} |g_2(x)|^2\, dx < \infty,$$

then the generalized Parseval Formula for the Fourier transformation

(4)
$$\int_{-\infty}^{+\infty} g_1(x)\, \overline{g_2(x)}\, dx = \frac{1}{2\pi}\int_{-\infty}^{+\infty} G_1(y)\, \overline{G_2(y)}\, dy,$$

holds. Eq. (4) *can also be written as follows:*

(5)
$$\int_{-\infty}^{+\infty} g_1(x)\, g_2(x)\, dx = \frac{1}{2\pi}\int_{-\infty}^{+\infty} G_1(y)\, G_2(-y)\, dy.$$

Remark: Observe that in formula (4), $\overline{G_2(y)}$ is certainly not the Fourier transform of $\overline{g_2(x)}$; actually, we have $\mathfrak{F}\{\overline{g_2(x)}\} = \overline{G_2(-y)}$.

Proof: The identity

$$4\,a\,\bar{b} = |a+b|^2 - |a-b|^2 + i\,|a+i\,b|^2 - i\,|a-i\,b|^2$$

can easily be verified by the introduction of the substitutions $|a+b|^2 = (a+b)(\bar{a}+\bar{b}), \ldots$ and the subsequent execution of all indicated multiplications. Using this identity, we demonstrate that $4g_1(x)\overline{g_2(x)}$ can be expressed as the sum of four exact squares; whence, the integral $4\int\limits_{-\infty}^{+\infty} g_1(x)\overline{g_2(x)}\,dx$ can be evaluated by integrals of these squares. Theorem 31.5 implies that these integrals are equal to integrals of the squares of the corresponding Fourier transforms, divided by 2π, as for instance shown by:

$$\int\limits_{-\infty}^{+\infty} |g_1(x) + g_2(x)|^2\,dx = \frac{1}{2\pi}\int\limits_{-\infty}^{+\infty} |G_1(y) + G_2(y)|^2\,dy.$$

Using the above identity once more, we produce by adding these integrals

$$4\,\frac{1}{2\pi}\int\limits_{-\infty}^{+\infty} G_1(y)\,\overline{G_2(y)}\,dy,$$

thus verifying formula (4).

Defining

$$\overline{g_2(x)} = k(x), \quad \text{and consequently} \quad g_2(x) = \overline{k(x)},$$

we find:

$$G_2(y) = \mathfrak{F}\{g_2\} = \mathfrak{F}\{\bar{k}\} = \int\limits_{-\infty}^{+\infty} e^{-iyx}\,\overline{k(x)}\,dx = \overline{\int\limits_{-\infty}^{+\infty} e^{iyx}\,k(x)\,dx} = \overline{K(-y)},$$

which implies that

$$\overline{G_2(y)} = K(-y).$$

Replacing in formula (4) the functions $\overline{g_2(x)}$ and $\overline{G_2(y)}$ by $k(x)$ and $K(-y)$ respectively, we produce a formula which differs from formula (5) only in notation.

The Parseval formula for the $\mathfrak{L}$-transformation can now readily be derived, using Theorem 31.5 and Theorem 31.6. Indeed, we obtain in this manner not only the Parseval formula for the one-sided $\mathfrak{L}$-transformation but also one for the two-sided $\mathfrak{L}$-transformation.

Theorem 31.7. *When, for some real x,*

$$\int\limits_{-\infty}^{+\infty} e^{-xt}\,|f(t)|\,dt < \infty, \qquad \int\limits_{-\infty}^{+\infty} e^{-2xt}\,|f(t)|^2\,dt < \infty,$$

then the Parseval Formula

$$(6)\qquad \int_{-\infty}^{+\infty} e^{-2xt}\,|f(t)|^2\,dt = \frac{1}{2\pi}\int_{-\infty}^{+\infty} |F(x+iy)|^2\,dy$$

is valid for the two-sided transformation $\mathfrak{L}_{\mathrm{II}}\{f\} = F(s)$. *When, with* $s_1 = x_1 + iy_1$ *and* $s_2 = x_2 + iy_2$,

$$\int_{-\infty}^{+\infty} e^{-x_1 t}\,|f_1(t)|\,dt < \infty, \qquad \int_{-\infty}^{+\infty} e^{-x_2 t}\,|f_2(t)|\,dt < \infty,$$

$$\int_{-\infty}^{+\infty} e^{-2x_1 t}\,|f_1(t)|^2\,dt < \infty, \qquad \int_{-\infty}^{+\infty} e^{-2x_2 t}\,|f_2(t)|^2\,dt < \infty,$$

then the generalized Parseval Formula:

$$(7)\qquad \int_{-\infty}^{+\infty} e^{-(s_1+\bar{s}_2)t}\,f_1(t)\,\overline{f_2(t)}\,dt = \frac{1}{2\pi}\int_{-\infty}^{+\infty} F_1(s_1+iy)\,\overline{F_2(s_2+iy)}\,dy$$

holds for the $\mathfrak{L}_{\mathrm{II}}$*-transformation.*

A corresponding theorem is true for the $\mathfrak{L}_{\mathrm{I}}$*-transformation; for this purpose, the integrals of the hypotheses, and of the left hand sides of formulae* (6) *and* (7) *are to be evaluated only between* 0 *and* ∞.

Proof: We substitute into formula (4) of Theorem 31.6

$$g_1(t) = e^{-s_1 t} f_1(t), \qquad g_2(t) = e^{-s_2 t} f_2(t),$$

hence

$$G_1(y) = F_1(s_1+iy), \quad G_2(y) = F_2(s_2+iy),$$

and we obtain formula (7). For $f_1 = f_2 = f$, and $s_1 = s_2 = x$, this yields formula (6). In the case of the $\mathfrak{L}_{\mathrm{I}}$-transformation, g_1 and g_2 are to be set equal to zero for $t < 0$.

In the theory of the one-sided $\mathfrak{L}_{\mathrm{I}}$-transformation, we may consider the function $F(s)$ in the Parseval formula not only on a single vertical line, but in an entire half-plane. This follows from the fact that convergence of

$$\int_0^{\infty} e^{-2x_0 t}\,|f(t)|^2\,dt$$

for some real x_0 implies convergence not only of

$$\int_0^{\infty} e^{-2xt}\,|f(t)|^2\,dt \quad \text{for } x > x_0,$$

but also of

$$\int_0^\infty e^{-xt} \, |f(t)| \, dt \quad \text{for } x > x_0.$$

The last conclusion follows from the Cauchy-Schwarz inequality:

$$\left(\int_0^\infty e^{-xt} \, |f(t)| \, dt\right)^2 = \left(\int_0^\infty e^{-(x-x_0)t} \, e^{-x_0 t} \, |f(t)| \, dt\right)^2$$

$$\leqq \int_0^\infty e^{-2(x-x_0)t} \, dt \int_0^\infty e^{-2x_0 t} \, |f(t)|^2 \, dt.$$

Thus, by Theorem 31.7, we may write the Parseval formula for $x > x_0$.

The above considerations suggest the introduction of the abscissa of convergence $\varkappa_2$ of the integral :

$$\int_0^\infty e^{-2xt} \, |f(t)|^2 \, dt.$$

For this purpose, we define $\varkappa_2$ as the Dedekind cut between those real x for which the integral converges, and those real x for which it diverges. Equivalently, we may define $\varkappa_2$ as the lower limit of those x for which the integral converges. We shall call $\varkappa_2$ the *second power abscissa* of the $\mathfrak{L}_{\mathrm{I}}$-transformation[4]. We can now formulate the following theorem.

Theorem 31.8. *The Parseval Formula of the* $\mathfrak{L}_{\mathrm{I}}$*-transformation*

$$\int_0^\infty e^{-2xt} \, |f(t)|^2 \, dt = \frac{1}{2\pi} \int_{-\infty}^{+\infty} |F(x+iy)|^2 \, dy \tag{8}$$

is valid for $x > \varkappa_2$.

The main interest of Theorem 31.8 lies in the following fact: The quadratic mean of $F(s)$ along a vertical line of abscissa x:

$$m(x) = \frac{1}{2\pi} \int_{-\infty}^{+\infty} |F(x+iy)|^2 \, dy$$

exists for $x > \varkappa_2$; it can be expressed as the $\mathfrak{L}_{\mathrm{I}}$-integral of the non-negative function $|f(t)|^2$, for the real argument $s = 2x$. Such an $\mathfrak{L}_{\mathrm{I}}$-integral exhibits a very perspicuous behaviour, as shown by

Theorem 31.9. *When* $\varphi(t) \geqq 0$, *then in the half-plane of convergence of* $\mathfrak{L}\{\varphi(t)\} = \Phi(s)$, *the function* Φ *restricted to real-valued* x *is completely monotonic and logarithmically convex, that is* $\log(\Phi(x))$ *is convex.*

[4] $\varkappa_2$ may, for some given function f, assume the values $+\infty$ or $-\infty$.

Proof: By Theorem 6.1, we have

$$\Phi^{(n)}(x) = (-1)^n \int_0^\infty e^{-xt}\, t^n\, \varphi(t)\, dt,$$

hence,

$$(-1)^n\, \Phi^{(n)}(x) \geqq 0.$$

A function which exhibits this property is called completely monotonic. It is monotonically decreasing, because of $\Phi'(x) \leqq 0$; it is convex, because of $\Phi''(x) \geqq 0$. The more specific property of logarithmic convexity is demonstrated by the following argumentation: We have, for $-\infty < y < +\infty$,

$$|\Phi(x + iy)| \leqq \int_0^\infty e^{-xt}\, \varphi(t)\, dt = \Phi(x),$$

thus,[5]

$$\sup_{-\infty < y < +\infty} |\Phi(x + iy)| = \Phi(x).$$

At this point we invoke a theorem, which is taken from the theory of functions:

Three Lines Theorem: *Suppose that the function $\Phi(s) = \Phi(x + iy)$ is analytic and bounded on the strip $a \leqq x \leqq b$. Let us define:*

$$\sup_{-\infty < y < +\infty} |\Phi(x + iy)| = M(x).$$

When $a \leqq x_1 < x_2 < x_3 \leqq b$, then we have for $\Phi(s) \not\equiv 0$:

$$\begin{vmatrix} x_1 & \log M(x_1) & 1 \\ x_2 & \log M(x_2) & 1 \\ x_3 & \log M(x_3) & 1 \end{vmatrix} \geqq 0.$$

This indicates that the triangle with the corners (x_ν, y_ν), $y_\nu = \log M(x_\nu)$ $(\nu = 1, 2, 3)$ has a positive area, which implies that the corner (x_2, y_2) lies beneath the straight line connection of the other corners (x_1, y_1) and (x_3, y_3); thus, the function $\log M(x)$ is convex.

In our application of this theorem, $M(x) = \Phi(x)$; thus the conclusion of Theorem 31.9 is verified.

Upon replacing $m(x)$ by the left hand side of (8), and then applying Theorem 31.9, we obtain:

Theorem 31.10. *When $\varkappa_2 < +\infty$, then there exists the quadratic mean $m(x)$ for*

[5] We use the symbol $\sup_{a \leqq y \leqq b} f(y)$ to designate the least upper bound of $f(y)$ in the interval $a \leqq y \leqq b$.

$x > \varkappa_2$; *it is an arbitrarily often differentiable, completely monotonic, logarithmically convex function.*

Formula (7) enables us to draw an important conclusion. Under the hypothesis $s_1 = s_2 = 0$ it has the form

$$\int_{-\infty}^{+\infty} f_1(t)\, \overline{f_2(t)}\, dt = \frac{1}{2\pi} \int_{-\infty}^{+\infty} F_1(iy)\, \overline{F_2(iy)}\, dy .$$

In the case that

$$\int_{-\infty}^{+\infty} f_1(t)\, \overline{f_2(t)}\, dt = 0,$$

that is when $f_1(t)$ and $f_2(t)$ are orthogonal, then it follows that

$$\int_{-\infty}^{+\infty} F_1(iy)\, \overline{F_2(iy)}\, dy = 0 .$$

We conclude that the corresponding functions $F_1(iy)$ and $F_2(iy)$ too are orthogonal. We summarize this conclusion in:

Theorem 31.11. *Suppose that the functions $f_n(t)$, $n = 0, 1, 2, \ldots$, have the following three properties:*

$$1.\ \int_{-\infty}^{+\infty} |f_n(t)|\, dt < \infty, \quad 2.\ \int_{-\infty}^{+\infty} |f_n(t)|^2\, dt < \infty, \ \textit{and}$$

$$3.\ \int_{-\infty}^{+\infty} f_n(t)\, \overline{f_m(t)}\, dt = 0 \quad \textit{for } n \neq m,$$

that is, the $f_n(t)$ constitute an orthogonal system of functions in the interval $(-\infty, +\infty)$. Then the corresponding image functions $F_n(s) = \mathfrak{L}_{II}\{f_n(t)\}$ on the imaginary axis, that is the functions $F_n(iy)$, also form an orthogonal system:

$$\int_{-\infty}^{+\infty} F_n(iy)\, \overline{F_m(iy)}\, dy = 0 \quad \textit{for } n \neq m .$$

Theorem 31.11 *is also true for the one-sided $\mathfrak{L}_I$-transformation, that is, when $f_n(t) = 0$ for $t < 0$.*

Example: Let $L_n(t)$ designate the nth Laguerre polynomial:

$$L_n(t) = \frac{e^t}{n!} \frac{d^n}{dt^n} (e^{-t} t^n) = \sum_{\nu=0}^{n} (-1)^\nu \binom{n}{\nu} \frac{t^\nu}{\nu!},$$

then, obviously, the functions

$$f_n(t) = e^{-t/2} L_n(t)$$

have the properties

$$\int_0^\infty |f_n(t)|\,dt < \infty, \text{ and } \int_0^\infty |f_n(t)|^2\,dt < \infty.$$

Moreover, these functions are orthogonal in $(0, +\infty)$, a fact which can readily be verified:

$$\int_0^\infty f_n(t)\,\overline{f_m(t)}\,dt = \int_0^\infty e^{-t} L_n(t)\, L_m(t)\,dt = 0 \quad \text{for } n \neq m.$$

The corresponding $\mathfrak{L}_{\mathrm{I}}$-transforms are given by:

$$F_n(s) = \frac{\left(s - \frac{1}{2}\right)^n}{\left(s + \frac{1}{2}\right)^{n+1}}.$$

Hence, we conclude that the functions

$$F_n(iy) = \frac{\left(iy - \frac{1}{2}\right)^n}{\left(iy + \frac{1}{2}\right)^{n+1}}$$

form an orthogonal system in the interval $(-\infty, +\infty)$.

When deriving the Parseval formula for the $\mathfrak{L}$-transformation, we started with the Parseval formula for the $\mathfrak{F}$-transformation in form of formula (4). Using the other form (5), we now derive an important result for the $\mathfrak{L}$-transformation of the product of original functions. The product was the one elementary combination of original functions for which we could not hitherto produce the corresponding $\mathfrak{L}$-transform. First we formulate the result for the $\mathfrak{L}_{\mathrm{II}}$-transform.

Theorem 31.12. *When for some given pair of fixed real values, x_1 and x_2, we have:*

$$\int_{-\infty}^{+\infty} e^{-x_1 t}\,|f_1(t)|\,dt < \infty, \qquad \int_{-\infty}^{+\infty} e^{-x_2 t}\,|f_2(t)|\,dt < \infty,$$

$$\int_{-\infty}^{+\infty} e^{-2x_1 t}\,|f_1(t)|^2\,dt < \infty, \quad \int_{-\infty}^{+\infty} e^{-2x_2 t}\,|f_2(t)|^2\,dt < \infty;$$

then, with $s_1 = x_1 + iy_1$ and $s_2 = x_2 + iy_2\,(-\infty < y_1, y_2 < +\infty)$, we find

(9) $$\int_{-\infty}^{+\infty} e^{-(s_1+s_2)t} f_1(t)\, f_2(t)\,dt = \frac{1}{2\pi}\int_{-\infty}^{+\infty} F_1(s_1 + iy)\, F_2(s_2 - iy)\,dy$$

and, with $s_1 + s_2 = s$,

$$\int_{-\infty}^{+\infty} e^{-st} f_1(t)\, f_2(t)\,dt = \frac{1}{2\pi i}\int_{x_1 - i\infty}^{x_1 + i\infty} F_1(\sigma)\, F_2(s - \sigma)\,d\sigma$$

(10) $$= \frac{1}{2\pi i}\int_{x_2 - i\infty}^{x_2 + i\infty} F_1(s - \sigma)\, F_2(\sigma)\,d\sigma \quad \textit{for } \Re s = x_1 + x_2.$$

Under the hypotheses of Theorem 31.12, the right hand sides of equation (10) define the "complex convolution" of $F_1(s)$ and $F_2(s)$, for which we use the symbol $F_1 \circ F_2$. With this notation, we present the above result as a theorem.

Theorem 31.13. *Suppose that the hypotheses of Theorem 31.12 are satisfied. Then $\mathfrak{L}_{II}\{f_1 \cdot f_2\}$ exists for the values of s on the vertical line $\Re s = x_1 + x_2$; it is, for these s, equal to $F_1 \circ F_2$:*

$$\mathfrak{L}_{II}\{f_1 \cdot f_2\} = F_1(s) \circ F_2(s) \quad (\Re s = x_1 + x_2).$$

Corresponding to the "Real Convolution Theorem" for the product of two image functions, we have the "*Complex Convolution Theorem*" for the product of two original functions.

Proof: Once again introducing into formula (5) the substitutions which were used during the initial steps of the proof of Theorem 31.7, we obtain formula (9). Setting $s_1 + s_2 = s$ and $s_1 + iy = \sigma$, so that $s_2 - iy = s - s_1 - iy = s - \sigma$, we have the first line of Eq. (10), although with the limits of integration $s_1 \pm i\infty$ which may, however, be replaced by $x_1 \pm i\infty$. The second line of Eq. (10) results when $s_2 + iy = \sigma$ is used instead of $s_1 + iy = \sigma$.

These Theorems become Theorems for the $\mathfrak{L}_I$-transformation, by the specifications: $f_1(t) = 0$ and $f_2(t) = 0$ for $t < 0$. In this case, all hypotheses are certainly satisfied for all x_1 and x_2 larger than the fixed values of Theorem 31.12. Moreover, absolute convergence of $\mathfrak{L}_I\{f_1\}$ for $\Re s \geqq x_1$, and of $\mathfrak{L}_I\{f_2\}$ for $\Re s \geqq x_2$, implies uniform convergence of $F_1(s)$ and of $F_2(s)$ towards zero in the respective right half-plane, as $y \to \pm\infty$. Thus, one may, similarly as on p. 159, shift the path of integration to any abscissa x, provided the path remains in the region of absolute convergence of *both* functions; that is, for instance, as long as, in the first line of (10), $x \geqq x_1$ and $\Re s - x \geqq x_2$. This conclusion is summarized in the following theorem:

Theorem 31.14. *When two real values x_1 and x_2 can be specified so that*

$$\int_0^\infty e^{-x_1 t}\, |f_1(t)|\, dt < \infty, \qquad \int_0^\infty e^{-x_2 t}\, |f_2(t)|\, dt < \infty,$$

$$\int_0^\infty e^{-2x_1 t}\, |f_1(t)|^2\, dt < \infty, \qquad \int_0^\infty e^{-2x_2 t}\, |f_2(t)|^2\, dt < \infty,$$

then, for all s with $\Re s \geqq x_1 + x_2$,

$$\int_0^\infty e^{-st} f_1(t)\, f_2(t)\, dt = \frac{1}{2\pi i} \int_{x-i\infty}^{x+i\infty} F_1(\sigma)\, F_2(s-\sigma)\, d\sigma \qquad (x_1 \leqq x \leqq \Re s - x_2)$$

$$(11) \qquad = \frac{1}{2\pi i} \int_{x-i\infty}^{x+i\infty} F_1(s-\sigma)\, F_2(\sigma)\, d\sigma \qquad (x_2 \leqq x \leqq \Re s - x_1);$$

that is, $\mathfrak{L}_I\{f_1 \cdot f_2\}$ exists for $\Re s \geqq x_1 + x_2$, and it is equal to the complex convolution $F_1 \circ F_2$ which is explicitly shown on the right hand side of (11). *When only the* *second line of the hypotheses is guaranteed, then* (11) *is valid for* $\Re s > x_1 + x_2$, *whereby* x *can vary in the interval* $x_1 < x < \Re s - x_2$ *or* $x_2 < x < \Re s - x_1$ *respectively.*

The latter conclusion is true since, as on p. 212, by the Cauchy-Schwarz inequality, convergence of

$$\int_0^\infty e^{-2x_\nu t}\, |f_\nu(t)|^2\, dt \quad (\nu = 1, 2)$$

implies convergence of

$$\int_0^\infty e^{-xt}\, |f_\nu(t)|\, dt \quad \text{for } x > x_\nu.$$

Example: The Gamma function $\Gamma(s)$ is defined for $\Re s > 0$ by

$$\Gamma(s) = \int_0^\infty z^{s-1} e^{-z}\, dz.$$

Replacement, in this formula, of the dummy variable of integration by αz with $\alpha > 0$, yields:

$$\int_0^\infty z^{s-1} e^{-az}\, dz = \frac{\Gamma(s)}{a^s} \quad (a > 0,\ \Re s > 0).$$

Substitution, in the last representation, of e^{-t} for z, produces the $\mathfrak{L}_{\mathrm{II}}$-transform:

$$\int_{-\infty}^{+\infty} e^{-st} \exp(-a\, e^{-t})\, dt = \frac{\Gamma(s)}{a^s}.$$

Defining

$$f_1(t) = \exp(-a_1 e^{-t}), \text{ and } f_2(t) = \exp(-a_2 e^{-t}) \quad (a_1, a_2 > 0),$$

we have

$$f_1(t)\, f_2(t) = \exp\left(-(a_1 + a_2)\, e^{-t}\right)$$

and

$$\int_{-\infty}^{+\infty} e^{-st} \exp\left(-(a_1 + a_2)\, e^{-t}\right) dt = \frac{\Gamma(s)}{(a_1 + a_2)^s}.$$

All hypotheses of Theorem 31.12 are satisfied for all $x_1 > 0$ and for all $x_2 > 0$; consequently, by (10),

$$(12) \qquad \frac{\Gamma(s)}{(a_1 + a_2)^s} = \frac{1}{2\pi i} \int_{x-i\infty}^{x+i\infty} \frac{\Gamma(\sigma)}{a_1^\sigma}\, \frac{\Gamma(s-\sigma)}{a_2^{s-\sigma}}\, d\sigma \quad (a_1 > 0,\ a_2 > 0,\ 0 < x < \Re s),$$

where x_1 is replaced by x, and $\Re s = x_1 + x_2$ may assume any value larger than x. Formula (12) provides a remarkable *transcendental addition theorem for the function* $\Gamma(s)/\alpha^s$.

32. The Concepts: Asymptotic Representation, Asymptotic Expansion

In previous Chapters we demonstrated how the $\mathfrak{L}$-transformation maps certain operations, like differentiation or convolution, in the original space to corresponding operations on the respective image functions in the image space. There is a further mapping property of a quite different kind which we now investigate. We shall show that the functional behaviour of some image function $F(s)$ as $s \to \infty$ is determined by the functional behaviour of the corresponding original function $f(t)$ near $t = 0$; one could say that the functional behaviour of $f(t)$ near $t = 0$ is mapped onto the functional behaviour of the $\mathfrak{L}$-transform $F(s)$ near $s = \infty$. Likewise, the behaviour of $f(t)$ as $t \to \infty$ is mapped onto the behaviour of $F(s)$ near a certain finite point.[1]

Asymptotic Representation

First, we must explain the term "behaviour of a function in the neighbourhood of some fixed point". On the real line, or in the complex plane, we designate as *neighbourhood* U of some given, fixed point z_0 an interval, or a region, which contains the given point z_0 in its interior or on its boundary; the point z_0 itself does not belong to the neighbourhood U.[2] For instance: the interval $0 < z < 1$ is a (one-dimensional) neighbourhood of $z_0 = 0$, on the real line; the angular region $|\arc z| < \pi/2$, $0 < |z| < 1$ is a (two-dimensional) neighbourhood of $z_0 = 0$, in the complex plane; the angular region $|\arc z| < \pi/4$, $0 < |z| < \infty$ is a (two-dimensional) neighbourhood of $z_0 = \infty$, in the complex plane. The behaviour of some function $\varphi(z)$ (which, in general, is a complicated function that is difficult to characterize) we shall describe in the neighbourhood U of some point z_0 by some "*comparison function*" $A\,\psi(z)$ (A is a constant, and $\psi(z)$ is an elementary, well known and understood function) which *asymptotically represents* the given function $\varphi(z)$ as z tends, in U, towards z_0, in the following sense:

$$(1) \qquad \lim_{z \to z_0,\, z \in U} \frac{\varphi(z)}{\psi(z)} = A\,.$$

We introduce the symbolic notation:

$$\varphi(z) \sim A\,\psi(z) \quad \text{as} \quad z \to z_0 \text{ in } U\,,$$

---

[1] The name "Abelian theorems" is used for those theorems which, for some given functional transformation, predict from the known functional behaviour of the original function the behaviour of the image function. This designation stems from the Abelian Continuity Theorem: Whenever the power series $\varphi(z) = \sum_{n=0}^{\infty} c_n z^n$ converges at $z = 1$ to the sum s, the limit $\lim_{z \to 1} \varphi(z)$ exists and equals s. In this case, one predicts from the behaviour of the "original sequence" c_n, as $n \to \infty$, $\lim_{n \to \infty} \sum_{\nu=0}^{n} c_\nu = s$, the behaviour of the image function $\varphi(z)$ near $z = 1$.

[2] In topology, one defines as a neighbourhood of some given point z_0 an open set which contains the point z_0. Here, in the sequel, we shall frequently encounter the situation that z_0 is a point of the boundary of U. Moreover, for our purpose, we must stipulate that the point z_0 does not belong to the neighbourhood.

which reads: "$\varphi(z)$ is asymptotically equal to $A\,\psi(z)$ as z tends, in U, towards z_0", or "$\varphi(z)$ behaves like $A_\psi(z)$ as $z \to z_0$, in U".

The definition (1) implies that

$$\lim_{z \to z_0,\, z \in U} \frac{\varphi(z) - A\,\psi(z)}{\psi(z)} \to 0\,, \tag{2}$$

that is, the "relative error" between φ and $A\,\psi(z)$ goes towards zero as z approaches z_0. Instead of (2), one could write

$$\varphi(z) = A\;\psi(z) + o(\psi) \quad \text{as } z \to z_0, \text{ in } U, \tag{3}$$

that is, the difference between φ and $A\,\psi$ is of smaller order (of magnitude) than ψ. The above shown expressions (1), (2), and (3) are equivalent; however, for applications, expression (3) is often preferred. The function $\psi(z)$ occurs in the denominator; hence, we must presume $\psi(z) \neq 0$ for z in U.

Asymptotic Expansion

In practical applications one is frequently not satisfied with a single comparison function; instead, one seeks a sequence of comparison functions which represent the given function with increasing accuracy. Suppose that a first comparison function $\psi_0(z)$ of $\varphi(z)$ is known:

$$\varphi(z) \sim \psi_0(z)\,, \text{ that is } \varphi(z) = \psi_0(z) + o(\psi_0)\,,$$

then one may attempt to represent the "error" $\varphi(z) - \psi_0(z)$ by a second comparison function $\psi_1(z)$:

$$\varphi(z) - \psi_0(z) \sim \psi_1(z)\,, \text{that is } \varphi(z) - \psi_0(z) = \psi_1(z) + o(\psi_1)$$

or

$$\varphi_0(z) = \psi_0(z) + \psi_1(z) + o(\psi_1)\,.$$

In all these expressions, the order symbol o refers to the limiting process $z \to z_0$. Proceeding in the above indicated manner, one generates a sequence of functions $\psi_\nu(z)$. We may formally, and without concern regarding convergence, form the series $\sum_{\nu=0}^{\infty} \psi_\nu(z)$. The partial sums of this series have the property:

$$\varphi(z) = \sum_{\nu=0}^{n} \psi_\nu(z) + o(\psi_n) \quad \text{as } z \to z_0\,. \tag{4}$$

Eq. (4) implies: *The difference between the given function $\varphi(z)$ and the partial sum is of smaller order (of magnitude) than the last included element.*

We say: $\varphi(z)$ has the *asymptotic expansion* $\sum_{\nu=0}^{\infty} \psi_\nu(z)$ as $z \to z_0$, provided that (4) is satisfied for $n = 0, 1, 2, \ldots$, and we write symbolically

$$\varphi(z) \approx \sum_{\nu=0}^{\infty} \psi_\nu(z) \quad \text{as } z \to z_0\,.$$

Rewriting the Eq. (4) in the following manner

$$\varphi(z) - \sum_{\nu=0}^{n-1} \psi_\nu(z) = \psi_n(z) + o(\psi_n)\,, \tag{5}$$

we produce an equation of the form of (3). Thus, one may also characterize an asymptotic expansion by the property: *The difference between the function and any partial sum behaves asymptotically like the next following element of the asymptotic expansion:*

$$\varphi(z) - \sum_{\nu=0}^{n-1} \psi_\nu(z) \sim \psi_n(z) \quad \text{as} \quad z \to z_0\,. \tag{6}$$

The expressions (4), (5), and (6) are equivalent.

Powers are most often employed as comparison functions; consequently, in most cases, the asymptotic expansions have the form of *power series*. Near some finite point z_0, these are powers of $(z - z_0)$ with increasing exponents:

$$\varphi(z) \approx \sum_{\nu=0}^{\infty} a_\nu (z - z_0)^{\lambda_\nu} \quad \text{as} \quad z \to z_0\,.$$

Near $z = \infty$, these are powers of z with decreasing exponents:

$$\varphi(z) \approx \sum_{\nu=0}^{\infty} \frac{a_\nu}{z^{\lambda_\nu}} \quad \text{as} \quad z \to \infty.$$

The λ_ν form a monotonically increasing sequence of real numbers, not necessarily integer-valued; a finite number of these may be negative:

$$\lambda_0 < \lambda_1 < \lambda_2 < \cdots \to \infty\,.$$

For some purposes one has to admit complex-valued λ_ν; in this case we require: $\Re\lambda_\nu < \Re\lambda_{\nu+1}$.

We call attention to the distinction between convergent series and asymptotic expansions. When dealing with a *convergent* series, we specify some fixed point z, and we consider the partial sums $\sum\limits_{\nu=0}^{n} \psi_\nu(z)$ and the limiting process $n \to \infty$. When using an *asymptotic* expansion, we study some partial sum for a fixed n, and the limiting process $z \to z_0$.

However, any power series with increasing exponents:

$$\varphi(z) = \sum_{\nu=0}^{\infty} a_\nu (z - z_0)^{\lambda_\nu}, \quad \lambda_0 < \lambda_1 < \cdots,$$

which converges absolutely[3] for $|z - z_0| \leqq \varrho$ is also an asymptotic expansion as $z \to z_0$. This is shown by

$$\left| \varphi(z) - \sum_{\nu=0}^{n} a_\nu (z - z_0)^{\lambda_\nu} \right| = |z - z_0|^{\lambda_{n+1}} \left| \sum_{\nu=n+1}^{\infty} a_\nu (z - z_0)^{\lambda_\nu - \lambda_{n+1}} \right|$$

$$\leqq |z - z_0|^{\lambda_{n+1}} \sum_{\nu=n+1}^{\infty} |a_\nu\, \varrho^{\lambda_\nu - \lambda_{n+1}}| = C\, |z - z_0|^{\lambda_{n+1}}\,.$$

[3] Attention is called to the remark in connection with Theorem 30.2.

The right hand side is $o(|z - z_0|^{\lambda_n})$ because of $\lambda_{n+1} > \lambda_n$; hence,

$$\varphi(z) \approx \sum_{\nu=0}^{\infty} a_\nu (z - z_0)^{\lambda_\nu} \quad \text{as} \quad z \to z_0$$

in the neighbourhood $0 < |z - z_0| \leqq \varrho$. Similarly, one can show that any absolutely converging power series with decreasing exponents is, in fact, an asymptotic expansion as $z \to \infty$.

33. Asymptotic Behaviour of the Image Function near Infinity

The relationship between the functional behaviour of some original function and the behaviour of the corresponding image function as mentioned in Chapter 32, is expressed by

Theorem 33.1. *Suppose that the two real-valued functions $f(t)$ and $\varphi(t)$, defined for $t > 0$, are continuous in some interval $0 < t < T$, where $\varphi(t) > 0$. Assume that the two corresponding $\mathfrak{L}$-transforms $\mathfrak{L}\{f\} = F(s)$ and $\mathfrak{L}\{\varphi\} = \Phi(s)$ exist in respective half-planes. Then we may conclude: When*

$$f(t) \sim A\,\varphi(t) \quad \textit{as real-valued } t \to 0 \ (A \textit{ an arbitrary, real constant}),$$

then

$$F(s) \sim A\,\Phi(s) \quad \textit{as real-valued } s \to \infty.$$

Proof: In the sequel we restrict s to the real axis of the common half-plane of convergence of both $\mathfrak{L}\{f\}$ and $\mathfrak{L}\{\varphi\}$. The function f/φ is continuous in every interval $0 < t \leqq \delta < T$, and $\varphi(t) > 0$; hence, we may invoke the first mean value theorem of integration, and we find:

$$\int_0^\delta e^{-st} f(t)\,dt = \int_0^\delta e^{-st} \frac{f(t)}{\varphi(t)}\,\varphi(t)\,dt = \frac{f(\vartheta)}{\varphi(\vartheta)} \int_0^\delta e^{-st} \varphi(t)\,dt,$$

where ϑ indicates some point with $0 < \vartheta < \delta$, which depends upon s. We define:

$$\int_0^\delta e^{-st} f(t)\,dt = F_1(s), \quad \text{and} \quad \int_0^\delta e^{-st} \varphi(t)\,dt = \Phi_1(s),$$

hence

$$\frac{F_1(s)}{\Phi_1(s)} = \frac{f(\vartheta)}{\varphi(\vartheta)}.$$

The property: $f(t)/\varphi(t) \to A$ as $t \to 0$ indicates that for every given $\varepsilon > 0$ one can choose a fixed value δ sufficiently small so that, for $0 < t < \delta$,

$$\left| \frac{f(t)}{\varphi(t)} - A \right| < \frac{\varepsilon}{2}.$$

Consequently, we also have

$$\left| \frac{f(\vartheta)}{\varphi(\vartheta)} - A \right| < \frac{\varepsilon}{2},$$

hence, for all s,

$$\left| \frac{F_1(s)}{\Phi_1(s)} - A \right| < \frac{\varepsilon}{2}. \tag{1}$$

Now we are going to show that for all large values of s the quotient $F(s)/\Phi(s)$ differs from $F_1(s)/\Phi_1(s)$ by less than $\varepsilon/2$. Then the Theorem will be verified.

Using the previously selected fixed value δ we define

$$\int_\delta^\infty e^{-st} f(t)\, dt = F_2(s), \text{ and } \int_\delta^\infty e^{-st} \varphi(t)\, dt = \Phi_2(s).$$

As $s \to \infty$ we have, by Theorem 23.2,

$$F_2(s) = e^{-\delta s} \int_0^\infty e^{-s\tau} f(\tau + \delta)\, d\tau = e^{-\delta s}\, o(1) = o(e^{-\delta s})$$

and, similarly,

$$\Phi_2(s) = o(e^{-\delta s}).$$

The function $\varphi(t)$ is continous and > 0 on the interval $\delta/2 \leqq t \leqq \delta$; thus it has a minimum $m > 0$ on this interval. For $s > 0$, we have:

$$\Phi_1(s) \geqq \int_{\delta/2}^{\delta} e^{-st} \varphi(t)\, dt \geqq m \frac{\delta}{2} e^{-\delta s};$$

hence

$$e^{-\delta s} \leqq \frac{2}{m\delta} \Phi_1(s).$$

The last expression enables us to replace the estimations for F_2 and for Φ_2 by

$$F_2(s) = o(\Phi_1), \text{ and } \Phi_2(s) = o(\Phi_1) \quad \text{as } s \to \infty.$$

Hence, one finds

$$\frac{F(s)}{\Phi(s)} = \frac{F_1(s) + F_2(s)}{\Phi_1(s) + \Phi_2(s)} = \frac{F_1(s) + o(\Phi_1)}{\Phi_1(s) + o(\Phi_1)} = \frac{\dfrac{F_1(s)}{\Phi_1(s)} + o(1)}{1 + o(1)} = \frac{F_1(s)}{\Phi_1(s)} + o(1)$$

as $s \to \infty$. For every given ε one can therefore find a s_0 so that, for $s \geqq s_0$,

$$\left| \frac{F(s)}{\Phi(s)} - \frac{F_1(s)}{\Phi_1(s)} \right| < \frac{\varepsilon}{2}. \tag{2}$$

Combining (1) with (2) we obtain:

$$\left|\frac{F(s)}{\Phi(s)} - A\right| = \left|\left(\frac{F(s)}{\Phi(s)} - \frac{F_1(s)}{\Phi_1(s)}\right) + \left(\frac{F_1(s)}{\Phi_1(s)} - A\right)\right|$$

$$\leqq \left|\frac{F(s)}{\Phi(s)} - \frac{F_1(s)}{\Phi_1(s)}\right| + \left|\frac{F_1(s)}{\Phi_1(s)} - A\right| < \varepsilon$$

for $s \geqq s_0$. This implies that

$$\frac{F(s)}{\Phi(s)} \to A \quad \text{as } s \to \infty.$$

An essential prerequisite for the above proof lies in the condition that both $f(t)$ and $\varphi(t)$ are real-valued functions,[1] and that s is restricted to real values. When f and φ can assume complex values, or when s tends towards ∞ in an angular region $|\operatorname{arc} s| \leqq \psi < \pi/2$,[2] then the above proof would not hold, and counter examples could be produced against a thus modified theorem.[3] A theorem which admits the thus extended conditions requires additional hypotheses.

Theorem 33.2. *Suppose that the two functions $f(t)$ and $\varphi(t)$ are real-valued or complex-valued for $t > 0$. Assume further that:*

I. $\varphi(t) \neq 0$ *and is continuous in some interval* $0 < t \leqq T$;

II. $f(t) \sim A\,\varphi(t)$ *as* $t \to 0$, A *an arbitrary, complex constant;*

III. $\mathfrak{L}\{f\} = F(s)$ *and* $\mathfrak{L}\{|\varphi|\} = \tilde{\Phi}(s)$ *exist in some half-plane* $x > x_0 > 0$ $(s = x + iy)$, *which also implies the existence of* $\mathfrak{L}\{\varphi\} = \Phi(s)$ *in the same half-plane;*

IV. *in the angular region* $\mathfrak{W}$, *which is defined by* $|\operatorname{arc} s| \leqq \psi < \pi/2$, *it is true that*

$$\frac{\tilde{\Phi}(x)}{|\Phi(s)|} < C \text{ when } \Re s = x > x_1.$$

---

[1] The requirement that these functions have but one sign in the neighbourhood of $t = 0$ poses *no further* restriction, for these functions are continuous, and $\varphi(t)$, and consequently also $f(t)$, must not vanish.

[2] Angular regions of the stated type are suggested since, in general, in such regions $F(s) \to 0$; this latter statement being, in fact, a statement concerning asymptotic behaviour.

[3] We have, for instance,

$$f(t) = \frac{1}{\sqrt{\pi t}} e^{i/t} + 1 \sim \varphi(t) = \frac{1}{\sqrt{\pi t}} e^{i/t} \quad \text{as } t \to 0,$$

and

$$F(s) = \frac{1}{\sqrt{s}} e^{-(1-i)\sqrt{2s}} + \frac{1}{s}, \quad \Phi(s) = \frac{1}{\sqrt{s}} e^{-(1-i)\sqrt{2s}}.$$

However, we cannot conclude that $F(s) \sim \Phi(s)$, instead, we have $F(s) \sim 1/s$ as $s \to \infty$, in $|\operatorname{arc} s| \leqq \psi < \pi/2$.
The real part, and the imaginary part, of $\varphi(t)$ and $\Phi(s)$ respectively, are employed as examples on several occasions (compare pp. 24, 226) to demonstrate remarkable situations. Here, we cannot use these since $(1/\sqrt{\pi t}) \cos 1/t$ and $(1/\sqrt{\pi t}) \sin 1/t$ change sign in the neighbourhood of $t = 0$.

Then

$$F(s) \sim A\,\Phi(s),$$

as s tends, two-dimensionally in $\mathfrak{W}$, towards ∞.[4]

Proof: Hypothesis II implies, for $t > 0$,

$$f(t) = A\,\varphi(t) + \varepsilon(t)\,\varphi(t), \quad \text{with } \varepsilon(t) \to 0 \text{ as } t \to 0. \tag{3}$$

Hence, for every given $\delta > 0$, for all s with $\Re s = x > x_0$,

$$F(s) = A\int_0^\infty e^{-st}\,\varphi(t)\,dt + \int_0^\delta e^{-st}\,\varepsilon(t)\,\varphi(t)\,dt + \int_\delta^\infty e^{-st}\,\varepsilon(t)\,\varphi(t)\,dt$$

$$= A\,\Phi(s) + \int_0^\delta e^{-st}\,\varepsilon(t)\,\varphi(t)\,dt + e^{-\delta s}\,I(s),$$

using

$$\int_0^\infty e^{-s\tau}\,\varepsilon(\tau+\delta)\,\varphi(\tau+\delta)\,d\tau = I(s).$$

Eq. (3) indicates that for every given $\varepsilon > 0$, one can select a (now fixed) value $\delta < T$, so that $|\varepsilon(t)| < \varepsilon$ for $0 < t \leqq \delta < T$. Thus, for $\Re s = x > x_0$, we may estimate as follows:

$$|F(s) - A\,\Phi(s)| \leqq \varepsilon \int_0^\delta e^{-xt}\,|\varphi(t)|\,dt + e^{-\delta x}\,|I(s)| \leqq \varepsilon\,\tilde{\Phi}(x) + e^{-\delta x}\,|I(s)|.$$

The fact that $|\varphi| > 0$ for $0 < t \leqq T$ implies that $\tilde{\Phi}(x) > 0$. Hypothesis IV indicates that $|\Phi(s)| \neq 0$ for all s in $\mathfrak{W}$ with $\Re s = x > x_1$. For thus restricted s, one can divide by $\Phi(s)$:

$$\left|\frac{F(s)}{\Phi(s)} - A\right| \leqq \varepsilon\,\frac{\tilde{\Phi}(x)}{|\Phi(s)|} + e^{-\delta x}\,\frac{|I(s)|}{|\Phi(s)|}. \tag{4}$$

The function $|\varphi(t)|$ is continuous and > 0 on $\delta/2 \leqq t \leqq \delta < T$; hence, $|\varphi(t)|$ has a minimum $m > 0$ on that interval. For $x > x_0 > 0$, it follows that

$$\tilde{\Phi}(x) = \int_0^\infty e^{-xt}\,|\varphi(t)|\,dt \geqq \int_{\delta/2}^\delta e^{-xt}\,|\varphi(t)|\,dt \geqq e^{-\delta x}\,m\,\frac{\delta}{2}$$

---

[4] This means: For every given $\varepsilon > 0$, there exists an $R > 0$, so that

$$\left|\frac{F(s)}{\Phi(s)} - A\right| < \varepsilon$$

for all s in $\mathfrak{W}$ with $|s| > R$. The situation $\psi = 0$ is permissible, in which case $\mathfrak{W}$ degenerates into the real axis of the s-plane.

or

$$e^{-\delta x} \leqq \frac{2}{m\,\delta}\,\tilde{\Phi}(x). \tag{5}$$

By Theorem 23.2, we can select an $R > 0$ so that[5]

$$|I(s)| < \frac{m\,\delta}{2}\,\varepsilon \quad \text{for all } s \text{ in } \mathfrak{W} \text{ with } |s| > R. \tag{6}$$

Moreover, we select R sufficiently large so that, for $|s| > R$ in $\mathfrak{W}$, $x > x_0$ and $x > x_1$. Because of (5) and (6), (4) yields:

$$\left|\frac{F(s)}{\Phi(s)} - A\right| \leqq 2\,\frac{\tilde{\Phi}(x)}{|\Phi(s)|}\,\varepsilon$$

or, because of Hypothesis IV,

$$\left|\frac{F(s)}{\Phi(s)} - A\right| \leqq 2\,C\,\varepsilon \quad \text{for all } s \text{ in } \mathfrak{W} \text{ with } |s| > R.$$

Thus, the Theorem is verified.

The most common comparison function is $\varphi(t) = t^\lambda$, with $\lambda > -1$. Application of Theorem 33.1 immediately yields the following property:

Suppose that $f(t)$ is a real-valued function, continuous in some neighbourhood of $t = 0$, and that $\mathfrak{L}\{f\} = F(s)$ has a half-plane of convergence; then the fact: $f(t) \sim A\,t^\lambda$, with $\lambda > -1$, as $t \to 0$ implies that $F(s) \sim \Gamma(\lambda+1)/s^{\lambda+1}$ as real-valued $s \to \infty$.

In order to make the conclusion useful for practical applications, we must extend the result in two directions: Firstly, we must provide for the situation that s tends towards ∞ in an angular region $\mathfrak{W}$, and secondly, we must admit a complex-valued exponent λ. To begin with, we verify quickly that $\mathfrak{L}\{t^\lambda\}$ converges (absolutely) for complex λ with $\Re\lambda > -1$ in the half-plane $\Re s > 0$, and that it is equal to $\Gamma(\lambda+1)/s^{\lambda+1}$. Indeed, with

$$\lambda = \lambda_1 + i\lambda_2 \quad \text{and} \quad \lambda_1 > -1, \quad \text{we have}$$

$$\left|\int_0^\infty e^{-st}\,t^\lambda\,dt\right| \leqq \int_0^\infty e^{-\Re s\,t}\,t^{\lambda_1}\,dt,$$

and the majorizing right hand side converges for $\Re s > 0$; clearly, $\mathfrak{L}\{t^\lambda\}$ diverges for $\Re s < 0$. The evaluation of $\mathfrak{L}\{t^\lambda\}$ is similar to the one employed for real-valued λ.

Prior to the application of Theorem 33.2, we must demonstrate compliance with Hypothesis IV. Indeed, for $\varphi(t) = t^\lambda$, with $\lambda = \lambda_1 + i\lambda_2$, we find that

$$\tilde{\Phi}(x) = \int_0^\infty e^{-xt}\,t^{\lambda_1}\,dt = \frac{\Gamma(\lambda_1+1)}{x^{\lambda_1+1}}$$

[5] We now realize why, especially, the angular region $\mathfrak{W}$ was chosen where the conclusion $F \sim A\,\Phi$ is to be verified.

and, using $s = r e^{i\vartheta}$,

$$|\Phi(s)| = \frac{|\Gamma(\lambda+1)|}{|s^{\lambda+1}|} = \frac{|\Gamma(\lambda+1)|}{|r^{\lambda_1+i\lambda_2+1} e^{i\varphi(\lambda_1+i\lambda_2+1)}|} = \frac{|\Gamma(\lambda+1)|}{r^{\lambda_1+1} e^{-\lambda_2\varphi}};$$

hence,

$$\frac{\tilde{\Phi}(x)}{|\Phi(s)|} = \frac{\Gamma(\lambda_1+1)}{|\Gamma(\lambda+1)|} \left(\frac{r}{x}\right)^{\lambda_1+1} e^{-\lambda_2\varphi}.$$

In every angular region $\mathfrak{W}$: $|\varphi| \leqq \psi$, for fixed $\psi < \pi/2$, one observes that

$$\frac{r}{x} = \frac{1}{\cos\varphi} \leqq \frac{1}{\cos\psi};$$

hence,

$$\frac{\tilde{\Phi}(x)}{|\Phi(s)|} \leqq \frac{\Gamma(\lambda_1+1)}{|\Gamma(\lambda+1)|} \frac{1}{\cos^{\lambda_1+1}\psi} e^{|\lambda_2|\psi} = C.$$

Application of Theorem 33.2 yields the more general conclusion which is expressed in

Theorem 33.3. *Suppose that the function $f(t)$ is real-valued or complex-valued for $t > 0$, and that $\mathfrak{L}\{f\} = F(s)$ has a half-plane of convergence. Then we conclude: When*

$f(t) \sim A\,t^{\lambda}$ *as $t \to 0$ (A an arbitrary, complex constant, λ complex, with $\Re\lambda > -1$), then*

$$F(s) \sim A\,\frac{\Gamma(\lambda+1)}{s^{\lambda+1}},$$

as s tends two-dimensionally in the angular region $\mathfrak{W}$: $|\text{arc}\,s| \leqq \psi < \pi/2$ towards ∞.

For $\lambda = 0$, we obtain the most often used special case:

Theorem 33.4. *Suppose that $\mathfrak{L}\{f\} = F(s)$ exists, and that $f(t)$ has the limit A when $t \to 0$; then $sF(s)$ has the same limit A when $s \to \infty$, in $|\text{arc}\,s| \leqq \psi < \pi/2$.*

Theorem 33.4 can alternatively be phrased as follows:

Theorem 33.5. *Suppose we know that the limit $f(0+)$ does exist, although its value is not known. Then, upon redefining $f(t)$ for large values of t in such a manner that $\mathfrak{L}\{f\}$ has a half-plane of convergence, we find that*

$$f(0+) = \lim_{s\to\infty} s\,F(s).$$

Obviously, Theorem 33.5 is particularly fruitful whenever $f(t)$ is a complicated function, and $F(s)$ is simple.

The order "hypotheses – conclusion" of Theorem 33.4 cannot be inverted; that is, the existence of $\lim sF(s)$ when $s \to \infty$, does not necessarily imply the existence of $\lim f(t)$ when $t \to 0$. This fact is demonstrated by the following example functions which were already employed on p. 24:

$$\mathfrak{L}\left\{\frac{1}{\sqrt{\pi t}} \begin{matrix}\cos\\ \sin\end{matrix} \frac{1}{t}\right\} = \frac{1}{\sqrt{s}} e^{-\sqrt{2s}} \begin{matrix}\cos\\ \sin\end{matrix} \sqrt{2s}.$$

Here, we have $\lim\limits_{s\to\infty} sF(s) = 0$; however, $\lim\limits_{t\to 0} f(t)$ does not exist. We emphasize this fact, since in the technical literature the improper inversion of Theorem 33.4 is often used instead of the above presented, correct Theorem 33.5.

After the power function, the most important comparison function is[6]

$$\varphi(t) = -t^\lambda \log t \quad (\Re \lambda > -1).$$

For this function, we find the corresponding image function

$$\Phi(s) = \frac{\Gamma(\lambda+1)}{s^{\lambda+1}} \left[\log s - \Psi(\lambda+1)\right],$$

where

$$\Psi(z) = \frac{\Gamma'(z)}{\Gamma(z)}.$$

This $\mathfrak{L}$-transformation is obtained by differentiation of the formula

$$\int_0^\infty e^{-st}\, t^\lambda \, dt = \frac{\Gamma(\lambda+1)}{s^{\lambda+1}}$$

with respect to λ. In order to prepare for the application of Theorem 33.2, we would have to estimate

$$\tilde{\Phi}(x) = -\int_0^1 e^{-xt}\, t^{\lambda_1} \log t \, dt + \int_1^\infty e^{-xt}\, t^{\lambda_1} \log t \, dt$$

for large values of x, an involved undertaking.[7] Thus, we restrain ourselves here to the conclusion indicated in Theorem 33.1.

Theorem 33.6. *Suppose that the real-valued function $f(t)$ is continuous in the neighbourhood of $t = 0$, and that $\mathfrak{L}\{f\} = F(s)$ converges in a right half-plane. Then we conclude: When*

$$f(t) \sim -A\, t^\lambda \log t \quad \textit{as } t \to 0 \ (A \textit{ an arbitrary real constant, } \lambda > -1),$$

then

$$F(s) \sim \frac{\Gamma(\lambda+1)}{s^{\lambda+1}} \left[\log s - \Psi(\lambda+1)\right] \quad \textit{as real } s \to +\infty.$$

In this case, one could replace the function $\Phi(s)$ by the asymptotically equivalent function $\Gamma(\lambda+1) \log s / s^{\lambda+1}$.

Employing these Theorems one can arrive at far-reaching conclusions concerning asymptotic expansions of functions which can be represented as $\mathfrak{L}$-transforms.

---

[6] The negative sign is included so that the function is positive in the neighbourhood of $t = 0$.

[7] Using asymptotic approximations for the so-called Incomplete Gamma Function

$$\Gamma(\lambda+1, x) = \int_x^\infty e^{-u}\, u^\lambda \, du,$$

one can demonstrate that condition IV of Theorem 33.2 is satisfied for every $\psi < \pi/2$.

Asymptotic Expansion of Image Functions

Here, we intend to develop asymptotic expansions of image functions $F(s)$ as $s \to \infty$, given the asymptotic expansions of the corresponding original functions $f(t)$ as $t \to 0$. In practical applications, the latter are often convergent power series, due to the mostly uncomplicated character of $f(t)$.

Theorem 33.7. *Suppose that $\mathfrak{L}\{f\} = F(s)$ has a half-plane of convergence. When $f(t)$ has the asymptotic expansion*

$$f(t) \approx \sum_{\nu=0}^{\infty} c_\nu t^{\lambda_\nu} \quad (-1 < \Re\lambda_0 < \Re\lambda_1 < \cdots) \quad \text{as real } t \to 0; \tag{7}$$

then $F(s)$ has the asymptotic expansion

$$F(s) \approx \sum_{\nu=0}^{\infty} c_\nu \frac{\Gamma(\lambda_\nu + 1)}{s^{\lambda_\nu+1}}, \tag{8}$$

as s tends two-dimensionally in the angular region $|\operatorname{arc} s| \leqq \psi < \pi/2$ towards ∞.

Proof: By definition (32.6), hypothesis (7) implies that

$$f(t) - \sum_{\nu=0}^{n-1} c_\nu t^{\lambda_\nu} \sim c_n t^{\lambda_n} \quad \text{as} \quad t \to 0.$$

The left hand side has a half-plane of convergence, hence, by Theorem 33.3,

$$\mathfrak{L}\left\{ f(t) - \sum_{\nu=0}^{n-1} c_\nu t^{\lambda_\nu} \right\} = F(s) - \sum_{\nu=0}^{n-1} c_\nu \frac{\Gamma(\lambda_\nu + 1)}{s^{\lambda_\nu+1}} \sim c_n \frac{\Gamma(\lambda_n + 1)}{s^{\lambda_n+1}},$$

as s tends two-dimensionally in $|\operatorname{arc} s| \leqq \psi < \pi/2$ towards ∞. Thus, Theorem 33.7 is verified.

For applications, one often employs the following specialization:

Theorem 33.8. *Suppose that $\mathfrak{L}\{f\} = F(s)$ has a half-plane of convergence. When $f(t)$ can be expressed in a neighbourhood of $t = 0$ by a convergent power series with fractional exponents of the form*

$$f(t) = \frac{1}{t} \sum_{\nu=1}^{\infty} a_\nu t^{\nu/m} \quad (m \text{ a natural number}),$$

then it follows that

$$F(s) \approx \sum_{\nu=1}^{\infty} a_\nu \frac{\Gamma\left(\frac{\nu}{m}\right)}{s^{\nu/m}}$$

as s tends two-dimensionally in $|\operatorname{arc} s| \leqq \psi < \pi/2$ towards ∞.

In the special case that (7) converges, one can interpret Theorem 33.7 as follows: A power series of $f(t)$ must, in general, not be transformed term by term in the process of forming $\mathfrak{L}\{f\} = F(s)$, although the series may converge for all values of t. For instance,

$$e^{-t^2} = \sum_{\nu=0}^{\infty} (-1)^\nu \frac{t^{2\nu}}{\nu!} \quad \text{for all } t,$$

and $\mathfrak{L}\{e^{-t^2}\}$ converges for all s; however, the series obtained by term-wise transformation

$$\sum_{\nu=0}^{\infty} (-1)^\nu \frac{(2\nu)!}{\nu!} \frac{1}{s^{2\nu+1}}$$

diverges for all s. Nevertheless, it is not useless, for it does provide an asymptotic expansion for $F(s)$. In this particular case, we have

$$F(s) = e^{(s/2)^2} \int_{s/2}^{\infty} e^{-x^2}\, dx,$$

a function closely related to the *Gaussian error function.* Thus, we have provided a first example of the application of Theorem 33.7. A further, important example is the following one:

The Bessel function

$$J_\alpha(z) = \sum_{\nu=0}^{\infty} \frac{(-1)^\nu}{\nu!\, \Gamma(\alpha+\nu+1)} \left(\frac{z}{2}\right)^{2\nu+\alpha},$$

when multiplied by $z^{-\alpha}$, can be expressed, for $\Re\alpha > -1/2$, by a finite Fourier integral[8]:

$$\sqrt{\pi}\, \Gamma\left(\alpha + \frac{1}{2}\right) \left(\frac{z}{2}\right)^{-\alpha} J_\alpha(z) = \int_{-1}^{+1} e^{izx} (1-x^2)^{\alpha-(1/2)}\, dx. \tag{9}$$

Using the substitutions: $z = is$, $x = t-1$, one obtains

$$\sqrt{\pi}\, \Gamma\left(\alpha + \frac{1}{2}\right) \left(\frac{is}{2}\right)^{-\alpha} e^{-s} J_\alpha(is) = \int_0^2 e^{-st} [t(2-t)]^{\alpha-(1/2)}\, dt, \tag{10}$$

which is a finite $\mathfrak{L}$-integral; it converges for all values of s, thus representing an entire function. The original function can be expanded in the neighbourhood of the origin into a convergent power series; the exponents of the latter are complex-valued whenever α is complex. This example demonstrates the need for the extension of Theorems 33.3 and 33.7 to include complex exponents. First consider the expansion:

$$[t(2-t)]^{\alpha-(1/2)} = (2t)^{\alpha-(1/2)} \left(1 - \frac{t}{2}\right)^{\alpha-(1/2)} = (2t)^{\alpha-(1/2)} \sum_{\nu=0}^{\infty} \binom{\alpha-\frac{1}{2}}{\nu} \left(-\frac{t}{2}\right)^\nu$$

$$= 2^{\alpha-(1/2)} \sum_{\nu=0}^{\infty} \frac{(-1)^\nu}{2^\nu} \binom{\alpha-\frac{1}{2}}{\nu} t^{\nu+\alpha-(1/2)}.$$

---

[8] Starting with $\mathfrak{L}\{t^\alpha J_\alpha(t)\}$ (compare p. 265), one derives the presented formula by a technique similar to the one employed in the derivation of the formula for J_0 shown on p. 57.

We have $\Re(\nu + \alpha - 1/2) > -1$, provided $\Re\alpha > -1/2$; hence, by Theorem 33.7,

$$\sqrt{2\pi}\,\Gamma\left(a+\frac{1}{2}\right)(i\,s)^{-a}\,e^{-s}\,J_a(i\,s) \approx \sum_{\nu=0}^{\infty} \frac{(-1)^\nu}{2^\nu}\binom{a-\frac{1}{2}}{\nu}\frac{\Gamma(\nu+a+\frac{1}{2})}{s^{\nu+a+(1/2)}}$$

as $s \to \infty$, in $|\operatorname{arc} s| \leqq (\pi/2) - \delta$, with $0 < \delta < \pi/2$. Substituting $z = i s$ and $i = e^{i(\pi/2)}$ yields:

$$\sqrt{2\pi}\,\Gamma\left(a+\frac{1}{2}\right) z^{-a}\,e^{iz}\,J_a(z) \approx \sum_{\nu=0}^{\infty} \frac{(-1)^\nu}{2^\nu}\binom{a-\frac{1}{2}}{\nu}\frac{\Gamma(\nu+a+\frac{1}{2})}{(e^{-i(\pi/2)}\,z)^{\nu+a+(1/2)}} \tag{11}$$

as $z \to \infty$, in $\delta < \operatorname{arc} z < \pi - \delta$. Since

$$\binom{a-\frac{1}{2}}{\nu} = \frac{(-1)^\nu\,\Gamma(\nu-a+\frac{1}{2})}{\nu!\,\Gamma(\frac{1}{2}-a)} \quad \text{and} \quad \frac{1}{\Gamma(\frac{1}{2}+a)\,\Gamma(\frac{1}{2}-a)} = \frac{\cos a\pi}{\pi}$$

we can rewrite the last statement as follows:

$$J_a(z) \approx \frac{\cos a\pi}{\sqrt{2}\,\pi^{3/2}}\,z^{-1/2}\,e^{-iz}\sum_{\nu=0}^{\infty}\frac{e^{[\nu+a+(1/2)]i\pi/2}}{2^\nu\,\nu!}\,\frac{\Gamma(\nu+a+\frac{1}{2})\,\Gamma(\nu-a+\frac{1}{2})}{z^\nu}. \tag{12}$$

Thus, we master the asymptotic behaviour of $J_a(z)$ in the upper half-plane, and also in the lower half-plane, since $z^{-a}J_a(z)$ is an even function, by (9). That is, we have control over the asymptotic behaviour of $J_a(z)$ in the entire plane, with the exception of the real axis. On the real axis, the asymptotic expansion has another form; it will be derived on pp. 267–269.

Employing Theorem 33.6 instead of Theorem 33.3, we can derive an analogue of Theorem 33.7:

Theorem 33.9. *Suppose that $\mathfrak{L}\{f\} = F(s)$ has a half-plane of convergence. Whenever $f(t)$ has the asymptotic expansion:*

$$f(t) \approx -\log t \sum_{\nu=0}^{\infty} c_\nu\,t^{\lambda_\nu} \quad (-1 < \lambda_0 < \lambda_1 < \cdots) \quad \textit{as real } t \to 0,$$

then the image function $F(s)$ has the asymptotic expansion:

$$F(s) \approx \sum_{\nu=0}^{\infty} c_\nu\,\frac{\Gamma(\lambda_\nu+1)}{s^{\lambda_\nu+1}}\left(\log s - \Psi(\lambda_\nu+1)\right) \quad \textit{as real } s \to \infty,$$

where $\Psi(z) = \Gamma'(z)/\Gamma(z)$.

Remark: The functions

$$\frac{\log s}{s^{\lambda_0+1}},\ \frac{1}{s^{\lambda_0+1}},\ \frac{\log s}{s^{\lambda_1+1}},\ \frac{1}{s^{\lambda_1+1}},\ \ldots$$

which occur in the expansion are of decreasing order (of magnitude) as $s \to \infty$.

34. Asymptotic Behaviour of the Image Function near a Singular Point on the Line of Convergence

In this Chapter we shall show that the functional behaviour of an original function $f(t)$ as $t \to \infty$, is reflected in the behaviour of the corresponding image function $F(s) = \mathfrak{L}\{f\}$ near some finite point s_0. At any point s_0 in the interior of the half-plane of convergence of $F(s)$, or on the line of convergence of $F(s)$, where the image function is holomorphic, the behaviour of $F(s)$ is trivial in the sense that $F(s)$ is continuous at s_0, and $F(s) \to F(s_0)$ as $s \to s_0$. Moreover, the $\mathfrak{L}$-integral cannot be called upon to provide information regarding the behaviour of $F(s)$ near such points s_0 outside the half-plane of convergence of $F(s)$, where $F(s)$ may happen to exist. Therefore, our interest concentrates upon singular points s_0 on the line of convergence.

We do not attempt here to develop an analogue to the general Theorem 33.1 for any comparison function; instead, we shall restrict our investigations to power functions as comparison functions, in the style of the specialized Theorem 33.3.

Theorem 34.1. *Suppose that the real-valued or complex-valued function $f(t)$ has the asymptotic property*

$$f(t) \sim A\, t^{\lambda} \quad \text{as } t \to \infty \ (A \text{ complex}, \ \Re\lambda > -1).$$

Then $\mathfrak{L}\{f\} = F(s)$ exists for $\Re s > 0$; it has, for $A \neq 0$, a singular point at $s = 0$, and it can be asymptotically represented in the following manner:

$$F(s) \sim A\,\frac{\Gamma(\lambda+1)}{s^{\lambda+1}},$$

as s tends two-dimensionally in the angular region $|\operatorname{arc} s| \leq \psi < \pi/2$ towards zero.

Proof: For $t \geqq 1$, we may write

$$f(t) = A\, t^{\lambda} + \varepsilon(t)\, t^{\lambda},$$

where $\varepsilon(t) \to 0$ as $t \to \infty$. Consequently, $\mathfrak{L}\{f\}$ converges in the right half-plane $\Re s > 0$. Using some as yet unspecified $T \geqq 1$, we have, for $\Re s > 0$,

$$\begin{aligned} F(s) &= \int_0^T e^{-st} f(t)\, dt + \int_T^\infty e^{-st}\,[A\, t^{\lambda} + \varepsilon(t)\, t^{\lambda}]\, dt \\ &= \int_0^T e^{-st}\,[f(t) - A\, t^{\lambda}]\, dt + \int_0^\infty e^{-st} A\, t^{\lambda}\, dt + \int_T^\infty e^{-st}\, \varepsilon(t)\, t^{\lambda}\, dt \\ &= A\,\frac{\Gamma(\lambda+1)}{s^{\lambda+1}} + \int_0^T e^{-st}\,[f(t) - A\, t^{\lambda}]\, dt + \int_T^\infty e^{-st}\, \varepsilon(t)\, t^{\lambda}\, dt. \end{aligned}$$

For any given, arbitrarily small $\varepsilon > 0$, we now select a fixed value $T \geqq 1$, sufficiently large so that $|\varepsilon(t)| \leqq \varepsilon$ for all $t \geqq T$. Then we have, with $\lambda = \lambda_1 + i\lambda_2$, the estimations:

$$\left|\int_T^\infty e^{-st}\, \varepsilon(t)\, t^{\lambda}\, dt\right| \leqq \varepsilon \int_T^\infty e^{-\Re s\cdot t}\, t^{\lambda_1}\, dt < \varepsilon \int_0^\infty e^{-\Re s\cdot t}\, t^{\lambda_1}\, dt = \varepsilon\,\frac{\Gamma(\lambda_1+1)}{(\Re s)^{\lambda_1+1}}$$

and

$$\left|\int_0^T e^{-st}\,[f(t) - A\,t^{\lambda}]\,dt\right| \leqq \int_0^T (|f(t)| + |A|\,t^{\lambda_1})\,dt = K,$$

where K designates a constant which is independent of the value of s. Thus, one concludes, for $\Re s > 0$, that

$$\left|F(s) - A\,\frac{\Gamma(\lambda+1)}{s^{\lambda+1}}\right| \leqq K + \varepsilon\,\frac{\Gamma(\lambda_1+1)}{(\Re s)^{\lambda_1+1}}$$

or

$$\left|F(s)\,\frac{s^{\lambda+1}}{\Gamma(\lambda+1)} - A\right| \leqq K\,\frac{|s^{\lambda+1}|}{|\Gamma(\lambda+1)|} + \varepsilon\,\frac{\Gamma(\lambda_1+1)}{|\Gamma(\lambda+1)|}\,\frac{|s^{\lambda+1}|}{(\Re s)^{\lambda_1+1}}\,.$$

With $s = |s|\,e^{i\vartheta}$ $(|\vartheta| < \pi/2)$, we have

$$s^{\lambda} = e^{\lambda\log s} = e^{(\lambda_1 + i\lambda_2)(\log|s| + i\vartheta)} = e^{\lambda_1\log|s| - \lambda_2\vartheta + i(\lambda_1\vartheta + \lambda_2\log|s|)},$$

hence

$$|s^{\lambda}| = e^{\lambda_1\log|s| - \lambda_2\vartheta} = |s|^{\lambda_1}\,e^{-\lambda_2\vartheta}\,.$$

Thus, one finds:

$$\left|F(s)\,\frac{s^{\lambda+1}}{\Gamma(\lambda+1)} - A\right| < \frac{K}{|\Gamma(\lambda+1)|}\,|s|^{\lambda_1+1}\,e^{-\lambda_2\vartheta} + \varepsilon\,\frac{\Gamma(\lambda_1+1)}{|\Gamma(\lambda+1)|}\left(\frac{|s|}{\Re s}\right)^{\lambda_1+1} e^{-\lambda_2\vartheta}\,.$$

We restrict s to the angular region $|\operatorname{arc} s| = |\vartheta| \leqq \psi < \pi/2$, so that

$$\frac{|s|}{\Re s} \leqq \frac{1}{\cos\psi} \qquad \text{and} \qquad e^{-\lambda_2\vartheta} \leqq e^{|\lambda_2|\,\psi}\,.$$

Also, we select ϱ sufficiently small so that, for $|s| < \varrho$,

$$|s|^{\lambda_1+1} < \varepsilon$$

because of $\lambda_1 > -1$. Then, we estimate for all $s \neq 0$ in the sector $|\operatorname{arc} s| \leqq \psi$ and $|s| < \varrho$:

$$\left|F(s)\,\frac{s^{\lambda+1}}{\Gamma(\lambda+1)} - A\right| < \varepsilon\left(\frac{K}{|\Gamma(\lambda+1)|}\,e^{|\lambda_2|\,\psi} + \frac{\Gamma(\lambda_1+1)}{|\Gamma(\lambda+1)|}\,\frac{e^{|\lambda_2|\,\psi}}{(\cos\psi)^{\lambda_1+1}}\right) = \varepsilon\cdot\text{const.}$$

The last statement implies that

$$F(s)\,\frac{s^{\lambda+1}}{\Gamma(\lambda+1)} \to A\,,$$

as s tends two-dimensionally in $|\operatorname{arc} s| \leqq \psi$ towards zero. This is the conclusion of the Theorem.

Using $\lambda = 0$, we produce a specialized case of the above Theorem:

Theorem 34.2. *Suppose that the function $f(t)$ has the limit A when $t \to \infty$. Then the corresponding image function $F(s) = \mathfrak{L}\{f\}$ exists for $\Re s > 0$, it has a singular point at $s = 0$, and it can be asymptotically represented by:*

$$F(s) \sim \frac{A}{s} \qquad \text{as } s \to 0, \quad \text{in } |\operatorname{arc} s| \leqq \psi < \pi/2.$$

This specialized Theorem can be used to find the limit $f(\infty)$.

Theorem 34.3. *When the limit of $f(t)$ as $t \to \infty$ exists, although its value is not known, then one can determine the value of the limit, using $\mathfrak{L}\{f\} = F(s)$, by means of*

$$\lim_{t\to\infty} f(t) = \lim_{s\to 0} s\,F(s).$$

This Theorem cannot be stated without the hypothesis regarding the existence of the limit $f(\infty)$. That is, the order "hypothesis – conclusion" of Theorem 34.2, and therefore of Theorem 34.1, cannot be reversed. This fact is demonstrated by the counterexample

$$f(t) \equiv \sin t, \qquad F(s) \equiv \frac{1}{s^2+1}.$$

For this example, $\lim\limits_{s\to 0} sF(s) = 0$; however, $f(t)$ does not have a limit as $t \to \infty$.

Conclusion: When $sF(s)$ has the limit A as $s \to 0$, then there are only two possibilities: Either $f(t)$ has a limit as $t \to \infty$, which equals A, or else the limit of $f(t)$ as $t \to \infty$ does not exist.

In the above Theorems which employ powers as comparison functions for the original function, $s = 0$ is the singular point, near which the behaviour of the image function is determined by the behaviour of the original function as $t \to \infty$. Other singular points could be encountered, when other functions are used as comparison functions of the original function.

Theorem 34.4. *When the original function $f(t)$ has the asymptotic property:*

$$f(t) \sim A e^{s_0 t} t^{\lambda} \quad \text{as } t \to \infty \quad (A \text{ and } s_0 \text{ complex}, \ \Re\lambda > -1),$$

then $\mathfrak{L}\{f\} = F(s)$ exists for $\Re s > \Re s_0$; it has, for $A \neq 0$, a singular point at s_0, and $F(s)$ can be asymptotically represented by:

$$F(s) \sim A \frac{\Gamma(\lambda+1)}{(s-s_0)^{\lambda+1}},$$

as s tends two-dimensionally in the angular region $|\operatorname{arc}(s-s_0)| \leq \psi < \pi/2$ towards s_0.

The proof results when Theorem 34.1 is applied to $\mathfrak{L}\{e^{-s_0 t} f(t)\} = F(s+s_0)$.

35. The Asymptotic Behaviour of the Original Function near Infinity, when the Image Function has Singularities of Unique Character

For most applicarions of the $\mathfrak{L}$-transformation, for instance when solving differential equations, one first derives the image function $F(s)$ of the sought solution $f(t)$, then one needs to determine the corresponding original function. Often, however, it is not possible to express $f(t)$ by means of known classical functions. Moreover, on many occasions one is less interested in the complete expression for $f(t)$ than in the asymptotic behaviour of $f(t)$ for large values of t; for instance, when investigating the stability of systems.

Thus, we are faced with the situation inverse to that of Chapter 34. The question is whether one could conclude from the (usually less complicated) behaviour of the image function $F(s)$ to the asymptotic behaviour of the corresponding (often more complicated) original function $f(t)$ as $t \to \infty$. A simple interchange of hypothesis and conclusion of Theorem 34.1 cannot be used to accomplish our aim, for we have shown in connection with Theorem 34.3 that the inverse of Theorem 34.1 is, in fact, incorrect.

The $\mathfrak{L}$-integral expresses $F(s)$ explicitly by $f(t)$.This fact explains how the knowledge of the behaviour of $f(t)$ enables us to predict the behaviour of $F(s)$, and why it is more difficult to draw the inverse conclusion. Obviously, if we want to describe the behaviour of $f(t)$ when the behaviour of $F(s)$ is given, we shall need an explicit expression of $f(t)$ involving $F(s)$: an inversion formula. The complex inversion formula is useful for the intended purpose; none of the other known inversion formulae has yielded results in this area of investigation.

From Chapter 24, we recall the formula:[1]

$$f(t) = \frac{1}{2\pi i} \int_{a-i\infty}^{a+i\infty} e^{ts} F(s)\, ds\,,$$

which represents the inverse of the $\mathfrak{L}_{\mathrm{I}}$- or the $\mathfrak{L}_{\mathrm{II}}$-transformation, provided certain hypotheses are satisfied. The abscissa a must be in the half-plane of holomorphy or in the strip of holomorphy respectively of $F(s)$. However, we can also interpret this formula as a transformation, in its own rights, with the original function $F(s)$ and the corresponding image function $f(t)$, regardless of the origin of $F(s)$ as a $\mathfrak{L}_{\mathrm{I}}$-transform or a $\mathfrak{L}_{\mathrm{II}}$-transform of $f(t)$. To this new transformation we assign the operational symbol $\mathfrak{B}$:

$$\text{(1)} \qquad \mathfrak{B}\{F\} \equiv \frac{1}{2\pi i} \int_{a-i\infty}^{a+i\infty} e^{ts} F(s)\, ds = f(t)\,.$$

However, we shall retain the original meaning of $\mathfrak{B}\{f\}$ insofar as $F(s)$ is presumed to be *analytic in some strip* which contains a. Thus, we exclude the case in which $F(s)$ is defined only on the line $\Re s = a$; for this case, $\mathfrak{B}\{f\}$ would coincide with

[1] We shall omit the symbol $V.P.$ hereafter.

the Fourier transformation, and we could not apply the method from the theory of functions which forms the principal technique in the subsequent considerations.

Nothing essential concerning the behaviour of $f(t)$ results as long as $F(s)$ is investigated in the strip of holomorphy only. We shall demonstrate that the behaviour of $f(t)$ as $t \to \infty$ depends upon the *singularities* of $F(s)$. In particular, the prediction of the behaviour of $f(t)$ as $t \to +\infty$ depends upon the behaviour of $F(s)$ near its singular points to the left of the strip of holomorphy(as $t \to -\infty$, upon those to the right of this strip). We presume that, when shifting the line of integration to the left, one encounters *isolated* singular points. The singularities could be single-valued, or else be of multi-valued character; this is, indeed, an essential distinction: in the first case we can move the line of integration beyond the singular point, provided the residue is properly accounted for; in the second case, this is not possible and other techniques are required (see Chapters 36, 37). For the remainder of this Chapter we restrict our investigation to the situation in which all singularities of $F(s)$ are single-valued.

We presume that the integral (1) converges, at least for $t > T$, and that $F(s)$ is an analytic function in the half-plane $\Re s \leq a$, with the possible exception of the poles $\alpha_0, \alpha_1, \alpha_2, \ldots$ with $a > \Re\alpha_0 > \Re\alpha_1 > \ldots$. (We could also admit isolated, essential singular points; accordingly, the residues would be represented not by finite sums but by infinite series.) We select the real point β_0 between $\Re\alpha_0$ and $\Re\alpha_1$ and we form a rectangle which contains the point α_0 using the vertical lines, through a and β_0 respectively, and the horizontal lines, at the heights $+\omega$ and $-\omega$. Then the integral of $e^{ts}F(s)$ along the boundary of the rectangle in the positive sense, divided by $2\pi i$, is equal to the residue $r_0(t)$ of $e^{ts}F(s)$ at the point α_0. If the principal part of the Laurent expansion of $F(s)$ at α_0 has the form

$$\frac{c_1^{(0)}}{s - a_0} + \cdots + \frac{c_{m_0}^{(0)}}{(s - a_0)^{m_0}},$$

then we have, by (26.5),

$$r_0(t) = \left(c_1^{(0)} + c_2^{(0)} \frac{t}{1!} + \cdots + c_{m_0}^{(0)} \frac{t^{m_0 - 1}}{(m_0 - 1)!}\right) e^{a_0 t}.$$

Next, we assume that $F(s)$ tends, uniformly with respect to $\Re s$ on the strip $\beta_0 \leq \Re s \leq a$, towards zero when $s \to \infty$. Thus, the values of the integrals along the horizontal sides tend towards zero when $\omega \to \infty$ (compare p. 159). The integral along the vertical line through a converges, by (1), towards $f(t)$. It follows that

$$r_0(t) = f(t) + \frac{1}{2\pi i} \int_{\beta_0 + i\infty}^{\beta_0 - i\infty} e^{ts} F(s)\, ds$$

or, upon rearranging,

$$f(t) = r_0(t) + \frac{1}{2\pi i} \int_{\beta_0 - i\infty}^{\beta_0 + i\infty} e^{ts} F(s)\, ds.$$

Iterating the above explained process, selecting the real points $\beta_1, \beta_2, \ldots$ so that

$$a > \Re a_0 > \beta_0 > \Re a_1 > \beta_1 > \cdots$$

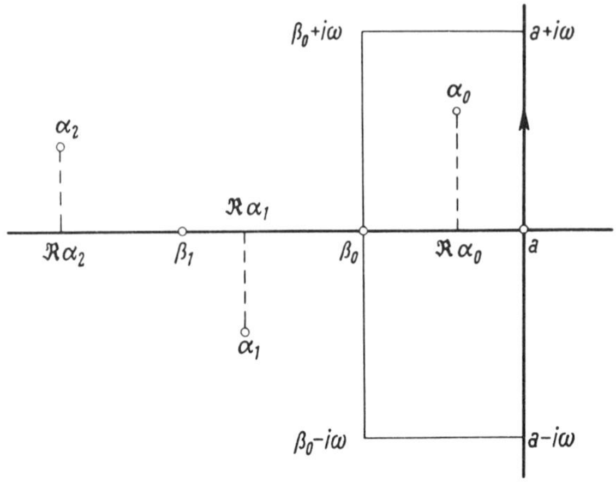

Figure 29

and assuming that $F(s)$ tends, uniformly on every strip: $\beta_\nu \leqq \Re s \leqq a$, towards zero when $s \to \infty$, we ultimately find:

$$f(t) = \sum_{\nu=0}^{n} r_\nu(t) + \frac{1}{2\pi i} \int_{\beta_n - i\infty}^{\beta_n + i\infty} e^{ts} F(s)\, ds, \tag{2}$$

where $r_\nu(t)$ designates the residue of $e^{st}F(s)$ at α_ν. When the principal part of $F(s)$ at α_ν has the form:

$$\frac{c_1^{(\nu)}}{s-\alpha_\nu} + \cdots + \frac{c_{m_\nu}^{(\nu)}}{(s-\alpha_\nu)^{m_\nu}},$$

then one obtains

$$r_\nu(t) = \left(c_1^{(\nu)} + c_2^{(\nu)} \frac{t}{1!} + \cdots + c_{m_\nu}^{(\nu)} \frac{t^{m_\nu-1}}{(m_\nu-1)!}\right) e^{\alpha_\nu t}.$$

All these steps and the results retrace those of pp. 170, 171, where a *convergent* expansion was developed for $f(t)$. We had then to select hypotheses so that the "remainder integral" of (2) would, for fixed t, tend towards 0 when $n \to \infty$. Here, we are satisfied with a less demanding hypothesis: The function $f(t)$ has the asymptotic expansion $\Sigma r_\nu(t)$, provided the remainder integral is, for a fixed n, of order $o(r_n(t))$, that is, of order $o(t^{m_n-1}e^{\alpha_n t})$ as $t \to \infty$. This condition can be fulfilled by means of a simple hypothesis. We write

$$\frac{1}{2\pi i} \int_{\beta_n - i\infty}^{\beta_n + i\infty} e^{ts} F(s)\, ds = \frac{1}{2\pi} e^{\beta_n t} \int_{-\infty}^{+\infty} e^{ity} F(\beta_n + iy)\, dy.$$

The integral

$$\int_{-\omega}^{+\omega} e^{ity} F(\beta_n + iy)\, dy \quad (\omega > 0 \text{ arbitrary, although fixed})$$

tends, according to the Riemann-Lebesgue Lemma (Theorem 23.3), towards zero when $t \to \infty$. Now, we suppose that the integral

$$\int_{-\infty}^{+\infty} e^{ity} F(\beta_n + iy)\, dy \tag{3}$$

converges uniformly for $t > T$. Then, from the beginning, we may choose ω sufficiently large so that the integrals

$$\int_{+\omega}^{+\infty} \quad \text{and} \quad \int_{-\infty}^{-\omega} e^{ity} F(\beta_n + iy)\, dy$$

are arbitrarily small for all $t > T$. Then the relation

$$\int_{-\infty}^{+\infty} e^{ity} F(\beta_n + iy)\, dy \to 0 \quad \text{when } t \to \infty,$$

is valid, that is

$$\frac{1}{2\pi i} \int_{\beta_n - i\infty}^{\beta_n + i\infty} e^{ts} F(s)\, ds = o(e^{\beta_n t}).$$

Since $\beta_n < \Re \alpha_n$, this implies that the remainder integral is certainly of order $o(t^{m_n - 1} e^{\alpha_n t})$ and, consequently, of order $o(r_n(t))$.

The requirement that the integral (3) converges uniformly for $t > T$ is certainly satisfied for an absolutely converging integral:

$$\int_{-\infty}^{+\infty} |F(\beta_n + iy)|\, dy < \infty.$$

However, for applications, the condition of uniform convergence is more advantageous, for even a simple integral like

$$\int_{Y}^{+\infty} e^{ity} \frac{1}{y^\lambda}\, dy \quad \text{with } 0 < \lambda \leqq 1 \quad (Y > 0)$$

does not converge absolutely, although it does converge uniformly for $t \geqq T > 0$. Integration by parts yields:

$$\left| \int_{\omega}^{\infty} e^{ity} \frac{1}{y^\lambda}\, dy \right| = \left| \left(\frac{e^{ity}}{it} \frac{1}{y^\lambda} \right)_{\omega}^{\infty} + \frac{\lambda}{it} \int_{\omega}^{\infty} e^{ity} \frac{1}{y^{\lambda+1}}\, dy \right|$$

$$\leqq \frac{1}{T} \frac{1}{\omega^\lambda} + \frac{\lambda}{T} \int_{\omega}^{\infty} \frac{dy}{y^{\lambda+1}} \quad \text{for } t \geqq T > 0,$$

and the last, majorizing expression is, independently of the value of t, arbitrarily small for sufficiently large values of ω.

The conclusions are summarized in

Theorem 35.1. *Suppose that $F(s)$ is analytic for $\Re s \leq a$, with the possible exception of the poles α_ν, with $a > \Re\alpha_0 > \Re\alpha_1 > \ldots$, where $F(s)$ has the respective principal parts*

$$\frac{c_1^{(\nu)}}{s-\alpha_\nu} + \cdots + \frac{c_{m_\nu}^{(\nu)}}{(s-\alpha_\nu)^{m_\nu}},$$

and that the integral

$$\mathfrak{V}\{F\} = \frac{1}{2\pi i}\int_{a-i\infty}^{a+i\infty} e^{ts} F(s)\,ds = f(t)$$

converges for $t > T$. Suppose, furthermore, that real points β_ν with $\Re\alpha_{\nu+1} < \beta_\nu < < \Re\alpha_\nu$ can be found so that:

1. *on every strip $\beta_\nu \leq \Re s \leq a$, $F(s)$ tends towards zero when $s \to \infty$, uniformly with regard to $\Re s$;*
2. *the integral*

$$\int_{-\infty}^{+\infty} e^{ity} F(\beta_\nu + iy)\,dy$$

converges uniformly for $t > T$.

Then we conclude that

$$f(t) \approx \sum_{\nu=0}^{\infty}\left(c_1^{(\nu)} + c_2^{(\nu)}\frac{t}{1!} + \cdots + c_{m_\nu}^{(\nu)}\frac{t^{m_\nu-1}}{(m_\nu-1)!}\right)e^{\alpha_\nu t} \quad \text{as } t \to +\infty.$$

The Hypotheses 1 and 2 are satisfied when, for instance,

$$|F(x+iy)| < \frac{C(\beta_\nu)}{|y|^{1+\delta}}\ (\delta > 0) \quad \text{for } \beta_\nu \leqq x \leqq a,\ |y| > Y(\beta_\nu).$$

The notation was chosen to indicate that the constants C and Y may depend upon the value of β_ν.

An application of Theorem 35.1 is shown on p. 297.

36. The Region of Convergence of the Complex Inversion Integral with Angular Path. The Holomorphy of the Represented Function

We now admit singularities of $F(s)$ to the left of a which are not all single-valued. In the case that the first encountered singularities to the left of a are single-valued, one can employ the previous method and thus separate from $f(t)$ the corresponding residues, until one finally encounters a many-valued singularity. Thus, we may, without loss of generality, assume that in the first encountered singular point to the left of a, that is *the singular point α_0 with largest real part $< a$, $F(s)$ has a many-valued singularity,* perhaps of the character $(s-\alpha_0)^{1/2}$ or $(s-\alpha_0)^{-1/2}$ or $\log(s-\alpha_0)$ or $(s-\alpha_0)^{1/2}\log(s-\alpha_0)$ etc. Possibly, one may encounter more than one singular point with identical largest real part $< a$; this particular situation will be discussed at the end of this Chapter.

The technique of Chapter 35 cannot here be employed, for the application of Cauchy's residue theorem requires single-valuedness of the function. When searching for another method, and thereby recalling the relative ease with which the asymptotic expansion of the $\mathfrak{L}$-integral could be developed in Chapter 33, one may ask whether the integral (35.1) could be reduced to $\mathfrak{L}$-integrals, perhaps by splitting it into two integrals, one along the ray from a to $(a+i\infty)$, the other from a to $(a-i\infty)$. However, as long as we employ vertical rays as paths of integration we shall not produce $\mathfrak{L}$-integrals, instead we obtain $\mathfrak{F}$-integrals:

$$\frac{1}{2\pi i}\int_a^{a+i\infty} e^{ts}F(s)\,ds = \frac{1}{2\pi}e^{at}\int_0^\infty e^{ity}\,F(a+iy)\,dy\,.$$

The only asymptotic property concerning $\mathfrak{F}$-integrals which is at our disposal is expressed in the Riemann-Lebesgue Lemma, when this lemma is extended to admit unbounded intervals of integration (compare p. 237); this lemma indicates that, in case of uniform convergence for large t, the $\mathfrak{F}$-integral is of order $o(1)$; it follows that integral (35.1) is of order $o(e^{at})$.

The situation is essentially altered when the *vertical ray can be replaced by one which is inclined towards the left;* such a move is permissible by Cauchy's theorem, provided certain conditions are satisfied. Along the inclined ray:

$$s = a + r\,e^{i\psi} \quad \text{with} \quad \frac{\pi}{2} < \psi < \pi\,,$$

we have

$$\int e^{ts}F(s)\,ds = e^{at}\int_0^\infty e^{tre^{i\psi}}F(a+r\,e^{i\psi})\,e^{i\psi}\,dr$$

$$= e^{at}\,e^{i\psi}\int_0^\infty e^{-te^{i(\psi-\pi)}r}F(a+r\,e^{i\psi})\,dr\,.$$

The last integral is an *ordinary $\mathfrak{L}$-integral* which employs r as its dummy variable of integration, and $te^{i(\psi-\pi)}$ assumes the position of the usual parameter s. Thus, one could invoke the Theorems of Chapter 33 and, upon applying these to the integral along the inclined ray, one could produce an asymptotic expansion for this integral. Analogous considerations apply to the integral along the other ray and, consequently, to the entire integral (35.1).

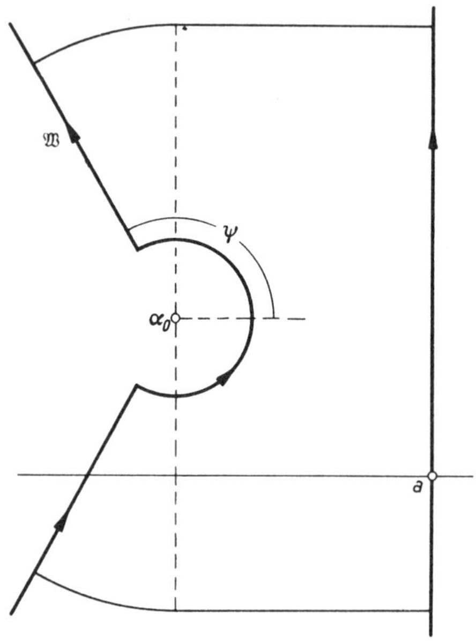

Figure 30

In the remarks following Theorem 37.1, we shall realize that a useful expansion can be obtained only after the path of integration of (35.1) has been moved from the quite arbitrary, holomorphic point a to the singular point α_0 where then the two vertical rays are inclined to the left. The function $F(s)$ may fail to be integrable at α_0, therefore we replace the path of integration near α_0 by a portion of a circle about α_0 (see Fig. 30). The new path of integration[1] is designated by $\mathfrak{W}$. We realize that the move of the straight line path of integration through a into the new, angular path is certainly permissible, provided $F(s)$ converges towards 0 when s tends two-dimensionally between the old and the new path towards ∞. This conclusion can easily be demonstrated by the following process: We insert between the old and the new path, above and beneath α_0, connecting curves, each composed of a portion of a circle centred at α_0 and a straight line section, as shown in Fig. 30. We apply Cauchy's theorem to the closed curve, and we consider the situation when the connecting curves are moved upwards and downwards respectively. The contribution to the integral along these connecting curves vanishes in the limit; for the above stipulated condition, this is obviously true along the bounded straight line sections; for the circular portions, it follows from Theorem 25.1.

Let us then assume that we managed indeed to move the path of integration

[1] We encountered the same contour during the discussion of the Hankel formula on pp. 165-168.

of (35.1) from the straight line through a to the contour $\mathfrak{W}$ with an half-angle of opening ψ, with $\pi/2 < \psi \leqq \pi$.[2] Henceforth we shall consider the integral:

$$\frac{1}{2\pi i} \int_{\mathfrak{W}} e^{ts} F(s)\, ds = f(t). \tag{1}$$

The integral (1) defines a correspondence between f and F; we shall interpret it as another transformation in its own right which we shall call the **$\mathfrak{W}$-transformation:**

$$\mathfrak{W}\{F(s)\} = f(t).$$

Our actual aim is to deduce the asymptotic behaviour of $f(t)$ from the asymptotic behaviour of $F(s)$. However, before we can pursue this aim, we must first investigate some fundamental properties of the new transformation.

Although we shall encounter in the subsequent applications situations as outlined above, that is, $F(s)$ is a $\mathfrak{L}$-transform and thus necessarily an analytic function, we require here merely that $F(s)$ is integrable on every finite portion of the contour $\mathfrak{W}$ (that is, locally integrable), and that the integral $\mathfrak{W}\{F\}$ converges for at least one real or complex value t.

It suffices to consider in the subsequent investigation the special case $\alpha_0 = 0$, for the substitution $s = \alpha_0 + \sigma$ in (1) yields

$$\frac{1}{2\pi i} e^{\alpha_0 t} \int e^{t\sigma} F(\alpha_0 + \sigma)\, d\sigma = f(t)\,; \tag{2}$$

the integral (2) is to be evaluated along a path of integration which is congruent to $\mathfrak{W}$, having its centre not at α_0 but at $\sigma = 0$. Knowing the properties of the integral (2), one can deduce the properties of the integral (1).

Thus, we are concerned with the transformation

$$\mathfrak{W}\{F\} \equiv \frac{1}{2\pi i} \int_{\mathfrak{W}} e^{ts} F(s)\, ds = f(t), \tag{3}$$

whereby $\mathfrak{W}$ is an angular contour which is composed of the circular arc $\mathfrak{K}$ having centre 0 and radius ϱ, and two rays through 0 at the respective angles $\pm\psi$, $\pi/2 < \psi \leqq \pi$, as shown in Fig. 31 (p. 242).

We shall observe that $\mathfrak{W}\{F\}$, in contrast to the inversion integral $\mathfrak{V}$ with a straight path of integration, converges not only for real-valued t but in a two-dimensional region of the complex t-plane; we want to discover this region.

The contribution to the contour integral, along the circular arc $\mathfrak{K}$ of finite length has a finite value for all complex t. The contribution along the ray in the $+\psi$-direction, where $s = re^{i\psi}$ with $\varrho \leqq r < \infty$, is given by:

$$\begin{aligned} \int_{\varrho e^{i\psi}}^{\infty(\psi)} e^{ts} F(s)\, ds &= \int_{\varrho}^{\infty} e^{tre^{i\psi}} F(re^{i\psi})\, e^{i\psi}\, dr \\ &= e^{i\psi} \int_{\varrho}^{\infty} e^{-te^{-i(\pi-\psi)}r} F(re^{i\psi})\, dr. \end{aligned} \tag{4}$$

[2] For $\psi = \pi$, we must locate the two horizontal rays on the opposite flanks of a branch cut.

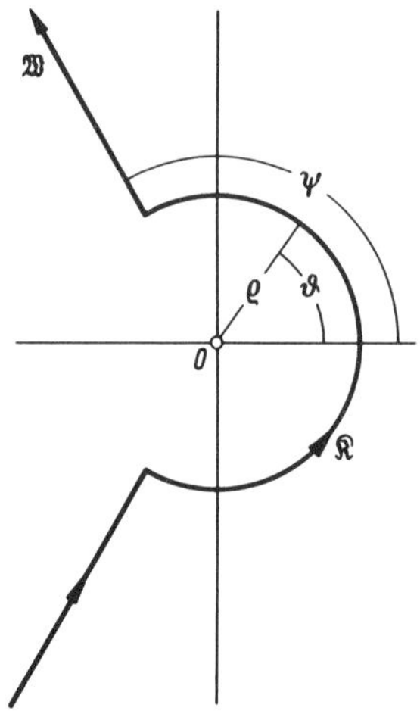

Figure 31

This is a $\mathfrak{L}$-integral with the dummy variable of integration r, the original function $F(re^{i\psi})$ and the image variable $te^{-i(\pi-\psi)}$. It converges, if anywhere, in some half-plane $\Re(te^{-i(\pi-\psi)}) > \beta_1$. The inner normal of this half-plane forms the angle $(\pi - \psi)$ with the positive real axis, as shown in Fig. 32. The contribution along the ray in the $-\psi$-direction, where $s = re^{-i\psi}$ with $\varrho \leqq r < \infty$, is given by

$$\int_{\varrho e^{-i\psi}}^{\infty(-\psi)} e^{ts} F(s)\, ds = \int_{\varrho}^{\infty} e^{tre^{-i\psi}} F(re^{-i\psi})\, e^{-i\psi}\, dr \tag{5}$$
$$= e^{-i\psi} \int_{\varrho}^{\infty} e^{-te^{i(\pi-\psi)}r} F(re^{-i\psi})\, dr .$$

This integral converges, if anywhere, in some half-plane $\Re(te^{i(\pi-\psi)}) > \beta_2$. The inner normal of this half-plane forms the angle $-(\pi - \psi) = (\psi - \pi)$ with the positive real axis. The entire integral (3) which is composed of the contribution along the circular arc $\mathfrak{K}$ plus (4) less (5) converges in the intersection of both half-planes of convergence. When t_0 designates the common point of the respective lines of convergence, then the integral (3) converges in the angular region with apex t_0, horizontal bisector, and half-angle $\psi - (\pi/2)$, as shown in Fig. 32.

The larger ψ, the larger the angular region of convergence. For $\psi = \pi$, its half-angle is $\pi/2$; the region of convergence is a right half-plane. (For the integral $\mathfrak{V}$, we find $\psi = \pi/2$; the corresponding region of convergence degenerates into a ray.)

When $F(s)$ designates the $\mathfrak{L}$-transform of a function $f(t)$ which is real-valued for all real t, then $F(\bar{s}) = \overline{F(s)}$. Conversely, when $F(s)$ has this property, then one can easily demonstrate for the $\mathfrak{W}$-transformation that the contribution (4) yields for $\bar{t}$ a value which is the conjugate of the contribution of (5) for t; similarly the contribution (5) for $\bar{t}$ is the conjugate of the contribution (4) for t. This implies that the half-planes of convergence of the respective contributions (4) and (5) are conjugates, that is $\beta_1 = \bar{\beta}_2$, and the difference between the two integrals for $\bar{t}$ is the conjugate of the difference for t. Analogous considerations hold for the integrals along the upper and lower halves of the circular arc $\mathfrak{K}$. We summarize thus: For a function $F(s)$ with the property $F(\bar{s}) = \overline{F(s)}$, the transform $\mathfrak{W}\{F\}$ assumes conjugate values for conjugate values of t; in particular, for real t, $\mathfrak{W}\{F\} = f(t)$ is real-valued. The region of convergence is symmetric with respect to the real axis.

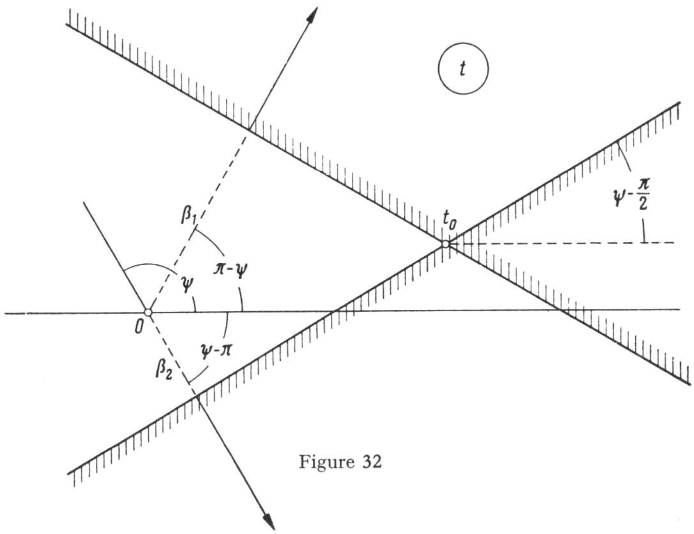

Figure 32

The $\mathfrak{L}$-integrals (4) and (5), evaluated along the rays in the respective directions $\pm\psi$, represent, by Theorem 6.1, analytic functions in the respective regions of convergence, and may be differentiated under the integral sign. This conclusion is also true for the integral along the circular arc $\mathfrak{K}$, for we may interchange summation and integration in:

$$\int_{\mathfrak{K}} e^{ts} F(s)\, ds = \int_{\mathfrak{K}} F(s) \sum_{n=0}^{\infty} \frac{t^n}{n!} s^n\, ds,$$

since the power series converges, for a fixed t, uniformly in s on $\mathfrak{K}$, and since $F(s)$ is Riemann integrable on $\mathfrak{K}$ and consequently bounded[3]:

$$\int_{\mathfrak{K}} e^{ts} F(s)\, ds = \sum_{n=0}^{\infty} \frac{t^n}{n!} \int_{\mathfrak{K}} F(s)\, s^n\, ds.$$

The power series in t may be differentiated term by term:

$$\frac{d}{dt} \int_{\mathfrak{K}} e^{ts} F(s)\, ds = \sum_{n=1}^{\infty} \frac{t^{n-1}}{(n-1)!} \int_{\mathfrak{K}} F(s)\, s^n\, ds = \sum_{n=0}^{\infty} \frac{t^n}{n!} \int_{\mathfrak{K}} F(s)\, s^{n+1}\, ds$$

$$= \sum_{n=0}^{\infty} \int_{\mathfrak{K}} F(s)\, s \frac{(ts)^n}{n!}\, ds.$$

[3] For a Lebesgue-integrable $F(s)$ one could also argue the legality of the interchange by Lemma 2 of Chapter 30.

Once again, using the same arguments as above, we can interchange summation and integration, hence:

$$\frac{d}{dt}\int_{\mathfrak{K}} e^{ts}F(s)\,ds = \int_{\mathfrak{K}} F(s)\,s\sum_{n=0}^{\infty}\frac{(ts)^n}{n!}\,ds = \int_{\mathfrak{K}} e^{ts}s\,F(s)\,ds.$$

It follows that the integral over $\mathfrak{K}$ may be differentiated under the integral sign for all complex t and represents, therefore, an entire function.

The above process may be iterated. This shows that, in fact, $f(t)$ represented by $\mathfrak{W}\{F\}$ in the angular region of convergence is analytic in this region, and that the derivatives have the form:

$$f^{(n)}(t) = \frac{1}{2\pi i}\int_{\mathfrak{W}} e^{ts}s^n F(s)\,ds. \tag{6}$$

The conclusions thus obtained can now be generalized by means of formula (2) to the case that the centre of the contour $\mathfrak{W}$ is no longer restrained to the origin; instead it can be any point α_0. Hence, we can formulate the following theorem.

Theorem 36.1. *Suppose that $\mathfrak{W}$ represents the angular contour with centre α_0 and the half-angle of opening ψ, with $\pi/2 < \psi \leqq \pi$, that $F(s)$ is locally integrable on $\mathfrak{W}$, and that the integral $\int_{\mathfrak{W}} e^{ts}F(s)\,ds$ converges for at least one real or complex value of t. Then it follows that this integral converges in an angular region of the complex t-plane with horizontal bisector and half-angle of opening $\psi - \pi/2$; i.e., in $|\operatorname{arc}(t - t_0)| < \psi - \pi/2$. The by means of the $\mathfrak{W}$-transformation:*

$$\mathfrak{W}\{F\} \equiv \frac{1}{2\pi i}\int_{\mathfrak{W}} e^{ts}\,F(s)\,ds = f(t) \tag{7}$$

generated function $f(t)$ represents in the angular region of convergence an analytic function the derivatives of which may be obtained by differentiation under the integral sign. In particular, when α_0 is real and when $F(\bar{s}) = \overline{F(s)}$, then the bisector of the angular region of convergence coincides with the real axis, and $f(t)$ is real-valued for real t.

The $\mathfrak{W}$-transform, the complex inversion integral with angular contour, has properties entirely different from those of the $\mathfrak{V}$-transform, the inversion integral with a straight path of integration. The latter is actually a $\mathfrak{F}$-integral, and it may, for real t, represent "arbitrary" functions, for instance functions with jumps.(Indeed, according to Theorem 24.3, it represents all functions which are locally of bounded variation and have an absolutely converging $\mathfrak{L}_{II}$-integral.) However, in general, we cannot differentiate the $\mathfrak{V}$-transform under the integral sign.

This is readily demonstrated by the following example:

$$\frac{1}{2\pi i}\int_{a-i\infty}^{a+i\infty} e^{ts}\frac{s}{s^2+1}\,ds = \cos t \qquad (t > 0,\ a > 0). \tag{8}$$

Interchange of differentiation and integration does not produce the known representation of $-\sin t$:

$$\frac{1}{2\pi i}\int_{a-i\infty}^{a+i\infty} e^{ts}\frac{-1}{s^2+1}\,ds = -\sin t;$$

instead, one obtains the diverging integral[4]

$$(9) \qquad \frac{1}{2\pi i} \int_{a-i\infty}^{a+i\infty} e^{ts} \frac{s^2}{s^2+1} \, ds.$$

By contrast, the $\mathfrak{W}$-integral can only represent those functions which are analytic in certain angular regions with horizontal bisectors. However, it offers the advantageous property that it can be differentiated under the integral sign, and it enables us to represent analytic functions which cannot be generated by the $\mathfrak{B}$-transformation. For instance, we have with $\alpha_0 = 0$, by (25.6),

$$(10) \qquad \mathfrak{W}\{s^{-\lambda}\} \equiv \frac{1}{2\pi i} \int_{\mathfrak{W}} e^{ts} s^{-\lambda} \, ds = \frac{t^{\lambda-1}}{\Gamma(\lambda)} \quad \text{for arbitrary complex } \lambda;$$

this formula, when considered as the inversion of $\mathfrak{L}\{t^{\lambda-1}/\Gamma(\lambda)\} = s^{-\lambda}$ with a straight line path of integration is valid only for $\Re\lambda > 0$. In Chapter 25 we derived the formula (10) for arbitrary ψ with $\pi/2 < \psi \leqq \pi$, only for $t > 0$. Theorem 36.1 shows that formula (10) is valid in the angular region $|\operatorname{arc} t| < \psi - \pi/2$. Selecting the largest permissible value $\psi = \pi$, we confirm it for $|\operatorname{arc} t| < \pi/2$, that is in the right half-plane.

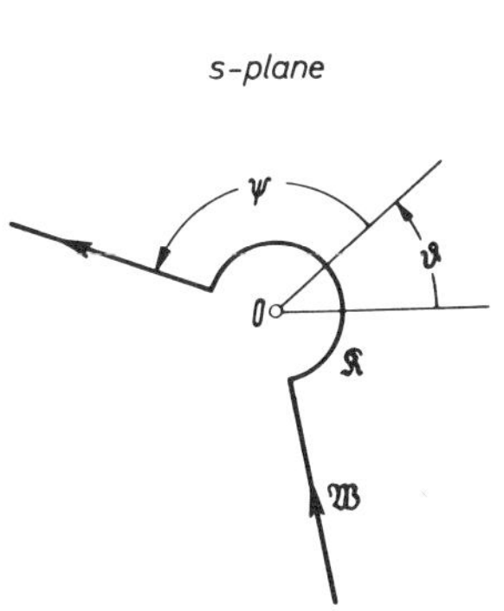

Figure 33a

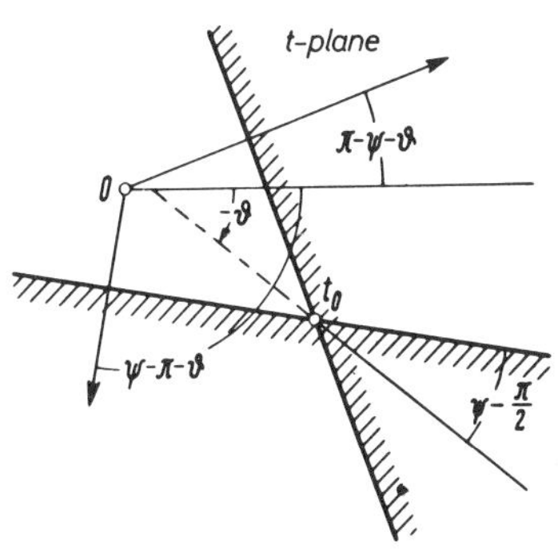

Figure 33b

In Chapter 37 we shall use integrals of the form (7) with a contour $\mathfrak{W}$ rotated about α_0 so that its bisector is not horizontal but forms some arbitrary angle ϑ with the real axis, as shown in Fig. 33a. Once again it suffices to use $\alpha_0 = 0$.

[4] We find the explanation for this in the observation that the integral (8) produces, for $t < 0$, the value zero (compare Theorem 24.4); the represented function, considered on the entire real axis, ought to be written $\cos t \cdot u(t)$. At $t = 0$ the latter has a jump of magnitude one. Consequently, it cannot have a conventional derivative. When looked upon as a distribution it has the distribution-derivative $-\sin t \cdot u(t) + \delta(t)$. The $\mathfrak{L}$-transform of the latter is

$$-\frac{1}{s^2+1} + 1 = \frac{s^2}{s^2+1}.$$

This is the function encountered in (9).

Naturally we inherit the conclusion that the integral along the circular arc $\mathfrak{K}$ converges for all t, thus producing an entire function. The ray of $\mathfrak{W}$, which forms the angle ψ with the bisector of $\mathfrak{W}$, now forms the angle $\psi + \vartheta$ with the positive real axis; hence, we find, instead of (4),

$$(11) \qquad e^{i(\psi+\vartheta)} \int_{\varrho}^{\infty} e^{-te^{-i(\pi-\psi-\vartheta)}r} F(re^{i(\psi+\vartheta)})\, dr.$$

Integral (11) converges in some half-plane $\Re(te^{-i(\pi-\psi-\vartheta)}) > \beta_1$; this is a half-plane, the inner normal of which forms the angle $\pi - \psi - \vartheta$ with the positive real axis. The other ray of $\mathfrak{W}$ now forms the angle $-\psi + \vartheta = -(\psi - \vartheta)$ with the positive real axis; hence, we obtain instead of (5),

$$(12) \qquad e^{-(\psi-\vartheta)} \int_{\varrho}^{\infty} e^{-te^{i(\pi-\psi+\vartheta)}r} F(re^{-i(\psi-\vartheta)})\, dr.$$

Integral (12) converges in some half-plane $\Re(te^{i(\pi-\psi+\vartheta)}) > \beta_2$; the inner normal of this half-plane forms the angle $-(\pi - \psi + \vartheta) = \psi - \pi - \vartheta$ with the positive real axis. The integral along the entire, rotated angular contour converges in the intersection of these half-planes; this is, as seen in Fig. 33b, an angular region with the half-angle of opening $\psi - \pi/2$ and a bisector which forms the angle $-\vartheta$ with the positive real axis. Observe that Fig. 33a is obtained from Fig. 31 by a rotation through the angle ϑ in the positive sense, whilst the corresponding Fig. 33b resembles Fig. 32 when rotated through the angle ϑ in the negative sense. This counter-behaviour clearly follows from the fact that in the exponential function of the integrand of (7) we encounter the product ts; hence, the behaviour of convergence of the integral remains unaltered when every increase of arc s (corresponding to a rotation of the path of integration) is counteracted by a decrease of equal magnitude of arc t (corresponding to a rotation of the region of convergence in the opposite sense).

Theorem 36.2. *Let $\mathfrak{W}$ designate an angular contour having its centre at α_0, and the half-angle of opening ψ, with $\pi/2 < \psi \leq \pi$. The bisector of $\mathfrak{W}$ forms the angle ϑ with the positive real axis. Suppose that $F(s)$ is locally integrable on $\mathfrak{W}$. Then we conclude that $\int_{\mathfrak{W}} e^{ts} F(s)\, ds$ converges in an angular region of the complex t-plane which has the half-angle of opening $\psi - \pi/2$, and a bisector which forms the angle $-\vartheta$ with the positive real axis. The $\mathfrak{W}$-transformation of the form (7) generates the function $f(t)$ which is analytic in the region of convergence; the derivatives of $f(t)$ are obtained by differentiation of (7) under the integral sign.*

When we use, with formula (10), an angular contour of the type described in Theorem 36.2, then the apex of the angular region of convergence in the t-plane is at $t = 0$; this is a consequence of the fact that the respective half-planes of convergence of the integrals along the rays of $\mathfrak{W}$ are obviously bounded by straight lines through the origin. Upon substituting $s = s' - \alpha_0$ in (10), the path of integration in the s'-plane is an angular contour having its centre at α_0; with this $\mathfrak{W}$ we obtain:

$$\frac{1}{2\pi i} \int_{\mathfrak{W}} e^{t(s'-\alpha_0)} (s' - \alpha_0)^{-\lambda}\, ds' = \frac{t^{\lambda-1}}{\Gamma(\lambda)}.$$

Thus, we produce the following generalization of (10):

The transformation

$$\text{(13)} \qquad \mathfrak{W}\{(s-\alpha_0)^{-\lambda}\} = \frac{t^{\lambda-1}}{\Gamma(\lambda)}\, e^{\alpha_0 t} \qquad (\lambda \text{ arbitrary, complex}^5)$$

is valid provided the following conditions are satisfied:

In the s-plane (see Fig. 34a): The angular contour $\mathfrak{W}$ has its centre at $s = \alpha_0$. Its half-angle of opening is ψ, with $\pi/2 < \psi \leqq \pi$. The bisector of the contour forms the angle ϑ with the positive real axis.

In the t-plane (see Fig. 34b): t is an interior point of the angular region which has its apex at $t = 0$. The bisector of this region forms the angle $-\vartheta$ with the positive real axis. Its half-angle of opening is $\psi - \pi/2$.

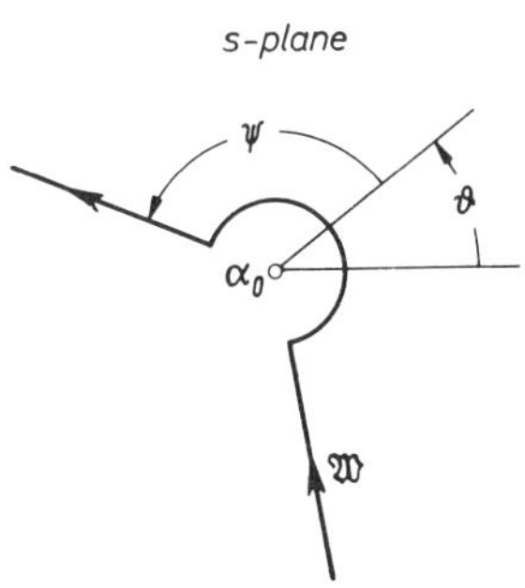

Figure 34 a

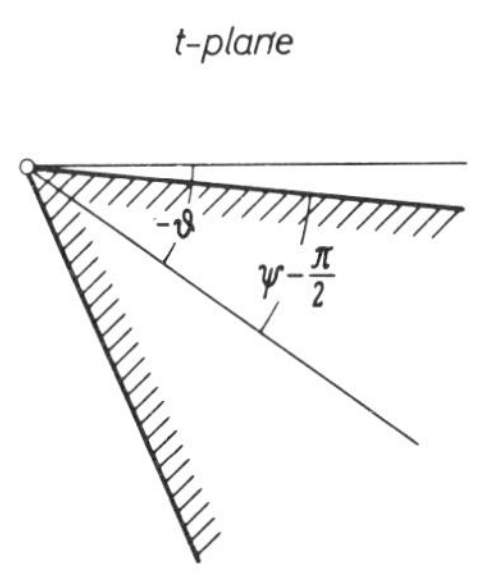

Figure 34 b

The Case that $F(s)$ is the $\mathfrak{L}$-transform of $f(t)$

For the previous Theorems we merely required that $F(s)$ is defined and locally integrable on the angular contour $\mathfrak{W}$. In applications, $F(s)$ often is the $\mathfrak{L}$-transform of some $f(t)$; that is, it is an analytic function in a right half-plane, and the $\mathfrak{W}$-transform results from the complex inversion formula with a straight line path of integration in the manner explained at the beginning of this Chapter. In order that the alteration of the straight line path of integration into an angular contour may be permissible, $f(t)$ must be analytic by Theorem 36.1, in an angular region with horizontal bisector. The for the $\mathfrak{L}$-transformation needed real axis $t \geqq 0$ certainly belongs to the angular region beyond a certain point T; however, it need not be entirely in this region, as shown in Fig. 35. For this latter situation, the function $f(t)$ is representable by a $\mathfrak{W}$-transform of its $\mathfrak{L}$-transform $F(s)$ only for $t > T$; the values of $f(t)$ for $0 \leqq t \leqq T$ are not included. The following simple example will help to elucidate the situation: The step function $f(t) \equiv u(t-1)$ has the $\mathfrak{L}$-transform:

$$\mathfrak{L}\{u(t-1)\} = \frac{e^{-s}}{s} \qquad \text{for} \qquad \Re s > 0.$$

---

[5] For $\lambda = 0, -1, -2, \cdots$, the right hand side is interpreted as zero (cf the remark to formula (25.6)).

According to the inversion formula we find[6]

$$\frac{1}{2\pi i}\ V.P. \int_{a-i\infty}^{a+i\infty} e^{ts}\frac{e^{-s}}{s}\,ds = \begin{cases} 1 & \text{for} \quad t > 1 \\ \frac{1}{2} & \text{for} \quad t = 1 \\ 0 & \text{for} \quad t < 1 \end{cases} \qquad (a > 0). \tag{14}$$

The function $1/s$ converges uniformly in every direction towards zero, when $s \to \infty$; consequently, we might firstly remove in the integral

$$\int e^{(t-1)s}\frac{1}{s}\,ds$$

the path of integration to the imaginary axis which we have to replace in the neighbourhood of $s = 0$ by a semicircle. By Theorem 25.1, we replace the straight line path of integration by an angular contour $\mathfrak{W}$ having the centre 0 and a half-angle of opening ψ, with arbitrary ψ in $\pi/2 < \psi \leqq \pi$; this can be done only for $t - 1 > 0$, since t of Theorem 25.1 must here be replaced by $t-1$.[7] Thus, we find:

$$\mathfrak{W}\left\{\frac{e^{-s}}{s}\right\} = 1 \quad \text{for} \quad t > 1\,. \tag{15}$$

This latter expression is actually correct for all t in the angular region $|\text{arc}\,(t-1)| < \psi - \pi/2$. For $0 \leqq t < 1$, we cannot alter the straight line path of integration into an angular contour. Obviously, the $\mathfrak{W}$-integral diverges for these values of t.[8]

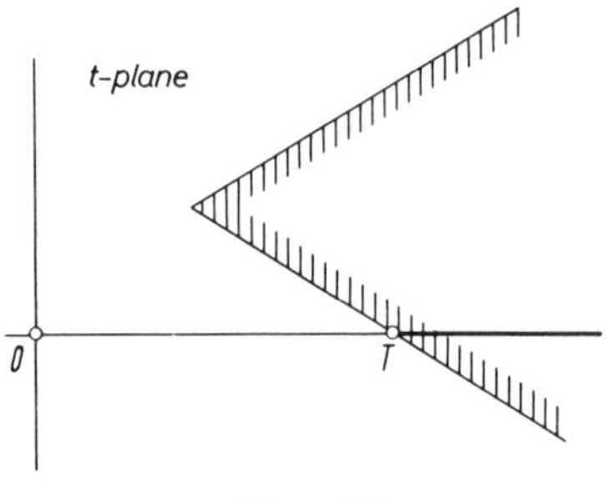

Figure 35

This example demonstrates what we must expect in view of the above development: The function $u(t-1)$ can, for $t > 1$, be embedded into the function 1 which is analytic in the entire plane; for $t > 1$ it may be represented as the $\mathfrak{W}$-transform of its $\mathfrak{L}$-transform. Along the entire positive axis $t \geqq 0$, the function $u(t-1)$

[6] The designation $V.P.$ is needed for $t = 1$ only; for $t \gtrless 1$, the integral converges in the conventional sense.

[7] In the left half-plane, e^{-s}/s does not converge to zero when $s \to \infty$; it is for this reason that e^{-s} must be joined with e^{ts} to form $e^{(t-1)s}$, otherwise Theorem 25.1 could not be invoked.

[8] When interpreting the $\mathfrak{W}$-integral in a manner analogous to the one employed with (14), as $V.P.$, then it does converge also for $t = 1$, yielding the value 1/2.

cannot be identical to an analytic function; hence, we cannot represent it as a $\mathfrak{W}$-transform. A similar, more complicated example will be explained on p. 299.

Any function $f(t)$, defined for $t \geqq 0$, which does not coincide beyond some fixed point T, in an angular region with an analytic function, cannot be represented by a $\mathfrak{W}$-transform. This is illustrated by the square wave function which is defined by: $f(t) = 0$ for $2n \leqq t < 2n+1$, and $f(t) = 1$ for $2n+1 \leqq t < 2n+2$ $(n = 0, 1, 2, \ldots)$. Its $\mathfrak{L}$-transform is given by $F(s) = 1/[s(1+e^s)]$; the latter has poles at $s = 0$, $s = (2k+1)\pi i (k = 0, \pm 1, \pm 2, \ldots)$. Indeed, these poles prevent a deformation of the vertical straight line path of integration into an angular contour $\mathfrak{W}$.

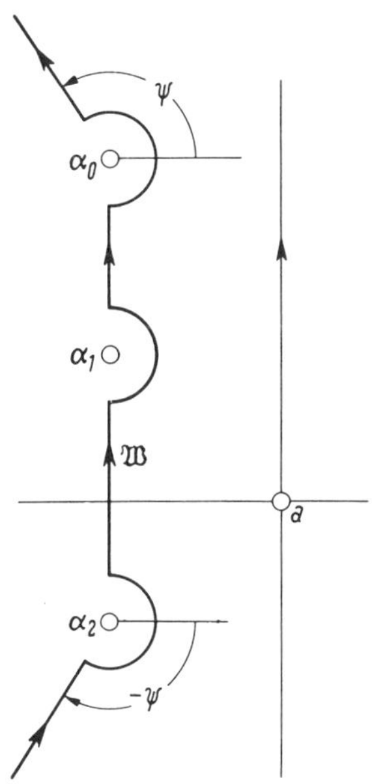

Figure 36

At the onset of this Chapter, we presumed that when moving the vertical path of integration to the left of a, exactly one many-valued singularity with maximum real part would be encountered. We now admit the situation that $F(s)$ has a finite number of singular points with identical, largest real part, say α_0, α_1, and α_2 as shown in Fig. 36. We now presume that the behaviour of $F(s)$ near infinity is such that the straight line path of integration through a may be replaced by an angular contour $\mathfrak{W}$ as shown in Fig. 36, at least for sufficiently large values $t > T$. The contributions to the $\mathfrak{W}$-transform along the several portions of circles and along the intermediate straight lines converge for all t; for the contributions along the rays at $\pm\psi$ respectively, the conclusions of Theorem 36.1 remain valid. Moreover, we may replace the straight line connections between the portions of circles by arbitrary, conveniently selected curves of finite length on which $F(s)$ is defined and integrable; we shall resort to such modifications in the sequel when applying the Theorem in Chapter 37.

37. The Asymptotic Behaviour of an Original Function near Infinity, when its Image Function is Many-Valued at the Singular Point with Largest Real Part

Having expounded the properties of the $\mathfrak{W}$-transformation, we now return to the task formulated at the beginning of Chapter 36. Some function $f(t)$ is given; its $\mathfrak{L}$-transform $F(s)$ has a many-valued singularity at its singular point α_0 with largest real part. Let us assume that $f(t)$ can be reproduced by means of the $\mathfrak{W}$-transformation of $F(s)$ employing an angular contour centred at α_0. Our objective is to deduce the asymptotic properties of $f(t)$ as $t \to \infty$ from the behaviour of $F(s)$ near α_0.

Once again, we presume here that $\alpha_0 = 0$, later on we can extend our conclusions to any arbitrarily located α_0, by means of formula (36.2).

Initially, we presume that the singularity of $F(s)$ at $s = 0$ is of such a nature that $F(s)$ may be represented by an asymptotic expansion of the form

(1) $$F(s) \approx \sum_{\nu=0}^{\infty} c_\nu \, s^{\lambda_\nu} \quad (-1 < \Re\lambda_0 < \Re\lambda_1 < \cdots)$$

in some sector $|\text{arc}\, s| \leq \psi$, with $\pi/2 < \psi \leqq \pi$. (In practical applications, we shall most often find $F(s)$ representable by an absolutely converging series of the indicated form in a surrounding neighbourhood of $s = 0$.) In this case that all $\Re\lambda_\nu > -1$, we may dispense with the circular arc of $\mathfrak{W}$ and we employ a simplified

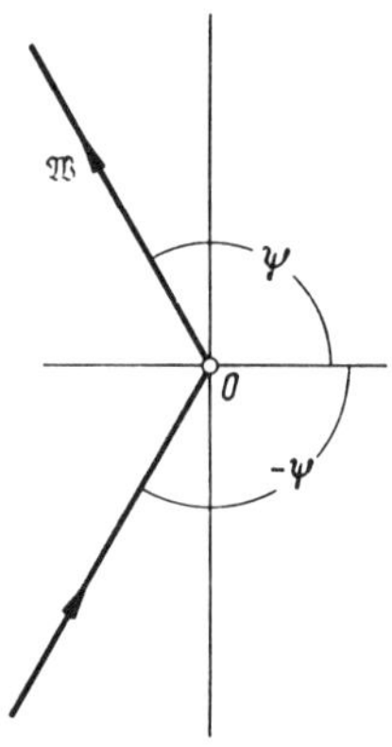

Figure 37

contour through $s = 0$ as shown in Fig. 37. This simplification is permissible according to the following argumentation. The integral $\int e^{ts} s^{\lambda_0} \, ds$, evaluated along the circular arc: $s = \varrho e^{i\vartheta}$, $-\psi \leqq \vartheta \leqq +\psi$, can be written, with $\lambda_0 = \mu + i\nu$, as follows:

$$\int_{-\psi}^{+\psi} e^{t\varrho e^{i\vartheta}} \varrho^{\mu+i\nu} e^{i(\mu+i\nu)\vartheta} \varrho \, e^{i\vartheta} \, i \, d\vartheta;$$

the absolute value of this integral is bounded above by

$$\int_{-\psi}^{+\psi} e^{t\varrho\cos\vartheta} \varrho^{\mu+1} e^{-\nu\vartheta} \, d\vartheta \leqq e^{t\varrho} \varrho^{\mu+1} e^{|\nu|\psi} \, 2\psi .$$

It follows that the integral tends towards zero when $\varrho \to 0$, since $\mu + 1 > 0$. We observe, furthermore, that $F(s) = c_0 s^{\lambda_0} + o(s^{\lambda_0})$; thus, the integral $\int e^{ts} F(s)\,ds$, evaluated along the circular arc, also tends towards zero, when $\varrho \to 0$.

Thus, we may write $f(t)$, using an easily understandable notation,

$$f(t) = \frac{1}{2\pi i}\left\{\int_0^{\infty(\psi)} e^{ts} F(s)\,ds - \int_0^{\infty(-\psi)} e^{ts} F(s)\,ds\right\}.$$

For the first integral, we obtain, by (36.4), now using $\varrho = 0$,

$$\int_0^{\infty(\psi)} e^{ts} F(s)\,ds = e^{i\psi} \int_0^{\infty} e^{-te^{i(\psi-\pi)}r} F(r\,e^{i\psi})\,dr\,. \tag{2}$$

We can apply Theorem 33.7 to the thus generated $\mathfrak{L}$-integral. According to (1), we have:

$$F(r\,e^{i\psi}) \approx \sum_{\nu=0}^{\infty} c_\nu\, e^{i\lambda_\nu\psi}\, r^{\lambda_\nu} \quad (-1 < \Re\lambda_0 < \Re\lambda_1 < \cdots) \quad \text{for } r \to 0\,,$$

hence

$$\int_0^{\infty(\psi)} e^{ts} F(s)\,ds \approx e^{i\psi} \sum_{\nu=0}^{\infty} c_\nu\, e^{i\lambda_\nu\psi} \frac{\Gamma(\lambda_\nu + 1)}{(t\,e^{i(\psi-\pi)})^{\lambda_\nu+1}} = \sum_{\nu=0}^{\infty} c_\nu\, e^{i\pi(\lambda_\nu+1)} \frac{\Gamma(\lambda_\nu+1)}{t^{\lambda_\nu+1}}\,,$$

as $te^{i(\psi-\pi)}$ tends two-dimensionally in the angular region

$$|\operatorname{arc}(t\,e^{i(\psi-\pi)})| \leqq \frac{\pi}{2} - \delta$$

(δ arbitrarily small) towards ∞. This implies that

$$-\frac{\pi}{2} + \delta \leqq \operatorname{arc} t + \psi - \pi \leqq \frac{\pi}{2} - \delta$$

or, equivalently,

$$\frac{\pi}{2} - \psi + \delta \leqq \operatorname{arc} t \leqq \frac{3}{2}\pi - \psi - \delta\,.$$

In an analogous manner one obtains for the second integral, by (36.5),

$$\int_0^{\infty(-\psi)} e^{ts} F(s)\,ds = e^{-i\psi} \int_0^{\infty} e^{-te^{i(\pi-\psi)}r}\, F(r e^{-i\psi})\,dr; \tag{3}$$

hence, since

$$F(r\,e^{-i\psi}) \approx \sum_{\nu=0}^{\infty} c_\nu\, e^{-i\lambda_\nu\psi}\, r^{\lambda_\nu} \quad \text{as } r \to 0,$$

by Theorem 33.7,

$$\int_0^{\infty(-\psi)} e^{ts} F(s)\,ds \approx e^{-i\psi} \sum_{\nu=0}^{\infty} c_\nu\, e^{-i\lambda_\nu\psi} \frac{\Gamma(\lambda_\nu+1)}{(t\,e^{i(\pi-\psi)})^{\lambda_\nu+1}} = \sum_{\nu=0}^{\infty} c_\nu\, e^{-i\pi(\lambda_\nu+1)} \frac{\Gamma(\lambda_\nu+1)}{t^{\lambda_\nu+1}}$$

as $te^{i(\pi-\psi)}$ tends in $|\operatorname{arc}(te^{i(\pi-\psi)})| \leq \pi/2 - \delta$ or, equivalently, as t tends in

$$-\frac{3}{2}\pi + \psi + \delta \leqq \operatorname{arc} t \leqq -\frac{\pi}{2} + \psi - \delta$$

two-dimensionally towards ∞.

The angular regions of the asymptotic representations of (2) and of (3) are portions of the respective half-planes:

$$\frac{\pi}{2} - \psi < \operatorname{arc} t < \frac{3}{2}\pi - \psi \quad \text{and} \quad -\frac{3}{2}\pi + \psi < \operatorname{arc} t < -\frac{\pi}{2} + \psi$$

as shown in Fig. 38. The boundary lines of these half-planes are parallel to the lines of convergence of the integrals (2) and (3) which were determined in Chapter 36, and shown in Fig. 32. This is elucidated by the angles of direction entered in Fig. 38.

Terms of the asymptotic expansions of the respective integrals (2) and (3), having the same index ν, are of the same order of magnitude; thus, one may subtract the two expansions term by term,[1] and in this manner one finds:

$$f(t) \approx \frac{1}{2\pi i} \sum_{\nu=0}^{\infty} c_\nu \left[e^{i\pi(\lambda_\nu+1)} - e^{-i\pi(\lambda_\nu+1)}\right] \frac{\Gamma(\lambda_\nu + 1)}{t^{\lambda_\nu+1}}$$

$$= \frac{1}{\pi} \sum_{\nu=0}^{\infty} c_\nu \sin \pi(\lambda_\nu + 1) \frac{\Gamma(\lambda_\nu + 1)}{t^{\lambda_\nu+1}}.$$

Using the well known expression:

$$\sin \pi(\lambda_\nu + 1) = \frac{\pi}{\Gamma(\lambda_\nu + 1)\,\Gamma(-\lambda_\nu)},$$

with the understanding that, for $\lambda_\nu = 0, 1, 2, \ldots,$

$$\frac{1}{\Gamma(-\lambda_\nu)} = 0,$$

we finally obtain

(4) $$f(t) \approx \sum_{\nu=0}^{\infty} \frac{c_\nu}{\Gamma(-\lambda_\nu)} \frac{1}{t^{\lambda_\nu+1}}.$$

This expression is valid provided t tends two-dimensionally in the intersection of both regions where the respective expansions are valid, that is in

(5) $$\frac{\pi}{2} - \psi + \delta \leqq \operatorname{arc} t \leqq -\frac{\pi}{2} + \psi - \delta$$

towards ∞. The larger an angle ψ we can select in the change from the $\mathfrak{L}$-transformation to the $\mathfrak{W}$-transformation, the larger is this angular region; for the limiting situation $\psi = \pi$, the angular region approaches a half-plane.

[1] The terms of an asymptotic expansion must be orderd by decreasing order of magnitude; it is for this reason that two asymptotic expansions, in general, cannot simply be superimposed term by term. More discussion concerning this is presented on p. 256.

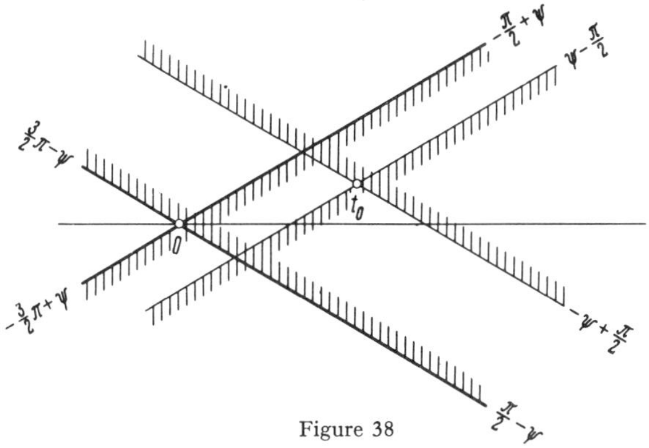

Figure 38

The angular region

(6) $$|\operatorname{arc} t| < -\frac{\pi}{2} + \psi,$$

which is arbitrarily closely approached by the region (5) is obtained by a parallel translation of the angular region of convergence, shifting the apex t_0 of the latter towards 0. The relative position shown in Fig. 38 is but one of four possibilities illustrated by Fig. 39. For every one of these situations we observe that any ray through the origin inside of the angular region (6) is, at least from some point onwards, inside the angular region of convergence, where $f(t)$ is defined. This suffices, since the asymptotic expansion refers only to large values of $|t|$.

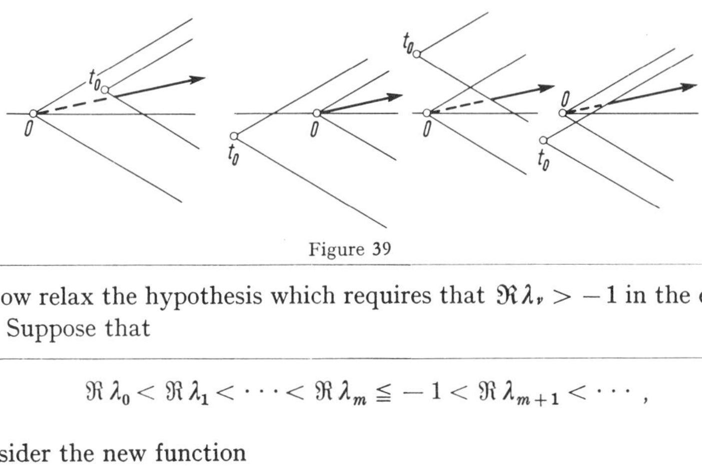

Figure 39

We now relax the hypothesis which requires that $\Re\lambda_\nu > -1$ in the expansion of $F(s)$. Suppose that

$$\Re\lambda_0 < \Re\lambda_1 < \cdots < \Re\lambda_m \leqq -1 < \Re\lambda_{m+1} < \cdots,$$

and consider the new function

$$F_1(s) = F(s) - c_0 s^{\lambda_0} - c_1 s^{\lambda_1} - \cdots - c_m s^{\lambda_m}.$$

Obviously, we have

$$F_1(s) \approx \sum_{\nu=m+1}^{\infty} c_\nu s^{\lambda_\nu} \quad (-1 < \Re\lambda_{m+1} < \cdots),$$

hence,

$$\mathfrak{W}\{F_1(s)\} \approx \sum_{\nu=m+1}^{\infty} \frac{c_\nu}{\Gamma(-\lambda_\nu)} \frac{1}{t^{\lambda_\nu+1}}.$$

We now invoke (36.10) which states that

$$\mathfrak{W}\{s^{\lambda_\nu}\} = \frac{1}{\Gamma(-\lambda_\nu)} \frac{1}{t^{\lambda_\nu+1}}$$

for arbitrary complex λ_ν, and every t in the angular region $|\operatorname{arc} t| < \psi - (\pi/2)$. We therefore obtain:

$$\mathfrak{W}\{F\} - \frac{c_0}{\Gamma(-\lambda_0)} \frac{1}{t^{\lambda_0+1}} - \cdots - \frac{c_m}{\Gamma(-\lambda_m)} \frac{1}{t^{\lambda_m+1}} \approx \sum_{\nu=m+1}^{\infty} \frac{c_\nu}{\Gamma(-\lambda_\nu)} \frac{1}{t^{\lambda_\nu+1}} .$$

From this, we conclude that

$$(7) \qquad f(t) \approx \sum_{\nu=0}^{\infty} \frac{c_\nu}{\Gamma(-\lambda_\nu)} \frac{1}{t^{\lambda_\nu+1}} \quad \text{as } t \to \infty, \text{ in } |\operatorname{arc} t| \leqq \psi - (\pi/2) - \delta .$$

For the case that the singularity is not at $s = 0$ but at $s = \alpha_0$, and that $F(s)$ has the asymptotic expansion

$$(8) \qquad F(s) \approx \sum_{\nu=0}^{\infty} c_\nu (s - \alpha_0)^{\lambda_\nu}$$

we find, by (36.2), the expansion for $f(t)$:

$$(9) \qquad f(t) \approx e^{\alpha_0 t} \sum_{\nu=0}^{\infty} \frac{c_\nu}{\Gamma(-\lambda_\nu)} \frac{1}{t^{\lambda_\nu+1}} .$$

The asymptotic expansion (9) for $f(t)$ is obtained simply by the termwise application of the $\mathfrak{W}$-transformation to the asymptotic expansion (8) for $F(s)$.

We have derived the following theorem.

Theorem 37.1. *Suppose that the function $f(t)$ can be presented as the $\mathfrak{W}$-transform of $F(s)$ for $t > T$, employing a contour $\mathfrak{W}$ centred at α_0 with the half-angle of opening ψ, $\pi/2 < \psi \leqq \pi$. This, in particular, is true when, in fact, we have initially*

$$f(t) = \frac{1}{2\pi i} \int_{a-i\infty}^{a+i\infty} e^{ts} F(s)\, ds \quad (a > \Re \alpha_0) ,$$

$F(s)$ being analytic in the region between the contour $\mathfrak{W}$ and the line $\Re s = a$, and tending towards zero when s tends two-dimensionally in this region towards ∞. Suppose, furthermore, that $F(s)$ has in $|\operatorname{arc}(s - \alpha_0)| \leqq \psi$ the asymptotic expansion

$$F(s) \approx \sum_{\nu=0}^{\infty} c_\nu (s - \alpha_0)^{\lambda_\nu} \quad (\Re \lambda_0 < \Re \lambda_1 < \cdots) \quad \text{as } s \to \alpha_0 .$$

Then we conclude that $f(t)$ has the asymptotic expansion

$$f(t) \approx e^{\alpha_0 t} \sum_{\nu=0}^{\infty} \frac{c_\nu}{\Gamma(-\lambda_\nu)} \frac{1}{t^{\lambda_\nu+1}} \quad \left(\frac{1}{\Gamma(-\lambda_\nu)} = 0 \text{ for } \lambda_\nu = 0, 1, 2, \cdots\right),$$

as t tends two-dimensionally in the angular region $|\operatorname{arc} t| \leqq \psi - (\pi/2) - \delta$ *towards* ∞. *The function* $f(t)$ *being a* $\mathfrak{W}$*-transform is eo ipso analytic in angular region* $|\operatorname{arc}(t - t_0)| < \psi - (\pi/2)$.

Attention is called to the remarkable fact that for possibly occuring exponents $\lambda_\nu = 0, 1, 2, \ldots$, the corresponding terms of the expansion of $f(t)$ vanish completely. This clearly results from the property that the terms 1, $(s - \alpha_0)^1$, $(s - \alpha_0)^2, \ldots$, which may occur in the expansion of $F(s)$, altogether form a function which is holomorphic at α_0 and does not contribute to the character of the singularity of

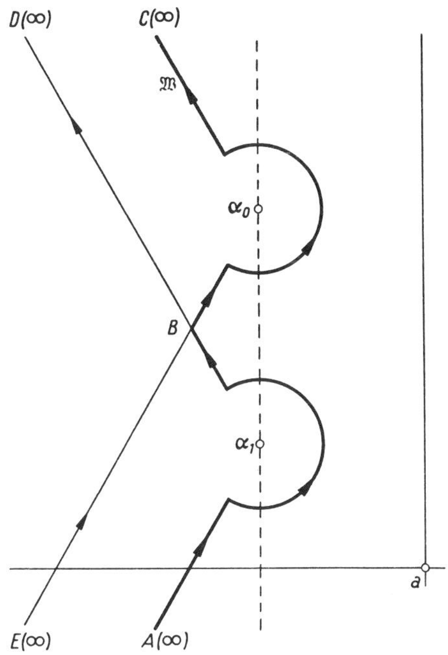

Figure 40

$F(s)$ at α_0. Thus, we now understand why the replacement of the straight line of integration through a by an angular contour with a as apex, as contemplated on p.239, would not yield a useful result. The function $F(s)$ being holomorphic at a implies that its power series expansion about the point a has exclusively non-negative integers as exponents; hence, all terms of the corresponding expansion of $f(t)$ would vanish. The only conclusion would be that $f(t) = o(e^{at})$, as already mentioned on p. 239; this conclusion, although correct, does not provide useful information. Indeed, by equation (9), $f(t)$ is actually of the order $o(e^{\alpha_0 t})$, with $\Re \alpha_0 < a$.

Remark: In the above proof of Theorem 37.1, we need not require the function $F(s)$ to be analytic on the two rays $\operatorname{arc}(s - \alpha_0) = \pm\psi$; the invoked Theorem 33.7 did not presume this property, it merely asks for convergence of the two integrals $\int_0^{\infty(\psi)}$ and $\int_0^{\infty(-\psi)}$. This fact is important for the intended extension of Theorem 37.1.

Also, Theorem 36.1 concerning the holomorphy of $f(t)$ merely presumes convergence of the two integrals.

In practical applications, one often encounters the situation in which, to the left of a, there are several singular points with largest real part; for instance, one might have two singular points α_0 and α_1, with $\Re\alpha_0 = \Re\alpha_1$, as shown in Figure 40. We presume that $f(t)$ can be presented as the $\mathfrak{W}$-transform of $F(s)$, using the in Fig. 40 heavily traced curve as contour $\mathfrak{W}$: it has two centres, at α_0 and at α_1; all finite and all unbounded straight line portions have the respective directions $+\psi$ or $-\psi$. According to the remark at the end of Chapter 36, it follows that $f(t)$ is analytic within an angular region $|\operatorname{arc}(t-t_0)| < \psi - (\pi/2)$.

Employing the notations of Fig. 40, we can write:

$$f(t) = \frac{1}{2\pi i}\int_{A(\infty)}^{B} e^{ts}F(s)\,ds + \frac{1}{2\pi i}\int_{B}^{C(\infty)} e^{ts}F(s)\,ds.$$

Extending the curve $A(\infty)\,B$ to form the curve $A(\infty)\,BD(\infty)$ which is of the original angular type $\mathfrak{W}$, and similarly extending $BC(\infty)$ into $E(\infty)\,BC(\infty)$, we find, upon setting $F(s)\equiv 0$ on the respective extensions,

$$f(t) = \frac{1}{2\pi i}\int_{A(\infty)}^{D(\infty)} e^{ts}F(s)\,ds + \frac{1}{2\pi i}\int_{E(\infty)}^{C(\infty)} e^{ts}F(s)\,ds.$$

When $F(s)$ has asymptotic expansions of the form (8) as both $s\to\alpha_1$ and $s\to\alpha_0$, then we may asymptotically develop the respective integrals of the last equation according to Theorem 37.1, since by the above remark, $F(s)$ need not be analytic on the rays. Both expansions are, for real t, of the same order of magnitude $e^{\Re\alpha_0 t}$, and we may combine the two expansions by superposition in such a manner that the orders of magnitude of the powers decrease.

Observe: The asymptotic expansion of a single $\mathfrak{W}$-transformation is valid in the angular region $|\operatorname{arc} t| < \psi - (\pi/2)$. By contrast, the superposition of several expansions is valid only for real-valued t.

Proof: The orders of magnitude of the encountered terms are determined by the factors $e^{\alpha_0 t}$ and $e^{\alpha_1 t}$. Using

$$\alpha_0 = \beta_0 + i\gamma_0, \quad \alpha_1 = \beta_1 + i\gamma_1; \quad \beta_0 = \beta_1, \ \gamma_0 \neq \gamma_1$$

we find, with $t = re^{i\varphi}$ ($|\varphi| < \psi - (\pi/2)$),

$$e^{\alpha_0 t} = e^{(\beta_0 + i\gamma_0)\,r(\cos\varphi + i\sin\varphi)},$$

hence

$$|e^{\alpha_0 t}| = e^{(\beta_0\cos\varphi - \gamma_0\sin\varphi)\,r}$$

and

$$|e^{\alpha_1 t}| = e^{(\beta_1\cos\varphi - \gamma_1\sin\varphi)\,r}.$$

Because of $\beta_0 = \beta_1$, we conclude that

$$\frac{|e^{\alpha_0 t}|}{|e^{\alpha_1 t}|} = e^{(\gamma_1 - \gamma_0)\sin\varphi\cdot r}.$$

$|e^{a_0 t}| = |e^{a_1 t}|$ holds only for $\varphi = 0$; in every other direction, that is for $\varphi \neq 0$, $|e^{a_0 t}|$ and $|e^{a_1 t}|$ differ by a factor of the form e^{cr}. For instance, when $(\gamma_1 - \gamma_0)\sin\varphi > 0$, then $e^{a_1 t}$ is of lesser order of magnitude than $e^{a_0 t}$, as $t \to \infty$. It follows that in the desired arrangement of the terms by order of magnitude, all terms resulting from the expansion at α_1 would follow behind the infinitely many terms resulting from the expansion at α_0; that is, the former terms would not at all be considered. The presented explanation also elucidates the fact that, in general, asymptotic expansions may not simply be added term by term, a process which is permissible for converging series, instead, one must insert the terms of one expansion amongst the terms of the other expansion with full attention to the proper position according to order of magnitude.

Theorem 37.2. *Suppose that the function $f(t)$ is, for $t > T$, representable as the $\mathfrak{W}$-transform of the function $F(s)$, the employed contour having the shape shown in Fig. 40: the two centres α_0 and α_1 have identical real parts, and all straight line sections have the directions $+\psi$ or $-\psi$, with $\pi/2 < \psi \leqq \pi$. Suppose, furthermore, that $F(s)$ has the asymptotic expansions:*

$$F(s) \approx \sum_{\nu=0}^{\infty} c_\nu (s - \alpha_0)^{\lambda_\nu} \quad (\Re\lambda_0 < \Re\lambda_1 < \cdots) \quad \text{as } s \to \alpha_0, \text{ in } |\operatorname{arc}(s - \alpha_0)| \leqq \psi,$$

$$F(s) \approx \sum_{\mu=0}^{\infty} d_\mu (s - \alpha_1)^{\varkappa_\mu} \quad (\Re\varkappa_0 < \Re\varkappa_1 < \cdots) \quad \text{as } s \to \alpha_1, \text{ in } |\operatorname{arc}(s - \alpha_1)| \leqq \psi$$

at α_0 and at α_1 respectively. Then the function $f(t)$ has, as real-valued $t \to \infty$, the asymptotic expansion which is constructed by superposition of the expansions

$$\sum_{\nu=0}^{\infty} \frac{c_\nu}{\Gamma(-\lambda_\nu)} \frac{e^{\alpha_0 t}}{t^{\lambda_\nu+1}} \quad \text{and} \quad \sum_{\mu=0}^{\infty} \frac{d_\mu}{\Gamma(-\varkappa_\mu)} \frac{e^{\alpha_1 t}}{t^{\varkappa_\mu+1}}$$

The function $f(t)$, being a $\mathfrak{W}$-transform, is not defined for real-valued $t > T$ only, but for all t in an angular region $|\operatorname{arc}(t - t_0)| < \psi - (\pi/2)$; it is analytic in the delineated region.

By contrast with the situation treated in Chapter 35, where $F(s)$ had single-valued singularities, we notice here, where $F(s)$ does have many-valued singularities, that only the singularities with the largest real part are utilized for the asymptotic expansion of $f(t)$. These singularities determine the behaviour of $f(t)$, as $t \to \infty$, in an angular region which lies symmetrically to the positive real axis; this region degenerates into a right hand portion of the real axis for several many-valued singularities with largest real part. The question is whether possible singularities of $F(s)$ *further to the left* would contribute to the asymptotic behaviour of $f(t)$. When this is the case then these would probably influence the behaviour of $f(t)$ in directions other than those mentioned above. This is, in fact, true for instance for *entire functions $f(t)$ of exponential type* and their $\mathfrak{L}$-transforms $F(s)$ which are holomorphic at $s = \infty$, assuming there the value zero, as considered in Theorem 30.3. A function $f(t)$ of exponential type submits to the same estimation along any ray through the origin; thus, one can easily demonstrate that its $\mathfrak{L}$-integral may be evaluated not only along the positive real axis but along any other ray starting at the origin (see Fig. 41a), and that all these integrals represent elements of the same function $F(s)$. Along a ray in the direction $-\vartheta$, where

$$t = e^{-i\vartheta}\tau \quad (0 \leqq \tau < \infty),$$

we find

$$(10) \qquad F(s) = \int_0^{\infty(-\vartheta)} e^{-st} f(t)\, dt = e^{-i\vartheta} \int_0^{\infty} e^{-se^{-i\vartheta}\tau} f(e^{-i\vartheta}\tau)\, d\tau.$$

Substituting

$$(11) \qquad f(e^{-i\vartheta}\tau) = f_1(\tau),$$

$$(12) \qquad \mathfrak{L}\{f_1(\tau)\} = F_1(\sigma),$$

whereby $\mathfrak{L}\{f_1\}$ is to be evaluated along the positive real axis (see Fig. 41b), we observe that $f_1(\tau)$ too is of exponential type, and that $F_1(\sigma)$ is holomorphic at $\sigma = \infty$, where it assumes the value zero. Eq. (10) implies that

$$(13) \qquad F(s) = e^{-i\vartheta} F_1(e^{-i\vartheta} s);$$

hence, with

$$\sigma = e^{-i\vartheta} s,$$

we find that

$$(14) \qquad F_1(\sigma) = e^{i\vartheta} F(e^{i\vartheta}\sigma).$$

Obviously, the complex inversion formula can be applied to the $\mathfrak{L}$-transformation (12); thus, one finds:

$$(15) \qquad f_1(\tau) = \frac{1}{2\pi i} \int_{v_1} e^{\tau\sigma} F_1(\sigma)\, d\sigma \quad \text{for} \quad 0 < \tau < \infty,$$

where v_1 designates a vertical line in the half-plane of absolute convergence of $\mathfrak{L}\{f_1\}$ (see Fig. 41c). We suppose that $F_1(\sigma)$ has only one singular point α_1 with largest real part. Then the vertical line v_1 is, necessarily, to the right of the point α_1. Replacing f_1 and F_1 in (15) according to (11) and (14) by f and F respectively, and σ by $e^{-i\vartheta}s$ yields:

$$f(e^{-i\vartheta}\tau) = \frac{1}{2\pi i} \int_{e^{i\vartheta}v_1} e^{\tau e^{-i\vartheta} s} e^{i\vartheta} F(s)\, e^{-i\vartheta}\, ds$$

or, with $e^{-i\vartheta}\tau = t$,

$$(16) \qquad f(t) = \frac{1}{2\pi i} \int_v e^{ts} F(s)\, ds \quad \text{for} \quad t = e^{-i\vartheta}\tau \ (0 < \tau < \infty),$$

evaluated along the straight line v which is obtained by a rotation of the plane through the angle ϑ (see Fig. 41d). The singular point α_1 of $F_1(\sigma)$ transfers to the singular point $\alpha_0 = e^{i\vartheta}\alpha_1$ of $F(s)$ which is furthest in the direction ϑ.

$F(s)$, and consequently also $F_1(\sigma)$, converge uniformly in all directions towards zero; hence, we may replace the path of integration v_1 in the σ-plane by an angular

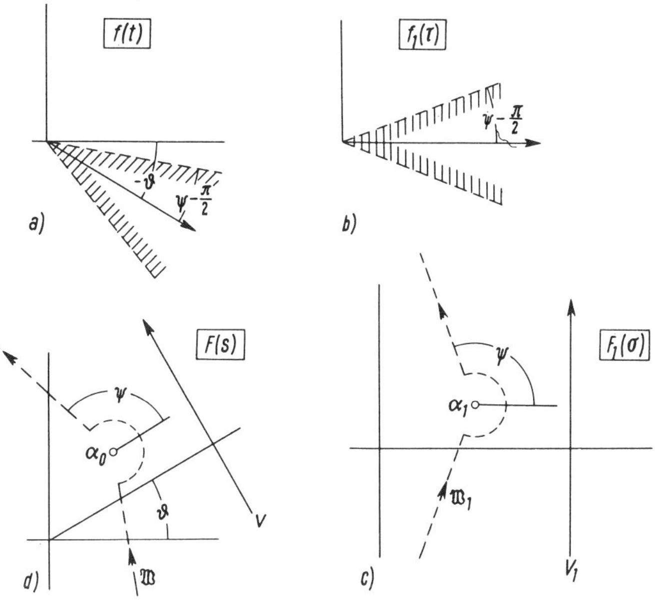

Figure 41

contour $\mathfrak{W}_1$ centred at α_1 having a fixed half-angle of opening $\psi > \pi/2$, chosen in such a manner that all other singular points of $F_1(\sigma)$ are to the left of $\mathfrak{W}_1$ (see Fig. 41c). In the s-plane this contour $\mathfrak{W}_1$ becomes the contour $\mathfrak{W}$ centred at α_0, its bisector forming the angle ϑ with the positive real axis (see Figure 41d). Theorem 36.2 guarantees the convergence of the integral (16) along the contour $\mathfrak{W}$ not only for the points on the ray in the direction $-\vartheta$ but actually in an angular region having a bisector which forms the angle $-\vartheta$ with the positive real axis, and the half-angle of opening $\psi - (\pi/2)$ (see Fig. 41a).

Let us suppose that $F(s)$ has the asymptotic expansion:

$$F(s) \approx \sum_{\nu=0}^{\infty} c_\nu (s - \alpha_0)^{\lambda_\nu} \qquad (\Re\lambda_0 < \Re\lambda_1 < \cdots) \tag{17}$$

as $s \to \alpha_0$ in the angular region $\vartheta - \psi \leqq \operatorname{arc}(s - \alpha_0) \leqq \vartheta + \psi$ about α_0; then we can derive from this an asymptotic expansion for $f(t)$. We shall accomplish this, using $F_1(\sigma)$ and $f_1(\tau)$.

From (17), using (14), because of

$$s = e^{i\vartheta}\sigma, \qquad \alpha_0 = e^{i\vartheta}\alpha_1,$$

we conclude that

$$F_1(\sigma) \approx e^{i\vartheta} \sum_{\nu=0}^{\infty} c_\nu \left(e^{i\vartheta}(\sigma - \alpha_1)\right)^{\lambda_\nu} = \sum_{\nu=0}^{\infty} c_\nu e^{i(\lambda_\nu + 1)\vartheta} (\sigma - \alpha_1)^{\lambda_\nu}$$

as $\sigma \to \alpha_1$ in $|\operatorname{arc}(\sigma - \alpha_1)| \leqq \psi$. Hence, by Theorem 37.1,

$$f_1(\tau) \approx e^{\alpha_1 \tau} \sum_{\nu=0}^{\infty} \frac{c_\nu e^{i(\lambda_\nu + 1)\vartheta}}{\Gamma(-\lambda_\nu)} \frac{1}{\tau^{\lambda_\nu + 1}} \qquad \text{as } \tau \to \infty \text{ in } \quad |\operatorname{arc} \tau| \leqq \psi - \pi/2 - \delta.$$

According to (11), it follows that

$$f_1(\tau) = f(e^{-i\vartheta}\tau) = f(t)\,;$$

since $\alpha_1 = e^{-i\vartheta}\alpha_0$ and $\tau = e^{i\vartheta}t$, we obtain

$$\alpha_1 \tau = \alpha_0 t, \qquad \frac{e^{i\vartheta}}{\tau} = \frac{1}{t}.$$

Thus, we finally find

$$(18) \quad f(t) \approx e^{\alpha_0 t} \sum_{\nu=0}^{\infty} \frac{c_\nu}{\Gamma(-\lambda_\nu)} \frac{1}{t^{\lambda_\nu+1}} \quad \text{as } t \to \infty \text{ in} \quad |\,(\operatorname{arc}(e^{i\vartheta}t)\,| \leqq \psi - \pi/2 - \delta,$$

$$\text{that is, in} \quad -\vartheta - \psi + \pi/2 + \delta \leqq \operatorname{arc} t \leqq -\vartheta + \psi - \pi/2 - \delta.$$

This expansion shows the same formal structure as the expansion of Theorem 37.1; however, here we have a different centre and the region of validity is rotated through the angle $-\vartheta$.

The process which we employed here for entire functions of exponential type can actually be used for other functions, provided these can be represented as $\mathfrak{W}$-transforms with some contour $\mathfrak{W}$ of the type shown in Fig. 41d. We summarize these results in a theorem.

Theorem 37.3. *Suppose that the function $f(t)$ can be represented, for $t > T$, as a $\mathfrak{W}$-transform of the function $F(s)$, the contour $\mathfrak{W}$ having its centre at α_0, and the half-angle of opening ψ, with $\pi/2 < \psi \leqq \pi$, the bisector of the contour forming the angle ϑ with the positive real axis. Suppose, furthermore, that $F(s)$ has the asymptotic expansion (17), as $s \to \alpha_0$ in the sector $\vartheta - \psi \leqq \operatorname{arc}(s - \alpha_0) \leqq \vartheta + \psi$. Then we conclude that $f(t)$ has the asymptotic expansion (18), as $t \to \infty$ in an angular region which is defined by its apex that is located at the origin, its bisector that forms the angle $-\vartheta$ with the positive real axis, and its half-angle of opening that equals $\psi - (\pi/2) - \delta$ ($\delta > 0$, arbitrarily small). The function $f(t)$ is analytic in a similar angular region, possibily having a different apex.*

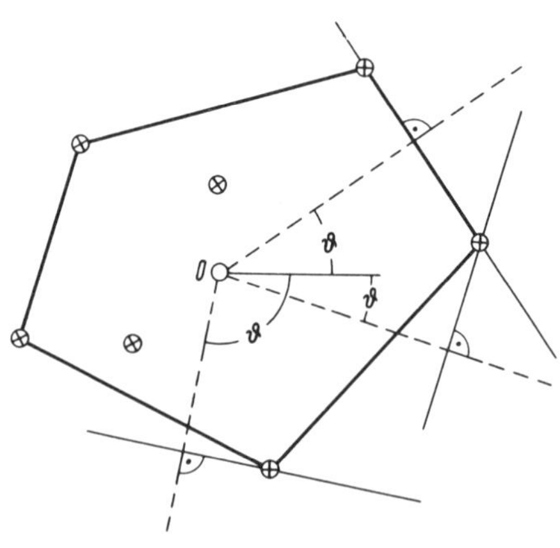

Figure 42

When the contour $\mathfrak{W}$ has several centres, as provided for in Theorem 37.2, then one must superimpose the respective expansions. The thus constructed expansion is not valid in an angular region, it is valid only on the ray in the direction $-\vartheta$.

For the special case that $F(s)$ is the $\mathfrak{L}$-transform of $f(t)$ and that $F(s)$ has merely a finite number of singular points, we can construct a smallest convex region which contains all singular points. Its boundary is a polygon, the corners of which are singular points. For every ray which starts at the origin and forms the angle ϑ with the positive real axis, we can find the corresponding "supporting line" of the polygon, the straight line orthogonal to the ray which touches the polygon. This supporting line passes through a single singular point, or through several singular points (Fig. 42).

We select that supporting line which corresponds to the angle ϑ and which contains the singular point α. We assume that $f(t)$ can be represented with $t = e^{-i\vartheta}\tau$ $(0 < \tau < \infty)$ by the complex inversion integral, employing the supporting line as the path of integration, excepting the point α which is circumvented by means of a semicircle as shown in Fig. 43. When this path can be replaced by an angular contour, and when $F(s)$ has an asymptotic expansion at α, then one finds, by

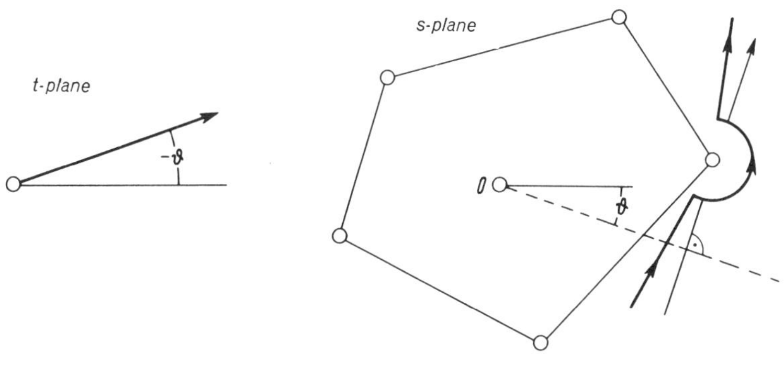

Figure 43

Theorem 37.3, an asymptotic expansion for $f(t)$ in the direction $-\vartheta$ (we disregard here the angular region within the asymptotic expansion of $f(t)$ may be valid). When continuously rotating the ray in the t-plane by varying the angle $-\vartheta$, we affect a simultaneous rotation in the s-plane of the corresponding supporting line in the opposite direction about the polygon, touching for most angles but *one* singular point, occasionally *several* singular points when it coincides with a side of the polygon. The described process clearly demonstrates for what directions in the t-plane one can expect to obtain a *single expansion*, and for what directions one obtains *superpositions of expansions*.

When the representation of $f(t)$ by a $\mathfrak{W}$-integral is not possible for all directions but only for a restricted region of directions, then one can find asymptotic expansions of $f(t)$ only for the corresponding directions.

The singularities in the interior of the polygon play no rôle for the asymptotic expansions of $f(t)$. They may be utilized though, when functions have been subtracted from $F(s)$ which remove the externally located singularities.

Examples of the presented method of asymptotic expansion will be presented in the next Chapter, and on pp. 291 and 297.

38. Ordinary Differential Equations with Polynomial Coefficients. Solution by Means of the Laplace Transformation and by Means of Integrals with Angular Path of Integration

The $\mathfrak{L}$-transformation serves to solve not only linear ordinary differential equations with constant coefficients, as explained in Chapter 15, but also linear differential equations with polynomial coefficients. We deferred the solution of the latter category, for it also provides examples for the application of the method of asymptotic expansion of a $\mathfrak{W}$-transform which was developed in Chapter 37.

An equation of nth order having polynomial coefficients of, at most, m^{th} degree may be written compactly as follows:

$$\sum_{\mu=0}^{m} \sum_{\nu=0}^{n} a_{\mu\nu}\, t^{\mu}\, y^{(\nu)}(t) = f(t). \tag{1}$$

The function $f(t)$ need not be a polynomial, it may be some arbitrary function. Presuming the existence of the $\mathfrak{L}$-transform of $y^{(n)}(t)$ and of $f(t)$, by application of Theorems 9.3 and 9.4 to Eq. (1), we find the image equation:

$$\sum_{\mu=0}^{m} \sum_{\nu=0}^{n} a_{\mu\nu}(-1)^{\mu} \frac{d^{\mu}}{ds^{\mu}} \left[s^{\nu}\, Y(s) - y(0+)\, s^{\nu-1} - \cdots - y^{(\nu-1)}(0+)\right] = F(s). \tag{2}$$

Each of the terms of (2) yields, since

$$\frac{d^{\mu}}{ds^{\mu}} \left[s^{\nu} Y(s)\right] = s^{\nu} Y^{(\mu)} + \binom{\mu}{1} \nu\, s^{\nu-1}\, Y^{(\mu-1)} + \cdots,$$

a sum of terms of the type $s^{\alpha} Y^{(\beta)}$, with $\alpha \leqq \nu$, $\beta \leqq \mu$. Transferring all terms of (2), which contain neither Y nor any of its derivatives, to the right hand side, we find for (2)

$$\sum_{\nu=0}^{n} \sum_{\mu=0}^{m} b_{\nu\mu}\, s^{\nu}\, Y^{(\mu)}(s) = F_1(s). \tag{3}$$

The resulting image equation is, again, a linear ordinary differential equation with polynomial coefficients; however, Eq. (3) has *the order of equation and the maximum degree of the polynomial coefficients interchanged* when compared with equation (1).

This observation immediately suggests the application of the $\mathfrak{L}$-transformation to equations with $m < n$, since this will result in a reduction of the order of the equation; the same method might be useful also for $m \geqq n$, in the case that the transformation leads to an equation with known solution.

Eq. (3) incorporates certain initial values of $y(t)$; some have been lost in the process of differentiation of (2); for $Y(s)$ we merely require one condition: it, and consequently all its derivatives, must be $\mathfrak{L}$-transforms. One obtains m linearly independent solutions of (3); from these we must select the functions which are $\mathfrak{L}$-transforms. For the latter, the corresponding original functions have to be determined.

For the case $m < n$ one, obviously, cannot obtain all linearly independent solutions of the original equation (1); one finds, at most, m of these. *At least* $(n - m)$ *of the linearly independent solutions of* (1) *must fail to have a* $\mathfrak{L}$*-transform of its* n^{th} *derivative.*

When the polynomial coefficients of (1) are of, at most, first degree, that is for $m = 1$, we generate an image equation (3) of the first order which can always be solved by quadratures, that is by definite integrals.

The method is first elucidated by the following example:

The Differential Equation of the Bessel Functions

The equation

(4) $$t^2 z'' + t z' + (t^2 - \alpha^2) z = 0 \quad (\alpha \text{ complex})$$

is known as Bessel differential equation. Having here $m = 2$ and $n = 2$, we must conclude that the application of the $\mathfrak{L}$-transformation would not reduce the order of the equation. However, the substitution

$$z(t) = t^{-\alpha} y(t),$$

followed by division by $t^{1-\alpha}$ leads to the equation

(5) $$t y'' - (2\alpha - 1) y' + t y = 0,$$

with $m = 1$ and $n = 2$; consequently, the image equation of the latter is a first order equation; it is

$$-\frac{d}{ds}[s^2 Y - y(0+)s - y'(0+)] - (2\alpha - 1)[sY - y(0+)] - Y' = 0$$

or

(6) $$(s^2 + 1) Y' + (2\alpha + 1) s Y = 2\alpha y(0+).$$

We seek the particular solution of (5) which satisfies

(7) $$\alpha y(0+) = 0;$$

this means that

$$y(0+) = \text{arbitrary value} \quad \text{for } \alpha = 0,$$
$$y(0+) = 0 \quad \text{for } \alpha \neq 0.$$

For this particular situation, equation (6) is reduced to the following form

$$\frac{Y'}{Y} = -(2a+1)\frac{s}{s^2+1}.$$

The general solution of the latter equation is

$$\log Y = -\frac{2a+1}{2}\log(s^2+1) + C$$

or

$$Y = c(s^2+1)^{-a-(1/2)}.$$

The value of α determines whether or not the last expression is a $\mathfrak{L}$-transform. For $\Re\alpha \leqq -1/2$ it cannot be a $\mathfrak{L}$-transform, for it fails to tend towards zero, when $s \to \infty$. It follows that Eq. (5) with $\Re a \leqq -1/2$ cannot have a solution, the second derivative of which has a $\mathfrak{L}$-transform. (Observe that we discuss here the equation for y in which α occurs in linear form, in contrast to the equation for z in which α^2 occurs, hence no distinction need be made between $+\alpha$ and $-\alpha$.) Theorem 28.3, and also the subsequent consideration, guarantees that, for $\Re\alpha > -1/2$, $Y(s)$ is indeed a $\mathfrak{L}$-transform. We can, for $|s| > 1$, expand $Y(s)$ into an absolutely converging power series:

$$Y(s) = c\, s^{-2a-1}(1+s^{-2})^{-a-(1/2)} = c\sum_{\nu=0}^{\infty}\binom{-a-\frac{1}{2}}{\nu} s^{-2\nu-2a-1}. \tag{8}$$

For real-valued α, by Theorem 30.2, one obtains the original function, for all complex $t \neq 0$, by termwise inverse transformation:

$$y(t) = c\sum_{\nu=0}^{\infty}\binom{-a-\frac{1}{2}}{\nu}\frac{t^{2\nu+2a}}{\Gamma(2\nu+2a+1)}.$$

Here, Theorem 30.2 may also be employed for complex values α. For one obtains in this case instead of the series $\sum|\alpha_\nu|/x_0^{\lambda_\nu}$, encountered in the proof on p. 195, when using $\alpha = \alpha_1 + i\alpha_2$,

$$\sum_{\nu=0}^{\infty}\frac{|a_\nu|}{x_0^{\Re\lambda_\nu}}\frac{\Gamma(\Re\lambda_\nu)}{|\Gamma(\lambda_\nu)|} = \sum_{\nu=0}^{\infty}\frac{|a_\nu|}{x_0^{2\nu+2a_1+1}}\frac{\Gamma(2\nu+2a_1+1)}{|\Gamma(2\nu+2a_1+1+2ia_2)|}. \tag{9}$$

The Stirling formula:

$$\log\Gamma(z) = \left(z-\frac{1}{2}\right)\log z - z + \log\sqrt{2\pi} + O\left(\frac{1}{|z|}\right)$$

is valid as $z \to \infty$ in the entire z-plane with the exception of the negative real axis. With the aid of this formula one demonstrates easily that, for fixed y, the quotient

$$\frac{\Gamma(x)}{\Gamma(x+iy)}$$

remains bounded as real-valued $x \to \infty$; thus the series (9) is majorized by a convergent series.

It follows that the series representation of $y(t)$ may be used for all α with $\Re\alpha > -1/2$. The binomial coefficient can be expressed by means of Γ-functions:

$$\binom{-\alpha-\frac{1}{2}}{\nu} = \frac{\left(-\alpha-\frac{1}{2}\right)\left(-\alpha-\frac{3}{2}\right)\cdots\left(-\alpha-\nu+\frac{1}{2}\right)}{\nu!}$$

$$= \frac{(-1)^\nu (2\alpha+1)(2\alpha+3)\cdots(2\alpha+2\nu-1)}{\nu!\,2^\nu}$$

$$= \frac{(-1)^\nu}{\nu!\,2^\nu}\,\frac{(2\alpha+1)(2\alpha+2)\cdots(2\alpha+2\nu)}{2^\nu(\alpha+1)\cdots(\alpha+\nu)}$$

(10) $$= \frac{(-1)^\nu}{\nu!\,2^{2\nu}}\,\frac{\Gamma(2\alpha+2\nu+1)\,\Gamma(\alpha+1)}{\Gamma(2\alpha+1)\,\Gamma(\alpha+\nu+1)};$$

hence,

$$y(t) = c\,\frac{\Gamma(\alpha+1)\,2^\alpha}{\Gamma(2\alpha+1)}\,t^\alpha \sum_{\nu=0}^{\infty} \frac{(-1)^\nu}{\nu!\,\Gamma(\alpha+\nu+1)}\left(\frac{t}{2}\right)^{2\nu+\alpha}.$$

The function $y(t)$ satisfies the requirement (7) only when either $\alpha = 0$ or else $\Re\alpha > 0$; hence, we must restrict α to these values.[1] Equation (5) is homogeneous, hence we can arbitrarily select any value for the constant c; for instance, so that

$$c\,\frac{\Gamma(\alpha+1)\,2^\alpha}{\Gamma(2\alpha+1)} = 1.$$

Returning to the initially presented Bessel differential equation (4), we find for $z(t)$ the following solution which is called *Bessel function* $J_\alpha(t)$:

(11) $$J_\alpha(t) = \sum_{\nu=0}^{\infty} \frac{(-1)^\nu}{\nu!\,\Gamma(\alpha+\nu+1)}\left(\frac{t}{2}\right)^{2\nu+\alpha} \quad (t \text{ arbitrary complex}).$$

This solution has been determined for $\alpha = 0$ or $\Re\alpha > 0$. One can easily verify that actually it satisfies Eq. (4) for all complex-valued α; this is a consequence of the fact that Eq. (4) contains only α^2, hence $+\alpha$ and $-\alpha$ produce the identical equation. It can be shown that $J_\alpha(t)$ and $J_{-\alpha}(t)$ are linearly independent, provided α is not an integer n; thus one has, for $\alpha \neq n$, two fundamental solutions of the Bessel equation (4). Another technique must be employed to construct the second fundamental solution for the case $\alpha = n$.

The above deduction also produces the $\mathfrak{L}$-*transform* of $t^\alpha J_\alpha(t)$ (notice the determination of the constant c):[2]

(12) $$\mathfrak{L}\{t^\alpha J_\alpha(t)\} = \frac{\Gamma(2\alpha+1)}{2^\alpha\,\Gamma(\alpha+1)}\,\frac{1}{(s^2+1)^{\alpha+(1/2)}} \quad \text{for } \Re\alpha > -\frac{1}{2}.$$

---

[1] For $\Re\alpha \leq 0$ ($\alpha \neq 0$), the initial value $y(0+)$ which is needed for the image equation (6) does not exist.

[2] Using the formula
$$\Gamma(2\alpha+1) = \frac{4^\alpha}{\sqrt{\pi}}\,\Gamma(\alpha+1)\,\Gamma\left(\alpha+\frac{1}{2}\right)$$
we can rewrite this briefly as follows:
$$\mathfrak{L}\{t^\alpha J_\alpha(t)\} = \frac{1}{\sqrt{\pi}}\,2^\alpha\,\Gamma\left(\alpha+\frac{1}{2}\right)\frac{1}{(s^2+1)^{\alpha+(1/2)}}.$$

It has a simpler structure than the $\mathfrak{L}$-transform of $J_\alpha(t)$, which may be obtained by termwise transformation of (11); we find thus:

$$\mathfrak{L}\{J_\alpha(t)\} = \frac{(\sqrt{s^2+1}-s)^\alpha}{\sqrt{s^2+1}} = \frac{1}{\sqrt{s^2+1}\,(\sqrt{s^2+1}+s)^\alpha} \quad \text{for } \Re\alpha > -1. \tag{13}$$

Hence, it is practical to start with (12) rather than with (13) when deriving properties of $J_\alpha(t)$ by means of the $\mathfrak{L}$-transformation.

In Eq. (8) we expanded $Y(s) = \mathfrak{L}\{t^\alpha J_\alpha(t)\}$ in a series of powers of s which converges when $|s| > 1$, that is outside the circle centred at $s = 0$ through the two singular points $s = \pm i$. Alternatively one can, for arbitrary s_0, expand $Y(s)$ in a series of powers of $(s - s_0)$; this series converges outside the smallest circular disk, centred at s_0, which contains the two singular points $\pm i$. For instance, with $s_0 = +i$ and $|s - i| > 2$, one finds:

$$Y(s) = c(s^2+1)^{-\alpha-(1/2)} = c(s-i)^{-\alpha-(1/2)}\,(s-i+2i)^{-\alpha-(1/2)}$$
$$= c(s-i)^{-2\alpha-1}\left(1+\frac{2i}{s-i}\right)^{-\alpha-(1/2)} = c\sum_{\nu=0}^{\infty}\binom{-\alpha-\frac{1}{2}}{\nu}\frac{(2i)^\nu}{(s-i)^{\nu+2\alpha+1}}.$$

This series is of the same type as (8); hence, it may be inversely transformed term by term. However, instead of s we now have $(s - i)$; this corresponds to a multiplication of the original function by e^{it}:

$$y(t) = c\,e^{it}\sum_{\nu=0}^{\infty}\binom{-\alpha-\frac{1}{2}}{\nu}(2i)^\nu\frac{t^{\nu+2\alpha}}{\Gamma(\nu+2\alpha+1)}.$$

Performing the following substitutions: c by its specified value, the binomial coefficient by the expression (10) and $y(t)$ by $t^\alpha J_\alpha(t)$, one finds:

$$J_\alpha(t) = e^{it}\sum_{\nu=0}^{\infty}\frac{(-i)^\nu}{\nu!}\,\frac{\Gamma(2\nu+2\alpha+1)}{\Gamma(\nu+\alpha+1)\,\Gamma(\nu+2\alpha+1)}\left(\frac{t}{2}\right)^{\nu+\alpha}, \tag{14}$$

and, for instance for $\alpha = 0$,

$$J_0(t) = e^{it}\sum_{\nu=0}^{\infty}\frac{(-i)^\nu\,(2\nu)!}{(\nu!)^3}\left(\frac{t}{2}\right)^\nu. \tag{15}$$

The series representations (11) and (14) with *increasing* powers of t converge well only for *small* values of $|t|$. For *large* values of $|t|$, *asymptotic expansions* involving *decreasing* powers of t are much better suited for use. Such asymptotic expansions can be developed by the method devised in Chapter 37. The hypotheses for the complex inversion formula are satisfied, hence by (12),

$$\frac{\sqrt{\pi}}{\Gamma(\alpha+\frac{1}{2})}\left(\frac{t}{2}\right)^\alpha J_\alpha(t) = \frac{1}{2\pi i}\int_{a-i\infty}^{a+i\infty} e^{ts}(s^2+1)^{-\alpha-(1/2)}\,ds, \tag{16}$$

provided $t > 0$, $\Re\alpha > -1/2$, and $a > 0$. In the entire plane, $F(s)$ has only the singular points $\pm i$ which are located to the left of a; these have equal real parts, hence both must be considered. The function $F(s)$ converges towards zero when s tends two-dimensionally in the entire plane towards ∞; hence, the path of integration of (16) may be replaced by a contour $\mathfrak{W}$ as shown in Fig. 44, having the two centres $\pm i$, and ψ may assume any value in $\pi/2 < \psi \leqq \pi$.

By Theorem 36.1, the integral evaluated along the contour $\mathfrak{W}$ converges not only for real-valued $t > 0$ but actually for all t in the angular region $|\operatorname{arc} t| < \psi - (\pi/2)$. (Here, the point t_0 is real-valued.) Moreover, the integral converges for all complex-valued α, thus providing a representation of $J_\alpha(t)$ for all values of

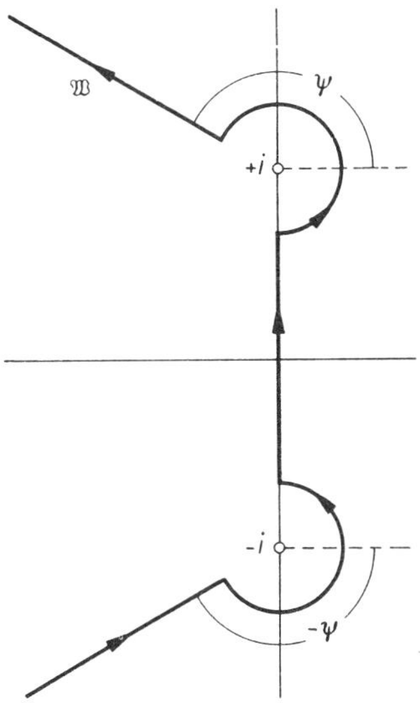

Figure 44

α with the exception of $\alpha = -1/2, -3/2, \ldots$; for these values of α, both sides of (16) yield 0. (For these values, the integrand is analytic in the whole plane; hence, the integral along a closed curve composed of $\mathfrak{W}$ together with a circular arc on the left side is zero, and the contribution to the integral along the circular arc tends towards zero when the radius grows.) We summarize this as follows:

$$(17) \qquad \frac{\sqrt{\pi}}{\Gamma(\alpha+\frac{1}{2})}\left(\frac{t}{2}\right)^{\alpha} J_\alpha(t) = \frac{1}{2\pi i}\int_{\mathfrak{W}} e^{ts}(s^2+1)^{-\alpha-(1/2)}\,ds$$

$$\text{for } |\operatorname{arc} t| < \psi - \frac{\pi}{2}, \quad \alpha \neq -\frac{1}{2}, -\frac{3}{2}, \ldots .$$

In order to obtain the asymptotic expansion which, according to the argumentation on p. 256, is valid for real-valued t only, we express $F(s)$ by power series at the two singular points $+i$ and $-i$. These power series may be $\mathfrak{W}$-transformed term by term; the resulting two expansions are then superimposed. We find, at $s = i$:

$$(s^2+1)^{-a-(1/2)} = (s-i)^{-a-(1/2)}\,(s-i+2i)^{-a-(1/2)}$$

$$= (2i)^{-a-(1/2)} \sum_{\nu=0}^{\infty} \binom{-a-\frac{1}{2}}{\nu} \frac{1}{(2i)^{\nu}}\,(s-i)^{\nu-a-(1/2)},$$

and at $s = -i$:

$$(s^2+1)^{-a-(1/2)} = (-2i)^{-a-(1/2)} \sum_{\nu=0}^{\infty} \binom{-a-\frac{1}{2}}{\nu} \frac{1}{(-2i)^{\nu}}\,(s+i)^{\nu-a-(1/2)}.$$

Hence, we obtain

$$\frac{\sqrt{\pi}}{\Gamma(a+\frac{1}{2})}\left(\frac{t}{2}\right)^{a} J_a(t)$$

$$\approx (2i)^{-a-(1/2)}\, e^{it} \sum_{\nu=0}^{\infty} \binom{-a-\frac{1}{2}}{\nu} \frac{1}{(2i)^{\nu}}\,\frac{1}{\Gamma(-\nu+a+\frac{1}{2})}\,\frac{1}{t^{\nu-a+(1/2)}}$$

$$+ (-2i)^{-a-(1/2)}\, e^{-it} \sum_{\nu=0}^{\infty} \binom{-a-\frac{1}{2}}{\nu} \frac{1}{(-2i)^{\nu}}\,\frac{1}{\Gamma(-\nu+a+\frac{1}{2})}\,\frac{1}{t^{\nu-a+(1/2)}},$$

as real-valued $t \to \infty$; and, with $\pm i = e^{\pm i\pi/2}$,

$$J_a(t) \approx \frac{\Gamma(a+\frac{1}{2})}{\sqrt{\pi}} \left\{ e^{it} \sum_{\nu=0}^{\infty} \binom{-a-\frac{1}{2}}{\nu} \frac{e^{-i(\pi/2)(\nu+a+(1/2))}}{\Gamma(a-\nu+\frac{1}{2})}\,\frac{1}{(2t)^{\nu+(1/2)}} \right. \tag{18}$$

$$\left. + e^{-it} \sum_{\nu=0}^{\infty} \binom{-a-\frac{1}{2}}{\nu} \frac{e^{i(\pi/2)(\nu+a+(1/2))}}{\Gamma(a-\nu+\frac{1}{2})}\,\frac{1}{(2t)^{\nu+(1/2)}} \right\}.$$

We compare (18) with (33.11) which is valid as $t \to \infty$ in the upper half-plane:

$$J_a(t) \approx \frac{e^{-it}}{\sqrt{\pi}\,\Gamma(a+\frac{1}{2})} \sum_{\nu=0}^{\infty} \binom{a-\frac{1}{2}}{\nu} e^{i(\pi/2)(\nu+a+(1/2))}\,\Gamma(a+\nu+\tfrac{1}{2})\,\frac{(-1)^{\nu}}{(2t)^{\nu+(1/2)}}. \tag{19}$$

Steps similar to the ones used in the derivation of (10) lead to the expressions:

$$\binom{-a-\frac{1}{2}}{\nu} = \frac{(-1)^{\nu}\,\Gamma(\nu+a+\frac{1}{2})}{\nu!\,\Gamma(\frac{1}{2}+a)}, \quad \text{and} \quad \binom{a-\frac{1}{2}}{\nu} = \frac{(-1)^{\nu}\,\Gamma(\nu-a+\frac{1}{2})}{\nu!\,\Gamma(\frac{1}{2}-a)}.$$

Also, we invoke two well known relations from the theory of the Γ-function:

$$\Gamma(\tfrac{1}{2}+a)\,\Gamma(\tfrac{1}{2}-a) = \frac{\pi}{\cos a\pi}, \quad \text{and} \quad \Gamma(\tfrac{1}{2}+\nu-a)\,\Gamma(\tfrac{1}{2}-\nu+a) = \frac{(-1)^{\nu}\pi}{\cos a\pi}$$

(integer-valued ν).

Using these four relations, one can easily demonstrate that the expansion (19) agrees with the second part of the expansion (18): Along the positive real axis, the expression (19) has to be augmented by another expansion.

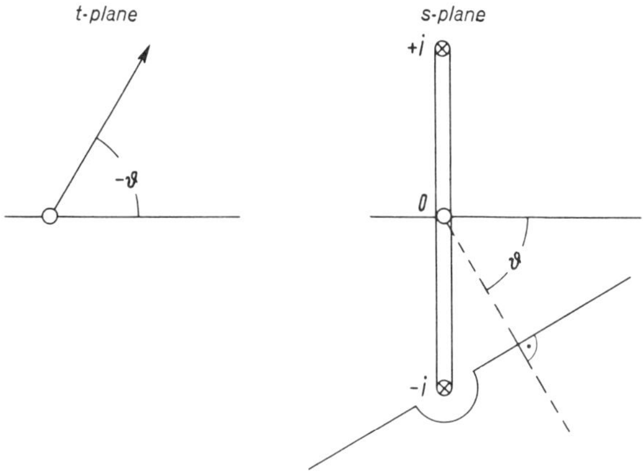

Figure 45

Theorem 37.3 and the remarks associated with this Theorem provide the explanation. In this case, the polygon of singular points is the line between $+i$ and $-i$, which must be considered twice to form a closed polygon. For every direction $-\vartheta$ in the upper t-half-plane without the borderline, that is for $0 < -\vartheta < \pi$, we have the corresponding direction ϑ in the s-plane, where $0 > \vartheta > -\pi$. The supporting line of the polygon of singular points, orthogonal to the direction ϑ, contains only the singular point $s = -i$. Consequently, for every direction in the upper t-half-plane we have only the asymptotic expansion due to the singularity at $s = -i$. For the positive real t-axis, we have $-\vartheta = 0$; the supporting line orthogonal to the direction $\vartheta = 0$ contains both singular points, $+i$ and $-i$. This necessitates the consideration of both asymptotic expansions (see Fig. 45). Also refer to Theorem 37.2.

Combining the two expressions of (18), and upon expressing the binomial coefficients by Γ-functions, one finds the expansion in the form ($\alpha \neq -1/2, -3/2, \ldots$):

$$(20) \qquad J_a(t) \approx \frac{2}{\sqrt{\pi}} \sum_{\nu=0}^{\infty} \frac{(-1)^\nu \Gamma\left(a + \nu + \frac{1}{2}\right)}{\nu!\, \Gamma\left(a - \nu + \frac{1}{2}\right)} \cos\left(t - \left(a + \nu + \tfrac{1}{2}\right) \frac{\pi}{2}\right) \frac{1}{(2\,t)^{\nu + (1/2)}}$$

as real-valued $t \to \infty$.

Having shown by this specific example how, by the $\mathfrak{L}$-transformation, one can obtain both converging and asymptotic representations of the solution, we now proceed to the general equation with polynomial coefficients. We shall restrict the following development to polynomial coefficients of first degree; this will enable us to execute the solutions in all detail. We shall modify the method and thus succeed in our search for all fundamental integrals of the differential equation.

The General Linear Homogeneous Differential Equation with Linear Coefficients

We shall consider differential equations of the form:

$$(a_n t + b_n)\, y^{(n)} + \cdots + (a_1 t + b_1)\, y' + (a_0 t + b_0)\, y = 0, \tag{21}$$

admitting real-valued as well as complex-valued coefficients a_ν, b_ν. We expressedly require that

$$a_n \neq 0.$$

When searching for *that particular solution, for which* $\mathfrak{L}\{y^{(n)}\}$ *exists, and which assumes the initial values*

$$y(0+) = y'(0+) = \cdots = y^{(n-1)}(0+) = 0,$$

then its $\mathfrak{L}$-transform must satisfy the image equation

$$\left(-a_n \frac{d}{ds} + b_n\right)(s^n Y) + \cdots + \left(-a_1 \frac{d}{ds} + b_1\right)(s\, Y) + \left(-a_0 \frac{d}{ds} + b_0\right) Y = 0$$

or

$$-Y'(a_n s^n + \cdots + a_1 s + a_0) + Y(b_n s^n - a_n n s^{n-1} + \cdots + b_1 s - a_1 + b_0) = 0. \tag{22}$$

We rewrite the last equation thus:

$$-p_1(s)\, Y' + p_0(s)\, Y = 0 \quad \text{or} \quad \frac{Y'}{Y} = \frac{p_0(s)}{p_1(s)}, \tag{23}$$

employing

$$\begin{aligned} p_1(s) &= a_n s^n + \cdots + a_1 s + a_0, \\ p_0(s) &= b_n s^n + (b_{n-1} - n\, a_n)\, s^{n-1} + \cdots + (b_0 - a_1). \end{aligned} \tag{24}$$

Because of $a_n \neq 0$, $p_1(s)$ is exactly of degree n, whilst $p_0(s)$ has degree n, or less. Let $\alpha_1, \ldots, \alpha_n$ designate the roots of $p_1(s)$ which are presumed to be distinct; then by partial fraction expansion, we have:

$$\frac{Y'}{Y} = \frac{p_0(s)}{p_1(s)} = d_0 + \frac{d_1}{s - \alpha_1} + \cdots + \frac{d_n}{s - \alpha_n},$$

where, similarly as in (15.9),

$$d_0 = \frac{b_n}{a_n}, \quad d_\nu = \frac{p_0(\alpha_\nu)}{p_1'(\alpha_\nu)} \quad (\nu = 1, \cdots, n).$$

Integration of the image equation yields:

$$\log Y = c + d_0 s + d_1 \log(s - \alpha_1) + \cdots + d_n \log(s - \alpha_n).$$

The constant c is indeterminate, since (23) and (21) are homogeneous equations; omitting c, we find:

$$Y(s) = e^{d_0 s}(s - \alpha_1)^{d_1} \cdots (s - \alpha_n)^{d_n}. \tag{25}$$

The exponential factor occurs only when $d_0 \neq 0$, that is when $b_n \neq 0$. The function (25) can be a $\mathfrak{L}$-transform only when d_0 is a negative real number or zero, since $Y(s)$ must tend towards zero when s tends towards ∞ in $|\operatorname{arc} s| < \pi/2$. When $d_0 \leqq 0$ then, by the Translation Theorem 7.2, it suffices that

$$Z(s) = (s - a_1)^{d_1} \cdots (s - a_n)^{d_n}$$

is a $\mathfrak{L}$-transform. The latter is true when and only when

$$\Re(d_1 + \cdots + d_n) < 0.$$

The corresponding original function $z(t)$ can be obtained by firstly expanding $Z(s)$ into a series of decreasing powers of s:

$$\begin{aligned} Z(s) &= s^{d_1 + \cdots + d_n}\left(1 - \frac{a_1}{s}\right)^{d_1} \cdots \left(1 - \frac{a_n}{s}\right)^{d_n} \\ &= s^{d_1 + \cdots + d_n} \sum_{k=0}^{\infty} \binom{d_1}{k} (-1)^k \left(\frac{a_1}{s}\right)^k \cdots \sum_{k=0}^{\infty} \binom{d_n}{k} (-1)^k \left(\frac{a_n}{s}\right)^k \\ &= \sum_{m=0}^{\infty} c_m\, s^{-m + d_1 + \cdots + d_n}, \end{aligned}$$

and then inversely transforming this series term by term, as permitted by Theorem 30.2, possibly requiring the extension of the latter to complex exponents as shown on p. 264. Thus, one develops a series for $z(t)$ with increasing powers of t which converges for all $t \neq 0$. The representation of $y(t)$ is then obtained, for all real $t > -d_0$, by

$$y(t) = z(t + d_0).$$

For $0 \leqq t < -d_0$, we have to set $y(t) = 0$.[3] Finally, we must determine if and under what hypotheses $y(t)$ is differentiable n times, and has vanishing initial values.

The method does yield a solution, only when $d_0 \leqq 0$ and $\Re(d_1 + \cdots + d_n) < 0$. For the above, special equation (5), we had

$$p_0(s) = -(2\alpha + 1)\, s, \quad p_1(s) = s^2 + 1$$

and

$$\frac{p_0(s)}{p_1(s)} = -\frac{2\alpha + 1}{2}\left(\frac{1}{s - i} + \frac{1}{s + i}\right),$$

hence

$$d_0 = 0, \quad d_1 = d_2 = -(2\alpha + 1)/2.$$

The first condition $d_0 \leqq 0$ is satisfied; the second condition $\Re(d_1 + d_2) < 0$ is satisfied only when $-\Re(2\alpha + 1) < 0$, that is when $\Re\alpha > -1/2$. This is exactly the

[3] $t = -d_0 = -b_n/a_n$ is the root of the polynomial coefficient $a_n t + b_n$ of $y^{(n)}$ in the differential equation; it is, in general, a singular point of the solutions as shown by function-theoretical considerations of linear differential equations.

same restriction which we discovered on p. 264. Moreover, we had to restrict α to the values 0 or $\Re\alpha > 0$ in order to comply with the specified initial value $\alpha Y(0) = 0$.

The effective application of the $\mathfrak{L}$-transformation to differential equations of the stated type is restricted by the required existence of $\mathfrak{L}\{y^{(n)}\}$ which presumes integrability of $y^{(n)}(t)$ in every interval $0 \leq t \leq T$, and the existence of the initial values $y(0+), \ldots, y^{(n-1)}(0+)$. However, in general, solutions do not satisfy these two conditions. Therefore, starting with the $\mathfrak{L}$-transformation, we intend to develop *another method* with further reaching applicability in the present case.

Upon determination of the $\mathfrak{L}$-transform $Y(s)$, one can represent the solution $y(t)$ by means of the complex inversion formula:

$$y(t) = \frac{1}{2\pi i}\int_{a-i\infty}^{a+i\infty} e^{ts}\, Y(s)\, ds\,. \tag{26}$$

Considering the fact that the solution $y(t)$ ultimately occurs in the form (26) one might wonder whether one could, *to begin with, postulate the solution in this form* and then determine the $Y(s)$ so that the $y(t)$ satisfies the differential equation.[4] For this purpose one must differentiate $y(t)$ n times. We differentiate under the integral sign:

$$y^{(\nu)}(t) = \frac{1}{2\pi i}\int_{a-i\infty}^{a+i\infty} e^{ts}\, s^{\nu}\, Y(s)\, ds \quad (\nu = 1, \cdots, n)\,, \tag{27}$$

and then substitute these terms into the differential equation. However, in general, formula (27) is wrong, as was already demonstrated in connection with Theorem 36.1. The cause of the discrepancy is clear from the point of view of the $\mathfrak{L}$-transformation, for we have the image function $s\,Y(s) - y(0^+)$ corresponding to $y'(t)$ and therefore, in fact,

$$y'(t) = \frac{1}{2\pi i}\int_{a-i\infty}^{a+i\infty} e^{ts}\,[s\,Y(s) - y(0^+)]\, ds\,,$$

in general,

$$y^{(\nu)}(t) = \frac{1}{2\pi i}\int_{a-i\infty}^{a+i\infty} e^{ts}\,[s^{\nu}\, Y(s) - y(0^+)\, s^{\nu-1} - \cdots - y^{(\nu-1)}(0^+)]\, ds. \tag{28}$$

We conclude that (27) is correct only when the first n initial values of $y(t)$ vanish.

Consequently, the *postulation* (26) *is useless* when the solution of the differential equation has non-zero initial values or has no initial values whatever. One may

[4] This is equivalent to the representation of $y(t)$ by a Fourier integral:

$$y(t) = \frac{1}{2\pi i}\, e^{at}\int_{-\infty}^{+\infty} e^{it\eta}\, Y(a + i\eta)\, d\eta\,,$$

in a manner similar to that employed for differential equations in a bounded region where the solution is postulated in the form of a Fourier series.

encounter this situation when solving differential equations with polynomial coefficients as, for instance, with solution (11) of Eq. (4), for $\Re\alpha < 0$.

However, when one replaces the straight line path of integration of (26) by an angular contour $\mathfrak{W}$, then, by Theorem 36.1, one may legitimately differentiate under the integral sign; that is, for a function of the form

(29) $$y(t) = \frac{1}{2\pi i}\int_{\mathfrak{W}} e^{ts}\, Y(s)\, ds,$$

one indeed finds

(30) $$y^{(\nu)}(t) = \frac{1}{2\pi i}\int_{\mathfrak{W}} e^{ts}\, s^{\nu}\, Y(s)\, ds.$$

We have to keep in mind that, according to Theorem 36.1, $y(t)$ must be analytic in some specified angular region and that, possibly, (29) may be correct for $t > T \geqq 0$ only.

When initially $y(t)$ is generated by means of the complex inversion formula using the $\mathfrak{L}$-transform $Y(s)$, then one can substantiate the correctness of (30) by an alternative process: Replacing in formula (28), which is valid under these circumstances, the straight line path of integration by an angular contour to the right of $s = 0$, the integrals involving the powers $s^{\nu-1}, \ldots, s, 1$ yield nought, according to formula (25.6), and one is left with (30).

We now seek to satisfy the differential equation (21) *in an interval* $t > T \geqq 0$ *by a function of the form* (29), leaving the actual position of the angular contour as yet unspecified. The derivatives $y^{(\nu)}(t)$ are then given by (30), and $Y(s)$ is to be specified so that

(31) $$\frac{1}{2\pi i}\sum_{\nu=0}^{n}(a_\nu t + b_\nu)\int_{\mathfrak{W}} e^{ts}\, s^{\nu}\, Y(s)\, ds = 0.$$

In order to permit the writing of the left hand side of (31) by an integral which contains the parameter t only in the factor e^{ts}, we alter $ty^{(\nu)}(t)$, invoking the rule of integration by parts:

$$2\pi i\, t\, y^{(\nu)}(t) = \int_{\mathfrak{W}} \left(t\, e^{ts}\right)\left(s^{\nu} Y(s)\right) ds = e^{ts}\, s^{\nu} Y(s)\,\Big|_{\mathfrak{W}} - \int_{\mathfrak{W}} e^{ts}\, \frac{d(s^{\nu}\, Y(s))}{ds}\, ds,$$

interpreting the first term of the last expression as the difference of the limits of $e^{ts} s^{\nu}\, Y(s)$ which are obtained when s moves on both rays of $\mathfrak{W}$ towards infinity. In the case that, for sufficiently large t, we have

(32) $$\lim_{\Re s \to -\infty} e^{ts}\, s^{\nu}\, Y(s) = 0 .$$

(the $Y(s)$ determined later on, indeed has this property), then we have:

$$2\pi i\, t\, y^{(\nu)}(t) = -\int_{\mathfrak{W}} e^{ts}\left[\nu\, s^{\nu-1}\, Y(s) + s^{\nu}\, Y'(s)\right] ds$$

and, instead of (31),

$$\int_{\mathfrak{W}} e^{ts}\left[-a_n(n\,s^{n-1}\,Y + s^n\,Y') + b_n\,s^n\,Y - \cdots - a_1\,(Y + s\,Y') + b_1\,s\,Y - a_0\,Y' + b_0\,Y\right] ds = 0.$$

This equation is certainly satisfied, when the expression within the square brackets vanishes identically,[5] that is

$$-Y'(a_n s^n + \cdots + a_1 s + a_0) + Y(b_n s^n - a_n n s^{n-1} + \cdots + b_1 s - a_1 + b_0) = 0.$$

This differential equation conforms with the equation (21), which was obtained by means of the $\mathfrak{L}$-transformation, in the case of vanishing initial values. This is obvious since for these conditions the derivatives of (26) are exactly equal to those of (29). When, as before, presuming distinct zeros for the polynomial $p_1(s)$ which is determined by (24), then $Y(s)$ has the form (25):

$$Y(s) = e^{d_0 s}(s - \alpha_1)^{d_1} \cdots (s - \alpha_n)^{d_n}.$$

This function satisfies, for $t > -\Re d_0$, the condition (32); hence, it is useful for our purpose. None of the constraints regarding the exponents $d_0, d_1, \ldots, d_n$, which we had to specify on p. 271, have been applied here. With this $Y(s)$ one obtains, according to the postulation (29):

(33) $$y(t) = \frac{1}{2\pi i} \int_{\mathfrak{W}} e^{(t+d_0)s}(s - \alpha_1)^{d_1} \cdots (s - \alpha_n)^{d_n}\, ds.$$

Theorem 36.1 guarantees the convergence of this integral in the angular region[6] $|\operatorname{arc}(t + d_0)| < \psi - (\pi/2)$ (see Fig. 32), where it represents an analytic function. Retracing our argumentation in the opposite direction, we obtain proof that $y(t)$ satisfies the differential equation (21).

Various positions can be selected for the angular contour $\mathfrak{W}$; it is this variability which enables us *to represent, with the integral (33), n linear independent solutions of* (21). First, we observe by Cauchy's theorem that the integral yields identical results when evaluated along two angular contours which do not enclose any of the roots α_ν that are, in general, branch points of the integrand. To generate distinct functions, we must select the various angular contours in such a manner that certain roots α_ν are located between the contours. Presuming that

$$\Re\alpha_1 > \Re\alpha_2 > \cdots > \Re\alpha_n,$$

---

[5] This is sufficient, although not necessary. When, for instance, $Y(s)$ is analytic in the region to the left of $\mathfrak{W}$ and we temporarily complete the contour $\mathfrak{W}$ by a circular arc $\mathfrak{B}$ to form a closed contour, then the integral along the closed contour yields the value zero. When $Y(s)$ behaves in such a manner that the value of the integral along $\mathfrak{B}$ tends towards zero, for increasing radius, then the integral along $\mathfrak{W}$ vanishes.

[6] In this case, convergence depends only upon the behaviour of the term $e^{(t+d_0)s}$. One can easily demonstrate that the exponent $(t + d_0)s$ tends, for $|\operatorname{arc}(t + d_0)| < \psi - (\pi/2)$, along the rays $\operatorname{arc} s = \pm\psi$ in the *left* half-plane towards ∞.

we can locate $\mathfrak{W}_1$ to the right of α_1, $\mathfrak{W}_2$ between α_1 and α_2, and so forth, and then select α_1 as the centre of $\mathfrak{W}_1$, α_2 as the centre of $\mathfrak{W}_2$, etc. The respective half-angles of opening ψ_ν need not be equal; we may select these so large that the next

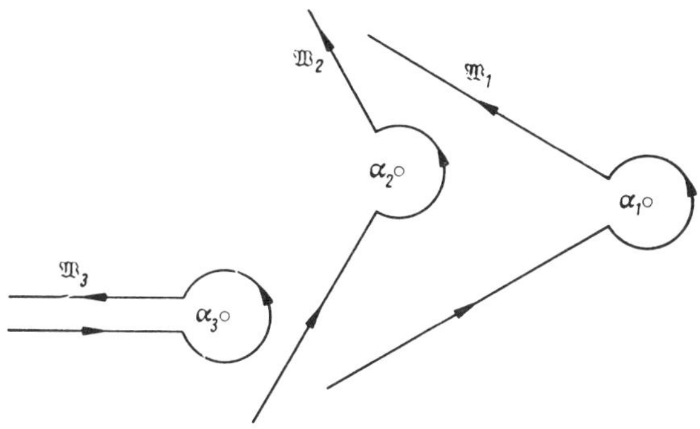

Figure 46

root $\alpha_{\nu+1}$ is to the left of $\mathfrak{W}_\nu$, as shown in Fig. 46. The functions generated in this manner

$$(34) \qquad y_\nu(t) = \frac{1}{2\pi i} \int\limits_{\mathfrak{W}_\nu} e^{(t+d_0)s} (s - a_1)^{d_1} \cdots (s - a_n)^{d_n} \, ds$$

are, indeed, linearly independent; this fact is demonstrated by comparison of the

asymptotic expansions

of these functions. The asymptotic expansions are obtained by means of the method of Theorem 37.1. For instance, when considering $y_1(t)$ we observe that the function $Y(s)/(s - \alpha_1)^{d_1}$ is holomorphic at α_1; thus, it can be expanded into a power series about α_1 with integer-valued exponents:

$$\frac{Y(s)}{(s - \alpha_1)^{d_1}} = e^{d_0 s}(s - a_2)^{d_2} \cdots (s - a_n)^{d_n} = \sum_{\nu=0}^{\infty} c_\nu (s - a_1)^\nu \quad (c_0 \neq 0)\,.$$

Hence, in some neighbourhood of α_1, we have the convergent expansion:

$$Y(s) = \sum_{\nu=0}^{\infty} c_\nu (s - a_1)^{\nu + d_1},$$

involving the generally complex constant d_1. The asymptotic expansion of $y_1(t)$ is obtained by means of Theorem 37.1; it is:

$$(35) \qquad y_1(t) \approx e^{a_1 t} \sum_{\nu=0}^{\infty} \frac{c_\nu}{\Gamma(-\nu - d_1)} t^{-\nu - d_1 - 1} \quad \text{as} \quad t \to \infty$$

in the angular region $|\operatorname{arc} t| < \psi_1 - (\pi/2)$. For the special case that d_1 is a positive integer, all terms of the expansion vanish, leading to the conclusion that $y_1(t) =$

$= o(e^{\alpha_1 t})$. When d_1 is a negative integer, then all terms with $\nu \geqq -d_1$ are zero. The case $d_1 = 0$ cannot occur.[7]

Similar asymptotic expansions can be found for the other integrals $y_2(t), \ldots, y_n(t)$.

The solutions $y_1(t), \ldots, y_n(t)$ are linearly independent provided none of the numbers $d_1, \ldots, d_n$ is a positive integer.

In our proof, we first recall that we presumed the real parts of all α_ν to be distinct, and to be arranged in the order $\Re\alpha_1 > \Re\alpha_2 \ldots > \Re\alpha_n$. Suppose there are constants $C_1, C_2, \ldots, C_n$ such that

$$(36) \qquad C_1 y_1(t) + \cdots + C_n y_n(t) \equiv 0,$$

then we would also have

$$(37) \qquad C_1 y_1(t)\, e^{-\alpha_1 t}\, t^{d_1+1} + C_2 y_2(t)\, e^{-\alpha_2 t}\, t^{d_2+1}\big(e^{-(\alpha_1-\alpha_2)t}\, t^{d_1-d_2}\big) + \cdots \equiv 0.$$

Using only the first term of each asymptotic expansion, one finds:

$$\lim_{t\to\infty} y_1(t)\, e^{-\alpha_1 t}\, t^{d_1+1} = \frac{c_0}{\Gamma(-d_1)} \neq 0,$$

$$\lim_{t\to\infty} y_2(t)\, e^{-\alpha_2 t}\, t^{d_2+1} \qquad\qquad \neq 0, \text{ and so forth.}$$

Moreover, we have $\Re(\alpha_1 - \alpha_2) > 0$, $\Re(\alpha_1 - \alpha_3) > 0, \ldots$, and in the limit as $t \to \infty$, (37) yields:

$$C_1 \frac{c_0}{\Gamma(-d_1)} = 0,$$

that is, $C_1 = 0$; and (36) becomes

$$C_2 y_2(t) + \cdots + C_n y_n(t) \equiv 0.$$

Iterating the above process, we similarly conclude that $C_2 = 0$, and so forth, and finally $C_n = 0$. Thus, we have shown that the relation (36) necessitates that all coefficients are zero. This implies that the $y_1, \ldots, y_n$ are linearly independent indeed.

Solutions of the type of asymptotic expansions having the form (35) are known as *Thomé's normal series* in the theory of differential equations.

Each solution $y_\nu(t)$ is analytic in some angular region $|\operatorname{arc}(t + d_0)| < \psi_\nu - (\pi/2)$, *which has its apex at* $-d_0 = -b_n/a_n$ *and extends towards the right.* When using angular contours $\mathfrak{W}_\nu$ with rays biased towards the right, then one would obtain solutions in angular regions with the same apex $-d_0$ which extend towards the left (see lower part of Fig. 47); accordingly, one would produce asymptotic expansions as t approaches infinity towards the left. *Rotation of the angular contours* $\mathfrak{W}_\nu$, so that the bisectors are no longer horizontal, *effects a counter-rotation of the angular regions of the $y_\nu(t)$* as explained in Theorem 37.3 and the subsequent remarks. Whenever, in the process of rotation, an angular contour contacts some

[7] When d_1 is a positive integer, then $Y(s)$ is holomorphic at α_1; when d_1 is a negative integer, then $Y(s)$ has a pole of order $-d_1$ at α_1.

root α_μ, one must stop there since α_μ is, in general, a singular point of $Y(s)$. Only when d_μ is a positive integer, that is when $Y(s)$ is holomorphic at α_μ, may the rotation be continued past α_μ; the root α_μ yields the same result as α_ν. This explains why one cannot obtain n linearly independent solutions when a positive integer is amongst the d_ν.

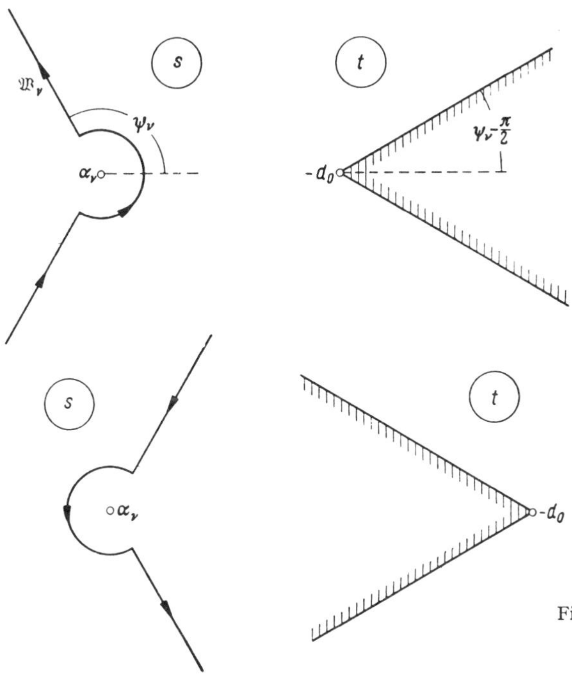

Figure 47

We presented the method as a kind of *reversal of the method of the $\mathfrak{L}$-transformation* insofar as the development *starts* here with a complex integral of the type of the complex inversion formula; the latter occurs *at the end* of the process when the $\mathfrak{L}$-transformation is used. Both methods overlap to some extent.

When one starts with the $\mathfrak{L}_{\mathrm{I}}$-transformation and one introduces the postulation (29) as another form of the complex inversion formula of this transformation, then one invariably selects the path $\mathfrak{W}$ of integration in the holomorphy half-plane of $Y(s)$; that is to the right of all singular points of $Y(s)$, as is proper for the inversion formula of the $\mathfrak{L}_{\mathrm{I}}$-transformation. The use of any path of integration amongst the singular points of $Y(s)$ is strange to the concepts of the $\mathfrak{L}_{\mathrm{I}}$-transformation. This choice seems natural, however, in view of the fact that the complex inversion formula is actually associated with the $\mathfrak{L}_{\mathrm{II}}$-transformation (Theorem 24.3). *One $\mathfrak{L}_{\mathrm{II}}$-transform may have distinct original functions in distinct vertical strips of the s-plane.* For instance, the function $Y(s) = 1/(s - \alpha)$ has, in the half-plane $\Re s > > \Re \alpha$, the corresponding original function:

$$y_1(t) = \begin{cases} e^{\alpha t} & \text{for } t \geqq 0 \\ 0 & \text{for } t < 0, \end{cases}$$

since

$$\int_{-\infty}^{+\infty} e^{-st}\, y_1(t)\, dt = \int_0^{\infty} e^{-(s-a)t}\, dt = \frac{1}{s-a} \quad \text{for } \Re(s-a) > 0;$$

whilst, in the half-plane $\Re s < \Re \alpha$, the corresponding original function is given by:

$$y_2(t) = \begin{cases} 0 & \text{for } t > 0 \\ -e^{at} & \text{for } t \leqq 0, \end{cases}$$

since

$$\int_{-\infty}^{+\infty} e^{-st}\, y_2(t)\, dt = -\int_{-\infty}^{0} e^{-(s-a)t}\, dt = \frac{1}{s-a} \quad \text{for } \Re(s-a) < 0.$$

This phenomenon is analogous to the fact that one and the same analytic function may in different annular regions, where it is analytic, be represented by different Laurent series (here, the analytic function is the image function, the sequence of the coefficients of the Laurent series is the original function); for instance,

$$\frac{1}{1-z} = \sum_{n=0}^{\infty} z^n = \sum_{n=-\infty}^{+\infty} a_n z^n \text{ with } a_n = \begin{cases} 1 & \text{for } n \geqq 0 \\ 0 & \text{for } n < 0 \end{cases} \quad \text{in } |z| < 1,$$

$$\frac{1}{1-z} = -\sum_{n=1}^{\infty} \frac{1}{z^n} = \sum_{n=-\infty}^{+\infty} a_n z^n \text{ with } a_n = \begin{cases} 0 & \text{for } n \geqq 0 \\ -1 & \text{for } n < 0 \end{cases} \quad \text{in } |z| > 1.$$

When $Y(s)$ is analytic in different vertical strips, then one obtains the original function which corresponds to some particular strip simply by selecting the path of integration in the inversion formula in that strip. When beginning the development of the solution $y(t)$ with the complex inversion integral of the $\mathfrak{L}_{II}$-transformation, then it is quite natural to select distinct paths of integration in the distinct strips of holomorphy, that is between two singular points.

39. Partial Differential Equations

The solution of ordinary differential equations with constant coefficients by means of the $\mathfrak{L}$-transformation is accomplished very easily, since the $\mathfrak{L}$-transformation removes the differentiation, a transcendental operation, and an algebraic image equation is obtained. When the original equation contains derivatives with respect to two variables, say x and t, that is it represents a *partial differential equation*, then the $\mathfrak{L}$-transformation applied to the variable t will remove differentiation with respect to t, and the image equation is an *ordinary differential equation*, with the variable x. Obviously, for this purpose we must presume that t varies in the interval $0 \leqq t < \infty$; the variable x may range in an interval which may be bounded or unbounded at one or both sides. Accordingly, we have in the xt-plane (see Fig. 48) as fundamental region of the partial differential equation either a half-strip or a quadrant or a half-plane.

To determine a definite solution $u(x, t)$ of the partial differential equation, one must specify u or certain derivatives of u or linear combinations of these on the boundary of the fundamental region: the *boundary values*. The boundary values which are specified along the boundary $t = 0$, usually are called *initial values*, since most often t represents time and $t = 0$ the initiation of the time scale of the problem. A partial differential equation together with the specified boundary values is a *boundary value problem*. When x varies between $-\infty$ and $+\infty$, in which case only initial values can be specified, one uses the designation *initial value problem*.

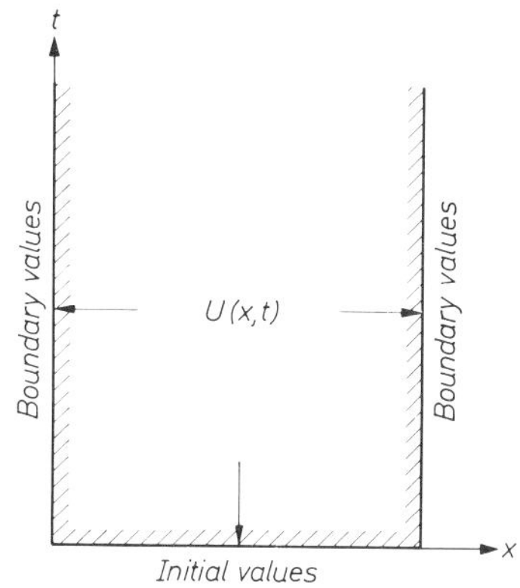

Figure 48

For most boundary value problems, the finished solution is meaningless on the points of the boundary of the fundamental region and, consequently, the solution cannot represent these boundary values. One can only request that the solution converges towards the specified boundary value, when the point (x, t) approaches some boundary point from the inside of the fundamental region, in general along a line orthogonal to the boundary. Thus, the term *boundary value* is to be understood as a *limit* (see Fig. 48).

There is no generally applicable rule which would enable us to specify type and number of necessary boundary conditions. Instead, one has to scrutinize each problem to discover what boundary conditions could or ought to be stipulated, to guarantee a solution, preferably a unique solution, of the problem.

To avoid unproductive generalities, we shall employ a few typical examples to demonstrate the technique of the $\mathfrak{L}$-transformation for boundary value problems, thus presenting all essential steps.

1. The Equation of Diffusion or Heat Conduction

Consider a recti-linear conductor of heat between $x = 0$ and $x = l$; let $u(x, t)$ designate the temperature at the location x and the time t. After appropriate normalization, we find the partial differential equation

$$\frac{\partial^2 u}{\partial x^2} = \frac{\partial u}{\partial t} \quad (0 < x < l, \ t > 0). \tag{1}$$

The identical equation is derived for other diffusion phenomena; we shall, however, for easier visualization, retain the terminology of heat transfer.

There are many possible *boundary conditions*. We shall select here the simplest ones the physical meaning of which suggests that they suffice to determine $u(x, t)$. We specify the *initial temperature* of the conductor, which may vary with x, that is we employ some function $u_0(x)$; we specify the *temperature at the end points* $x = 0$ and $x = l$, which may be functions of t, that is we use $a_0(t)$ and $a_1(t)$. According to the above remarks, these boundary conditions are understood as limits when (x, t) approaches the respective boundary along a line orthogonal to it; in mathematical terms we write the boundary conditions as follows:

(2)
$$\lim_{t \to +0} u(x, t) = u_0(x) \quad (0 < x < l);$$

(3)
$$\lim_{x \to +0} u(x, t) = a_0(t), \qquad \lim_{x \to l-0} u(x, t) = a_1(t) \quad (t > 0).$$

The application of the $\mathfrak{L}$-transformation to $u(x, t)$ with respect to the variable t, does not involve the other variable x which assumes the rôle of a parameter which re-appears in the image function; the latter is written accordingly as $U(x, s)$; it is

$$\mathfrak{L}\{u(x, t)\} = \int_0^\infty e^{-st} u(x, t)\, dt = U(x, s) \quad (0 < x < l).$$

When seeking the image equation to Eq. (1), that is when applying the $\mathfrak{L}$-transformation to the latter, Theorem 9.1 is to be used for the derivative with respect to t. Hence, we find:

$$\mathfrak{L}\left\{\frac{\partial u}{\partial t}\right\} = s\,U(x, s) - u(x, 0+) = s\,U(x, s) - u_0(x),$$

obviously employing the hypothesis:

W_1: *$\mathfrak{L}\{\partial u/\partial t\}$ does exist.*

In order to express $\mathfrak{L}\{\partial^2 u/\partial x^2\}$ in terms of $U(x, s)$, we require another hypothesis:

W_2: *$\mathfrak{L}$-transformation may be interchanged with differentiation of $u(x, t)$ with respect to the parameter x:*

$$\mathfrak{L}\left\{\frac{\partial^2 u}{\partial x^2}\right\} = \frac{\partial^2}{\partial x^2}\,\mathfrak{L}\{u\} = \frac{\partial^2 U(x, s)}{\partial x^2}.$$

With these hypotheses we can find the *image equation* of (1); it is

$$\frac{\partial^2 U(x, s)}{\partial x^2} = s\,U(x, s) - u_0(x).$$

In this image equation only a derivative with respect to the parameter x occurs; we can treat this equation as an ordinary differential equation, replacing $\partial^2/\partial x^2$ by d^2/dx^2:

(4)
$$\frac{d^2 U(x, s)}{dx^2} - s\,U(x, s) = -u_0(x).$$

The boundary condition (2), which is called *initial condition*, is already incorporated in (4) and thus automatically accounted for. Next, we investigate the effects of the $\mathfrak{L}$-transformation upon the other boundary conditions (3).

Writing rather carelessly

$$u(0, t) = a_0(t), \quad u(l, t) = a_1(t),$$

one could argue: When transforming $u(x, t)$ for every value of the parameter x in the interval $0 < x < l$, we produced $U(x, s)$; similarly, we $\mathfrak{L}$-transform $u(x, t)$ for the values of parameter x, $x = 0$ and $x = l$:

$$\mathfrak{L}\{u(0, t)\} = \mathfrak{L}\{a_0(t)\} \quad \text{or} \quad U(0, s) = A_0(s),$$

$$\mathfrak{L}\{u(l, t)\} = \mathfrak{L}\{a_1(t)\} \quad \text{or} \quad U(l, s) = A_1(s).$$

The boundary values of $u(x, t)$ produce, upon transformation, the boundary values of $U(x, s)$.

Actually, one does need a further hypothesis to establish the above presented result. This is due to the limit character of the boundary conditions of $u(x, t)$:

W_3: *$\mathfrak{L}$-transformation may be interchanged with the limiting operations $x \to +0$ and $x \to l - 0$.*

With Hypothesis W_3, we indeed find:

$$\mathfrak{L}\{a_0(t)\} = \mathfrak{L}\{\lim_{x \to +0} u(x, t)\} = \lim_{x \to +0} \mathfrak{L}\{u(x, t)\} = \lim_{x \to +0} U(x, s),$$

$$\mathfrak{L}\{a_1(t)\} = \mathfrak{L}\{\lim_{x \to l-0} u(x, t)\} = \lim_{x \to l-0} \mathfrak{L}\{u(x, t)\} = \lim_{x \to l-0} U(x, s);$$

that is, with $\mathfrak{L}\{a_0\} = A_0(s)$ and $\mathfrak{L}\{a_1\} = A_1(s)$,

$$\lim_{x \to +0} U(x, s) = A_0(s), \quad \lim_{x \to l-0} U(x, s) = A_1(s). \tag{5}$$

Thus, we obtain as "*the image*" *of the boundary value problem* involving a partial differential equation in two independent variables a *boundary value problem involving an ordinary differential equation.* Upon solving the latter, one finds the solution of the former original problem through inverse $\mathfrak{L}$-transformation.

Scheme:

Original space { partial differential equation / plus initial condition / plus boundary conditions } → Solution

↓ $\mathfrak{L}$-transformation ↑ $\mathfrak{L}^{-1}$-transformation

Image space { ordinary differential equation / plus boundary conditions } ⟶ Solution

We ought to be aware of the fact that the presented method (like any other method) requires certain hypotheses. Possibly, there might be solutions which do not comply with these hypotheses and, therefore, cannot be obtained by our method. Further considerations regarding this question will be presented on p. 287.

In the image space, we have to solve the ordinary differential equation (4) with the boundary conditions (5). We have solved this problem in Chapter 16. We begin the further development for the special case $l = \infty$ which, in the inverse transformation step, does not require higher transcendental functions.

The Case of Infinite Length

Complying with the hypotheses of Theorem 16.1, we assume that the initial temperature $u_0(x)$ is continuous and has a finite limit $u_0(\infty)$. By (16.8), for equation (4) we find the only consistent boundary value:

$$U(\infty, s) = \frac{u_0(\infty)}{s};$$

the corresponding boundary value in the original space is:

$$u(\infty, t) = u_0(\infty) = \text{const},$$

a physically plausible requirement.

The solution of the boundary value problem in the image space is, by (16.10),

$$U(x, s) = A_0(s)\, e^{-x\sqrt{s}} + \int_0^\infty \gamma_\infty(x, \xi; s)\, u_0(\xi)\, d\xi, \tag{6}$$

with

$$\gamma_\infty(x, \xi; s) = \begin{cases} \dfrac{1}{\sqrt{s}}\, e^{-x\sqrt{s}} \sinh \xi\sqrt{s} & \text{for } 0 \leqq \xi \leqq x \\ \dfrac{1}{\sqrt{s}}\, e^{-\xi\sqrt{s}} \sinh x\sqrt{s} & \text{for } x \leqq \xi < \infty, \end{cases} \tag{7}$$

where, as emphasized on p. 89, $\sqrt{s}$ represents the root with positive real part, the principle branch of $\sqrt{s}$.

For the purpose of inverse transformation we employ the correspondences:

$$e^{-k\sqrt{s}} \circ\!\!-\!\!\bullet\ \psi(k, t) = \frac{k}{2\sqrt{\pi}\, t^{3/2}}\, e^{-k^2/(4t)} \quad (k > 0),$$

$$\frac{1}{\sqrt{s}}\, e^{-k\sqrt{s}} \circ\!\!-\!\!\bullet\ \chi(k, t) = \frac{1}{\sqrt{\pi t}}\, e^{-k^2/(4t)} \quad (k \geqq 0).$$

Thus, we find:

$$\frac{1}{\sqrt{s}}\, e^{-x\sqrt{s}} \sinh \xi\sqrt{s} = \frac{1}{2\sqrt{s}}\left(e^{-(x-\xi)\sqrt{s}} - e^{-(x+\xi)\sqrt{s}}\right) \circ\!\!-\!\!\bullet\ \frac{1}{2}\left[\chi(x-\xi, t) - \chi(x+\xi, t)\right]$$
$$(0 \leqq \xi \leqq x),$$

$$\frac{1}{\sqrt{s}}\,e^{-\xi\sqrt{s}}\sinh x\sqrt{s}=\frac{1}{2\sqrt{s}}\left(e^{-(\xi-x)\sqrt{s}}-e^{-(\xi+x)\sqrt{s}}\right)\bullet\!\!-\!\!\circ\,\frac{1}{2}\left[\chi(\xi-x,t)-\chi(\xi+x,t)\right]$$
$$(x\leqq\xi<\infty).$$

The right hand sides of these two correspondences are, in fact, identical, since $\chi(x-\xi,t)=\chi(\xi-x,t)$, and we find for γ_∞ the single expression

$$\text{(8)}\qquad \frac{1}{2}\left[\chi(\xi-x,t)-\chi(\xi+x,t)\right].$$

When $\int_0^\infty \ldots d\xi$ may be interchanged with the $\mathfrak{L}^{-1}$-transformation, then we find the original function[1] of (6):

$$\text{(9)}\qquad u(x,t)=a_0(t)*\psi(x,t)+\frac{1}{2}\int_0^\infty\left[\chi(\xi-x,t)-\chi(\xi+x,t)\right]u_0(\xi)\,d\xi.$$

When, as presupposed above, $u_0(x)$ does have a limit when $x\to\infty$, then the integral converges absolutely and, indeed, we find that

$$\mathfrak{L}\left\{\int_0^\infty\cdots\right\}=\int_0^\infty\mathfrak{L}\{\cdots\}.$$

We may invoke the principle of extension (see p. 74) and confirm that $u(x,t)$ represents the solution of the boundary value problem (1), (2), (3) for $l=\infty$ irrespective of the introduced hypotheses, provided only that $a_0(t)$ and $u_0(x)$ are continuous functions. To begin with, one quickly confirms that $\psi(x,t)$, $\chi(x-\xi,t)$ and $\chi(x+\xi,t)$ satisfy the differential equation (1). The function ψ as well as its first and second derivative with respect to x are continuous, hence

$$\frac{\partial^2}{\partial x^2}\,a_0(t)*\psi(x,t)=a_0(t)*\frac{\partial^2\psi(x,t)}{\partial x^2}.$$

By Theorem 10.5, since $\psi(x,0+)=0$, one finds that

$$\frac{\partial}{\partial t}\,a_0(t)*\psi(x,t)=a_0(t)*\frac{\partial\psi(x,t)}{\partial t};$$

the first term of (9) satisfies (1). The second term can be differentiated under the integral sign with respect to either x or t, since all functions are continuous and the by differentiation created integrals converge uniformly in a neighbourhood of $x>0$ as well as in a neighbourhood of $t>0$.This shows that the second term,too, satisfies equation (1).

[1] On p. 57 we derived the formula $\psi(x_1,t)*\psi(x_2,t)=\psi(x_1+x_2,t)$; according to (9) we read from this formula the following physical implication: For the case $u_0(t)\equiv 0$, one can calculate the temperature $u(x,t)$ at the location $x=x_1+x_2$ by either of two methods: One can directly determine $u(x_1+x_2,t)=a_0(t)*\psi(x_1+x_2,t)$, or one can employ two steps, first evaluating $u(x_1,t)=a_0(t)*\psi(x_1,t)$ and then, using this function as a new excitation at x_1, once more employ (9) and calculate $u(x_1+x_2,t)=$ $=a_0(t)*\psi(x_1,t)*\psi(x_2,t)$.

With regard to the boundary conditions, one observes easily that both

$$u_1(x, t) = a_0(t) * \psi(x, t)$$

with fixed $x > 0$ as $t \to 0$, and

$$u_2(x, t) = \frac{1}{2} \int_0^\infty [\chi(\xi - x, t) - \chi(\xi + x, t)] \, u_0(\xi) \, d\xi$$

with fixed $t > 0$ as $x \to 0$, tend towards zero. The verification of the other conditions is more involved. We start with the integral

$$v(x) = \int_0^\infty \varphi(\tau) \, \psi(x, \tau) \, d\tau ,$$

which we presume to converge, when $x > 0$. The function $\psi(x, \tau)$ has, for fixed x, as a function of τ, the graph illustrated in Fig. 49; it assumes its maximum value C/x^2 ($C = 3\sqrt{6}e^{-3/2}/\sqrt{\pi}$) at $\tau = x^2/6$. One can easily verify that the area under the

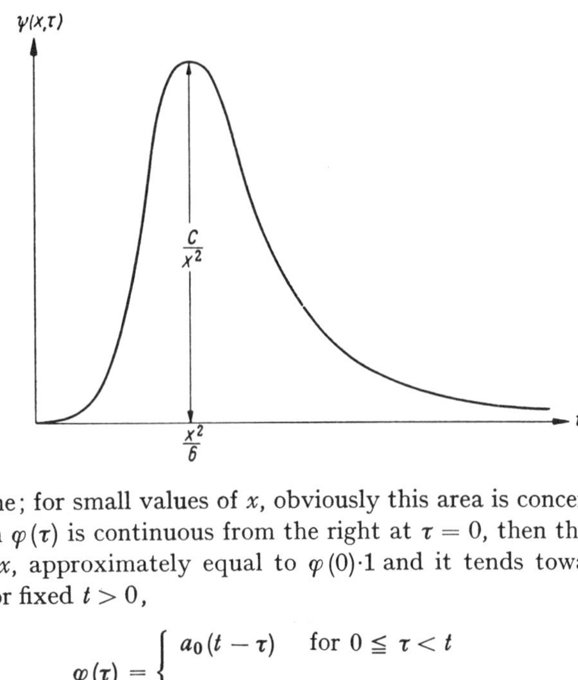

Figure 49

graph has size one; for small values of x, obviously this area is concentrated near the origin. When $\varphi(\tau)$ is continuous from the right at $\tau = 0$, then the $v(x)$ is, for small values of x, approximately equal to $\varphi(0) \cdot 1$ and it tends towards $\varphi(0)$ as $x \to 0$. Setting, for fixed $t > 0$,

$$\varphi(\tau) = \begin{cases} a_0(t - \tau) & \text{for } 0 \leqq \tau < t \\ 0 & \text{for } \tau \geqq t, \end{cases}$$

we find that $u_1(x, t) = v(x)$ and we arrive at the following conclusion:

When $a_0(t)$ is continuous from the left in $t > 0$,[2] then we find that

$$(10) \qquad u_1(x, t) = a_0(t) * \psi(x, t) \to a_0(t) \quad \text{as } x \to 0.$$

---

[2] When, at t, $a_0(t)$ is continuous from the left, then $a_0(t - \tau)$, considered as a function of τ, at $\tau = 0$ is continuous from the right.

Thus, we have here the typical situation in which the solution of a boundary value problem does not actually assume the boundary values; however, it does have the boundary values as its limits. In the expression

$$u_1(x, t) = \frac{x}{2\sqrt{\pi}} \int_0^t a_0(t-\tau) \frac{e^{-x^2/4\tau}}{\tau^{3/2}} d\tau,$$

the integral diverges for $x = 0$; moreover, even if it had a finite value, then due to the factor x, $u_1(0, t)$ would be zero instead of $a_0(t)$.

In order to find the limit of $u_2(x, t)$ for fixed $x > 0$, when $t \to 0$, we perform the following modification:

$$\begin{aligned} u_2(x, t) &= \frac{1}{2\sqrt{\pi t}} \left\{ \int_0^\infty e^{-(\xi-x)^2/4t} u_0(\xi)\, d\xi - \int_0^\infty e^{-(\xi+x)^2/4t} u_0(\xi)\, d\xi \right\} \\ &= \frac{1}{2\sqrt{\pi t}} \left\{ \int_0^\infty e^{-(\xi-x)^2/4t} u_0(\xi)\, d\xi - \int_{-\infty}^0 e^{-(\xi-x)^2/4t} u_0(-\xi)\, d\xi \right\}. \end{aligned}$$

We define a function $w(\xi)$ by

$$w(\xi) = \begin{cases} u_0(\xi) & \text{when } \xi \geqq 0 \\ -u_0(-\xi) & \text{when } \xi < 0, \end{cases}$$

and we find[3]

$$u_2(x, t) = \frac{1}{2\sqrt{\pi t}} \int_{-\infty}^{+\infty} e^{-(\xi-x)^2/4t} w(\xi)\, d\xi.$$

Setting

$$\xi - x = -\sqrt{\tau} \quad \text{for } -\infty < \xi < x,$$

$$\xi - x = +\sqrt{\tau} \quad \text{for } x \leqq \xi < +\infty,$$

and $1/(4t) = s$, one obtains

$$\begin{aligned} u_2(x, t) &= \sqrt{\frac{s}{\pi}} \left\{ \int_0^\infty e^{-s\tau} w(x-\sqrt{\tau}) \frac{d\tau}{2\sqrt{\tau}} + \int_0^\infty e^{-s\tau} w(x+\sqrt{\tau}) \frac{d\tau}{2\sqrt{\tau}} \right\} d\tau \\ &= \sqrt{\frac{s}{\pi}} \int_0^\infty e^{-s\tau} \frac{w(x-\sqrt{\tau}) + w(x+\sqrt{\tau})}{2\sqrt{\tau}} d\tau. \end{aligned}$$

[3] This integral defines a transformation which transforms the function $w(\xi)$ into the function $u_2(x,t)$ of x, involving the parameter t. It is known as *Weierstrass Transformation* since Weierstrass has shown for the first time that $u_2(x,t) \to w(x)$ as $t \to 0$, at every point where $w(x)$ is continuous. We shall show this property as a consequence of an earlier theorem.

At all points x where w is continuous we have

$$\frac{w(x-\sqrt{\tau})+w(x+\sqrt{\tau})}{2} \to w(x), \quad \text{hence} \quad \frac{w(x-\sqrt{\tau})+w(x+\sqrt{\tau})}{2\sqrt{\tau}} \sim w(x)\,\tau^{-1/2},$$

as $\tau \to 0$. Thus, by Theorem 33.3,

$$\sqrt{\frac{\pi}{s}}\, u_2\left(x, \frac{1}{4s}\right) \sim w(x)\, \frac{\Gamma(\frac{1}{2})}{s^{1/2}} \quad \text{as } s \to \infty \quad (\Gamma(1/2) = \sqrt{\pi});$$

that is

$$u_2(x, t) \to w(x) \quad \text{as } t \to 0.$$

Restricting $x > 0$, we find that

$$u_2(x, t) \to u_0(x) \quad \text{as } t \to 0. \tag{11}$$

Finally, we show that, for fixed t, $u(x, t) \to u_0(\infty)$ when $x \to \infty$. Obviously, we have

$$a_0(t) * \psi(x, t) \to 0, \quad \frac{1}{2\sqrt{\pi t}} \int_0^\infty e^{-(\xi+x)^2/4t}\, u_0(\xi)\, d\xi \to 0 \quad \text{as } x \to \infty.$$

Consequently, one merely need consider

$$u_3(x, t) = \frac{1}{2\sqrt{\pi t}} \int_0^\infty e^{-(\xi-x)^2/4t}\, u_0(\xi)\, d\xi = \frac{1}{2\sqrt{\pi t}} \int_{-x}^\infty e^{-\xi^2/4t}\, u_0(\xi+x)\, d\xi.$$

Using

$$\frac{1}{2\sqrt{\pi t}} \int_{-\infty}^{+\infty} e^{-\xi^2/4t}\, d\xi = 1,$$

we construct

$$u_3(x, t) - u_0(\infty) = \frac{1}{2\sqrt{\pi t}} \int_{-x}^\infty e^{-\xi^2/4t} \left[u_0(\xi+x) - u_0(\infty)\right] d\xi$$

$$- u_0(\infty)\, \frac{1}{2\sqrt{\pi t}} \int_{-\infty}^{-x} e^{-\xi^2/4t}\, d\xi.$$

The function u_0 is continuous, and has a limit when $x \to \infty$; hence, it follows that

$$|u_0(\xi+x) - u_0(\infty)| < M \quad \text{for all } \xi + x \geqq 0.$$

For every given $\varepsilon > 0$, we can find an X so that for all $x > X$ we have:

$$\frac{1}{2\sqrt{\pi t}} \int_{-x}^{-x/2} e^{-\xi^2/4t}\, d\xi < \varepsilon,$$

$$|u_0(\xi + x) - u_0(\infty)| < \varepsilon \text{ with } -\frac{x}{2} \leqq \xi < \infty, \quad \text{that is} \quad \frac{x}{2} \leqq \xi + x < \infty,$$

$$\frac{1}{2\sqrt{\pi t}} \int_{-\infty}^{-x} e^{-\xi^2/4t}\, d\xi < \varepsilon .$$

It follows that, for $x > X$,

$$|u_3(x, t) - u_0(\infty)| \leqq M\varepsilon + \varepsilon \frac{1}{2\sqrt{\pi t}} \int_{-x/2}^{\infty} e^{-\xi^2/4t}\, d\xi + |u_0(\infty)|\,\varepsilon < (M + 1 + |u_0(\infty)|)\,\varepsilon;$$

this shows that $u_3(x, t) \to u_0(\infty)$ as $x \to \infty$.

The conclusions can be summarized in the following theorem.

Theorem 39.1. *Consider in the fundamental region $x > 0$ and $t > 0$, the following boundary value problem: In the interior of the fundamental region, the function $u(x, t)$ satisfies the partial differential equation*

$$\frac{\partial^2 u}{\partial x^2} = \frac{\partial u}{\partial t} .$$

On the boundary $t = 0$, we specify, for $x \geqq 0$, the continuous boundary value function $u_0(x)$, that is

$$\lim_{t \to +0} u(x, t) = u_0(x) \quad \textit{for } x > 0.$$

We assume that $\lim_{x \to \infty} u_0(x) = u_0(\infty)$ exists. At both boundaries, $x = 0$ and $x = \infty$, we specify the boundary function $a_0(t)$ which is continuous for $t > 0$, and the constant value $u_0(\infty)$ respectively:

$$\lim_{x \to +0} u(x, t) = a_0(t), \quad \lim_{x \to \infty} u(x, t) = u_0(\infty) \quad \textit{for } t > 0.$$

It follows that (9)' *is a solution of the problem.*

We now ask if (9) is the only solution. Presuming the existence of two distinct solutions, we conclude that the difference between these solutions is also a solution, which has boundary values zero along all boundaries. This difference must fail to comply with at least one of the hypotheses W_1, W_2, and W_3. Indeed, such "*singular solutions*" do exist as shown, for instance, by

$$u(x, t) = \psi(x, t), \quad u(x, t) = \frac{\partial^n \psi(x, t)}{\partial t^n}, \text{ and } u(x, t) = \begin{cases} 0 & \text{for } 0 < t \leqq t_0 \\ \psi(x, t - t_0) & \text{for } t > t_0 . \end{cases}$$

One can easily demonstrate that each of these functions (including the last one) satisfies the differential equation in the interior of the fundamental region; also, each has *zero boundary values*. They fail to comply with hypothesis W_3, for the corresponding $\mathfrak{L}$-transforms are:

$$U(x, s) = e^{-x\sqrt{s}}, \quad U(x, s) = s^n e^{-x\sqrt{s}}, \quad U(x, s) = e^{-t_0 s} e^{-x\sqrt{s}},$$

with the respective limits, as $x \to 0$:

$$1, \qquad s^n, \qquad e^{-t_0 s};$$

none has the boundary value $\mathfrak{L}\{0\} = 0$.

These singular solutions are included in formula (9), provided we replace functions by distributions. Using the boundary functions $u_0(x) \equiv 0$ and, respectively, for $a_0(t)$:

$$\delta(t),\ \delta^{(n)}(t),\ \delta(t - t_0),$$

then (9) yields, in view of the properties (14.8) and (14.10) of the δ-distribution:

$$\delta(t) * \psi(x, t) = \psi(x, t), \quad \delta^{(n)}(t) * \psi(x, t) = \frac{\partial^n \psi(x, t)}{\partial t^n}, \text{ and}$$

$$\delta(t - t_0) * \psi(x, t) = \begin{cases} 0 & \text{for } 0 < t \leqq t_0 \\ \psi(x, t - t_0) & \text{for } t > t_0. \end{cases}$$

These singular solutions describe temperature distributions due to impulses and multi-impulses at $t = 0$ and $t = t_0$ at the point $x = 0$ in a heat conductor of initial temperature zero. In physics, we can interpret these boundary values as *heat explosions*.

We must conclude that the solution of the boundary value problem is not unique, unless further conditions are stipulated. Necessarily, the proofs of uniqueness presented in some text books must be faulty. Indeed, when scrutinizing these proofs one discovers the use of several not explicitly formulated hypotheses.

The Case of Finite Length

In the image space, we are concerned with the solution of the ordinary differential equation (4) with the boundary conditions (5); this problem has been solved in Chapter 16. As was done there, we split the problem into two parts, to clarify the development.

1. *Thermal Conductor with Vanishing Initial Temperature*

With $u_0(x) \equiv 0$, the Eq. (4) becomes a homogeneous equation; the boundary values may be chosen arbitrarily. The solution is, by (16.2),

(12) $$U(x, s) = A_0(s)\, U_0(x, s) + A_1(s)\, U_1(x, s)$$

with

$$U_0(x, s) = \frac{\sinh(l-x)\sqrt{s}}{\sinh l\sqrt{s}}, \quad U_1(x, s) = \frac{\sinh x\sqrt{s}}{\sinh l\sqrt{s}},$$

where $\sqrt{s}$ represents the principal branch. These functions are analytic in $\Re s > 0$, since the characteristic values of the differential equation (compare p. 88), where the denominator $\sinh l\sqrt{s}$ vanishes, are exclusively located at 0 and on the negative real axis.

The corresponding original functions are determined by means of Eq. (30.10). These are:

$$u_0(x,t) = -\frac{1}{2l^2}\left[\frac{\partial\vartheta_3\left(v, \frac{t}{l^2}\right)}{\partial v}\right]_{v=\frac{x}{2l}} \quad \text{for } 0 < x < 2l,$$

$$u_1(x,t) = -\frac{1}{2l^2}\left[\frac{\partial\vartheta_3\left(v, \frac{t}{l^2}\right)}{\partial v}\right]_{v=\frac{l-x}{2l}} \quad \text{for } -l < x > l.$$

We employ the Convolution Theorem to find the original function of (12); it is:

(13) $$u(x,t) = a_0(t) * u_0(x,t) + a_1(t) * u_1(x,t) \quad \text{for } 0 < x < l.$$

The explicit presentations are (compare (30.8.)):

(14) $$u_0(x,t) = \frac{1}{2\sqrt{\pi}\,t^{3/2}} \sum_{\nu=-\infty}^{+\infty} (2\nu l + x)\, e^{-(2\nu l+x)^2/4t} = \sum_{\nu=-\infty}^{+\infty} \psi(2\nu l + x, t),$$

(15) $$u_1(x,t) = \sum_{\nu=-\infty}^{+\infty} \psi(2\nu l + l - x, t).$$

2. *Thermal Conductor with Vanishing End Temperatures*

With $a_0(t) \equiv a_1(t) \equiv 0$, and arbitrary $u_0(x)$, the Eq. (4) is an inhomogeneous equation with vanishing boundary values. The solution is, by (16.5),

(16) $$U(x,s) = \int_0^l \Gamma(x,\xi;s)\, u_0(\xi)\, d\xi,$$

with

$$\Gamma(x,\xi;s) = \begin{cases} \dfrac{\sinh \xi\sqrt{s}\, \sinh(l-x)\sqrt{s}}{\sqrt{s}\,\sinh l\sqrt{s}} & \text{for } 0 \leqq \xi \leqq x \\[2ex] \dfrac{\sinh x\sqrt{s}\, \sinh(l-\xi)\sqrt{s}}{\sqrt{s}\,\sinh l\sqrt{s}} & \text{for } x \leqq \xi \leqq l. \end{cases}$$

To facilitate the inverse transformation we replace the products in the numerators by differences thus

$$\Gamma(x,\xi;s) = \begin{cases} \dfrac{\cosh(x-\xi-l)\sqrt{s} - \cosh(x+\xi-l)\sqrt{s}}{2\sqrt{s}\,\sinh l\sqrt{s}} & \text{for } 0 \leqq \xi \leqq x \\[2ex] \dfrac{\cosh(\xi-x-l)\sqrt{s} - \cosh(\xi+x-l)\sqrt{s}}{2\sqrt{s}\,\sinh l\sqrt{s}} & \text{for } x \leqq \xi \leqq l. \end{cases}$$

The individual fractions are $\mathfrak{L}$-transforms of certain ϑ_3-functions, for we can demonstrate that

$$\frac{\cosh(y-l)\sqrt{s}}{\sqrt{s}\sinh l\sqrt{s}} \bullet\!\!-\!\!\circ \frac{1}{l}\,\vartheta_3\left(\frac{y}{2l}, \frac{t}{l^2}\right) \quad \text{for } 0 \leqq y \leqq 2l$$

in the manner that was employed for formula (30.10). When inversely transforming the function Γ, we obtain in the first position of the first line $\vartheta_3(x - \xi/2l, t/l^2)$ and of the second line $\vartheta_3(\xi - x/2l, t/l^2)$. The definition (30.9) clearly shows that $\vartheta_3(-v, t) = \vartheta_3(v, t)$; consequently, the two resulting inverse transforms are identical, and we obtain the resulting single expression for the original function of Γ

$$\gamma(x, \xi; t) = \frac{1}{2l}\left[\vartheta_3\left(\frac{x-\xi}{2l}, \frac{t}{l^2}\right) - \vartheta_3\left(\frac{x+\xi}{2l}, \frac{t}{l^2}\right)\right]$$

$$(17) \qquad = \frac{1}{2\sqrt{\pi t}} \sum_{\nu=-\infty}^{+\infty} \left(e^{-(x-\xi+2\nu l)^2/4t} - e^{-(x+\xi+2\nu l)^2/4t}\right)$$

$$= \frac{1}{2} \sum_{\nu=-\infty}^{+\infty} \left(\chi(x-\xi+2\nu l, t) - \chi(x+\xi+2\nu l, t)\right).$$

Upon interchanging the $\mathfrak{L}$-transformation with $\int_0^\infty \ldots d\xi$, one finds the original function of (16); it is:

$$(18) \qquad u(x, t) = \int_0^l \gamma(x, \xi; t)\, u_0(\xi)\, d\xi.$$

Superposition of the solutions (13) and (18) yields the solution of the boundary value problem (1) with (2) and (3).

Similarly as for the case $l = \infty$, one can now verify that the derived result is a solution irrespective of the hypotheses W_1, W_2, and W_3. We can base this verification upon the result of the case $l = \infty$; in the difficult limiting process for $u_0(x, t)$ and $\gamma(x, \xi; t)$ the middle terms ($\nu = 0$) are the only important ones, and these are identical with the single terms occurring in the solution for $l = \infty$. Thus, one finds

Theorem 39.2. *The boundary value problem* (1), (2) *and* (3) *has the solution which is obtained by superposition of* (13) *and* (18), *provided the functions* $a_0(t)$ *and* $a_1(t)$ *are continuous for* $t > 0$, *and the function* $u_0(x)$ *is continuous for* $0 < x < l$.

The solution for finite l too is not unique; for one can find solutions which have boundary values zero on all boundaries as, for instance,

$$u_0(x, t), \quad \frac{\partial^n u_0(x, t)}{\partial t^n}, \quad u(x, t) = \begin{cases} 0 & \text{for } 0 < t \leqq t_0 \\ u_0(x, t) & \text{for } t > t_0. \end{cases}$$

These fail to comply with Hypothesis W_3.

Asymptotic Expansion of the Solution

The presented solutions of the heat conduction problem are quite clear from a purely mathematical point of view; however, they are difficult to evaluate in practical applications, and it is not at all obvious how specific boundary conditions influence the solution in the interior of the fundamental region. For these inquiries one can successfully employ the method of asymptotic expansion as explained in Chapter 37. There is another reason why we indicate this method with emphasis: When engaged with complicated boundary value problems, it may happen that the solution in the image space can be determined with relative ease; however, the corresponding original function cannot explicitly be expressed in terms of classical transcendental functions. When such is the case, then the development of an asymptotic expansion based upon the image function may be *the only resort to gain insight into the sought original function.*

We shall show the method of the asymptotic expansion using the example of heat conduction in the *infinite* x-interval. When the initial temperature vanishes, the temperature at $x = 0$ is described by $a_0(t)$; and the temperature at $x = \infty$ remains at zero, then, by Theorem 39.1, we have the solution

$$u(x,t) = a_0(t) * \psi(x,t).$$

In order to develop an asymptotic expansion, we must be given $a_0(t)$ explicitly. We select specifically for the latter the complex oscillation $a_0(t) = e^{i\omega t}$, and we shall designate the associated specific solution by $u_\omega(x,t)$.

We start with the corresponding image function:

$$U_\omega(x,s) = A_0(s)\, e^{-x\sqrt{s}} = \frac{1}{s - i\omega}\, e^{-x\sqrt{s}}$$

and we first represent $u_\omega(x,t)$ by the complex inversion integral

$$u_\omega(x,t) = \frac{1}{2\pi i} \int_{a-i\infty}^{a+i\infty} e^{ts} \frac{1}{s - i\omega}\, e^{-x\sqrt{s}}\, ds. \tag{19}$$

The integrand has a pole at $s = i\omega$ and a branch point at $s = 0$. These two singular points have identical real components, hence both must be considered. Theorem 37.1 may be invoked, provided we can replace the straight line path of integration in (19) by an angular contour with the two centres 0 and $i\omega$. $\sqrt{s}$ means principal branch; hence, for $x > 0$, $e^{-x\sqrt{s}}$ is bounded in the entire sheet $-\pi < \operatorname{arc} s \leqq +\pi$, since $\Re \sqrt{s} \geqq 0$ in this sheet. The factor $1/(s - i\omega)$ tends, uniformly on the entire plane, towards zero when $s \to \infty$; consequently, this is true for $U_\omega(x,s)$. It follows that (19) may be evaluated along an angular contour having the half-angle of opening ψ, with $\pi/2 < \psi \leqq \pi$.

We must expand $U_\omega(x,s)$ in a series of powers of $(s - i\omega)$ and of s respectively. The factor $e^{-x\sqrt{s}}$ is holomorphic at $s = i\omega$; hence, it may be expanded thus

$$e^{-x\sqrt{s}} = \sum_{n=0}^{\infty} c_n (s - i\omega)^n \quad \left(c_0 = e^{-x\sqrt{i\omega}}\right).$$

Consequently, we have

$$U_\omega(x, s) = \sum_{n=0}^{\infty} c_n (s - i\omega)^{n-1}.$$

All terms with non-negative integer exponents vanish in the process of inverse transformation, and we are left with

$$c_0 e^{i\omega t} = e^{-x\sqrt{i\omega} + i\omega t}.$$

At the point $s = 0$, we have

$$\frac{1}{s - i\omega} = -\sum_{n=0}^{\infty} \frac{s^n}{(i\omega)^{n+1}},$$

and

$$e^{-x\sqrt{s}} = \sum_{n=0}^{\infty} \frac{(-x)^n}{n!} s^{n/2};$$

hence,

$$U_\omega(x, s) = -\frac{1}{i\omega} + \frac{x}{i\omega} s^{1/2} - \left(\frac{1}{(i\omega)^2} + \frac{x^2}{2i\omega}\right) s + \left(\frac{x^3}{3!\,i\omega} + \frac{x}{(i\omega)^2}\right) s^{3/2} - \cdots.$$

This produces, upon inverse transformation, the contribution

$$\frac{x}{i\omega} \frac{1}{\Gamma(-1/2)} \frac{1}{t^{3/2}} + \left(\frac{x^3}{3!\,i\omega} + \frac{x}{(i\omega)^2}\right) \frac{1}{\Gamma(-3/2)} \frac{1}{t^{5/2}} + \cdots.$$

Involving the formula

$$\frac{1}{\Gamma(\frac{1}{2} - \nu)} = \frac{(-1)^\nu (2\nu)!}{4^\nu \nu! \sqrt{\pi}},$$

we find altogether for the asymptotic expansion

$$u_\omega(x, t) \approx e^{-x\sqrt{i\omega} + i\omega t} + \frac{x i}{2\sqrt{\pi}\,\omega} \frac{1}{t^{3/2}} - \left(\frac{x^3 i}{3!\,\omega} + \frac{x}{\omega^2}\right) \frac{3}{4\sqrt{\pi}} \frac{1}{t^{5/2}} + \cdots \tag{20}$$

as $t \to \infty$. All terms after the first one tend towards zero when $t \to \infty$, hence, for sufficiently large values of t, the first term alone need be considered. We conclude that the first term represents the *stationary state* $\tilde{u}_\omega(x, t)$:

$$\tilde{u}_\omega(x, t) = e^{-x\sqrt{i\omega} + i\omega t}. \tag{21}$$

In complete analogy to the concepts developed for the solution of ordinary differential equations. (p. 96) one could call this function the *frequency response*; the modulus of the factor of $e^{i\omega t}$ tells the amplitude of $\tilde{u}_\omega$, its arc indicates the phase shift of $\tilde{u}_\omega$, relative to the boundary excitation $e^{i\omega t}$.

The frequency response can immediately be read from the image function of the general solution

$$U(x, s) = A_0(s)\, e^{-x\sqrt{s}}$$

Designating the factor of $A_0(s)$ by $G(s)$,

$$G(s) = e^{-x\sqrt{s}},$$

we can write the frequency response (or stationary state) in the form

$$\tilde{u}_\omega(x, t) = G(i\,\omega)\, e^{i\omega t}.$$

We observe that the latter *may, bypassing the process of inverse transformation, directly be obtained from the solution in the image space.* All this is patently analogous to the circumstances in Chapter 17.2.

Notice that $\sqrt{i} = (1 + i)/\sqrt{2}$, hence

$$G(i\,\omega) = e^{-x\sqrt{\omega/2}\,(1+i)}$$

and

$$\tilde{u}_\omega(x, t) = e^{-x\sqrt{\omega/2}}\, e^{i\omega(t - x/\sqrt{2\omega})}.$$

Both the amplitude $e^{-x\sqrt{\omega/2}}$ and the phase shift $x/\sqrt{2\omega}$ tend, for any fixed location $x > 0$, monotonically towards zero when ω grows.

We separate the *real part* and the *imaginary part* of $u_\omega(x, t)$, and in this manner we derive expansions for the states corresponding to the respective boundary excitations $a_0(t) = \cos\omega t$, and $a_0(t) = \sin\omega t$; these are:

$$e^{-x\sqrt{\omega/2}}\cos\left(\omega t - x\sqrt{\omega/2}\right) - \frac{3}{4\sqrt{\pi}}\,\frac{x}{\omega^2}\,\frac{1}{t^{5/2}} + \cdots,$$

$$e^{-x\sqrt{\omega/2}}\sin\left(\omega t - x\sqrt{\omega/2}\right) + \frac{1}{2\sqrt{\pi}}\,\frac{x}{\omega}\,\frac{1}{t^{3/2}} - \frac{1}{8\sqrt{\pi}}\,\frac{x^3}{\omega}\,\frac{1}{t^{5/2}} + \cdots.$$

These asymptotic expansions permit far easier numerical evaluation when compared with the evaluation of $u_\omega(x, t)$ by means of a convolution integral; moreover, they provide more intimate insight into the characteristic behaviour of the solution $u_\omega(x, t)$.

The equation of heat conduction is the normal form of the second order partial differential equation of the *parabolic type*; its solutions are represented by integrals. Next, we shall investigate a specifie equation of *hyperbolic type*, the solution of which has a fundamentally different form; therefore, a new element also appears in the method of the $\mathfrak{L}$-transformation. The normal form of the equation of the hyperbolic type is the *wave* equation:

$$\frac{\partial^2 u}{\partial x^2} = a^2\,\frac{\partial^2 u}{\partial t^2}.$$

We shall consider a more general form of the equation.

2. The Telegraph Equation

One encounters this equation in the theory of electric transmission lines and in other branches of the sciences where media capable of oscillation are investigated. Consider an electric double line which extends between $x = 0$ and $x = l$, and which has, per unit length of line, the following invariant electric characteristics:

Resistance R, Inductance L, Capacitance C, Leakance G.

We use t to designate the time variable and find the differential equation

$$\frac{\partial^2 u}{\partial x^2} = L\,C\,\frac{\partial^2 u}{\partial t^2} + (R\,C + L\,G)\,\frac{\partial u}{\partial t} + R\,G\,u$$

for both the current in the line and the potential difference between the lines. Upon introducing $L\,C = a,\quad R\,C + L\,G = b,\quad R\,G = c,$

the partial differential equation becomes

$$\frac{\partial^2 u}{\partial x^2} - a\,\frac{\partial^2 u}{\partial t^2} - b\,\frac{\partial u}{\partial t} - c\,u = 0.$$

From the physical point of view, all constants, a, b and c, are inherently positive; for the subsequent mathematical investigations we shall merely require that $a > 0$.

Starting with a double line which is initially at rest in which neither current nor voltage is recorded at $t = 0$, we have the initial conditions:[4]

$$u(x, 0+) = 0, \quad u_t(x, 0+) = 0.$$

Also, the voltage (or the current) at the end points of the line is presumed known; that is, we know the boundary conditions:

$$u(0+, t) = a_0(t), \quad u(l-, t) = a_1(t).$$

Employing hypotheses analogous to those introduced in Chapter 39.1, we have, for the mixed boundary and initial value problem, the corresponding *boundary value problem in the image space*:

$$\frac{d^2 U}{dx^2} - (a\,s^2 + b\,s + c)\,U = 0,$$

$$U(0+, s) = A_0(s), \quad U(l-, s) = A_1(s).$$

We have solved this problem on p. 86; here, we have specifically $f(x) \equiv 0$. Observe that here, because of a, b, and $c > 0$, the expression $as^2 + bs + c$ cannot be negative real-valued for $\Re s > 0$; hence, the characteristic values cannot occur (compare p. 88). We shall restrict the subsequent considerations to the special case $l = \infty$. The practical consequence of the supposition $l = \infty$ is that reflections originated at the right boundary need not be considered. In the terminology of

[4] We write briefly $u(x, 0+)$ for $\lim_{t \to +0} u(x,t)$, and we shall write $u(0+,t)$ and $u(l-,t)$ instead of $\lim_{x \to +0} u(x,t)$ and $\lim_{x \to l-0} u(x,t)$ respectively. u_t indicates $\partial u/\partial t$.

Theorem 16.1, we have $f(\infty) = 0$; hence, zero is the only admissible value for $U(\infty, s)$ and, consequently, also for $u(\infty, t)$, and we have

$$U(x, s) = A_0(s)\, e^{-x\sqrt{a s^2 + b s + c}}. \tag{22}$$

Here, we encounter a remarkable situation which *precludes inverse transformation by means of the Convolution Theorem*: the exponential function cannot be a $\mathfrak{L}$-transform. This fact already becomes obvious for the special case $b = c = 0$, that is for the wave equation (concerning e^{-as}, compare p. 25). In this case, we have

$$U(x, s) = A_0(s)\, e^{-x\sqrt{a}\, s},$$

which indicates that here the Translation Theorem 7.2 is to be applied. In this manner, we find

$$u(x, t) = \begin{cases} 0 & \text{for } t < x\sqrt{a} \\ a_0(t - x\sqrt{a}) & \text{for } t \geqq x\sqrt{a}. \end{cases}$$

Strictly the same process can be employed to find the inverse transform in the case that $a s^2 + b s + c$ is the exact square of a linear function. This is true when and only when the discriminant

$$d = a c - \left(\frac{b}{2}\right)^2$$

vanishes; in this special case, we have

$$a s^2 + b s + c = \left(\sqrt{a}\, s + \frac{b}{2\sqrt{a}}\right)^2,$$

hence

$$U(x, s) = A_0(s)\, e^{-x\sqrt{a}(s + (b/2a))}$$

and

$$u(x, t) = e^{-(b/2\sqrt{a})x}\, a_0(t - x\sqrt{a}) \quad \text{with } a_0(t) = 0 \text{ when } t < 0. \tag{23}$$

Using the electric characteristics of the line, we find that

$$d = L C R G - \frac{1}{4}(R C + L G)^2 = -\frac{1}{4}(R C - L G)^2.$$

Consequently, the condition $d = 0$ is equivalent with

$$R C = L G \quad \text{or} \quad \frac{R}{L} = \frac{G}{C}. \tag{24}$$

A line having characteristics which satisfy the condition (24) is said to be *distorsionless*, a designation which will be properly understood only after the solution of the general case has been produced. The "signal" $a_0(t_0)$ is imposed at the location

$x = 0$ at the time t_0; it arrives at some point $x > 0$ at the time t; location x and arrival time t are interrelated by

(25) $$t - x\sqrt{a} = t_0, \text{ that is } t = t_0 + x\sqrt{a}\,.$$

The signal requires a time interval $x\sqrt{a}$ to arrive at x; that is, the *speed of the signal* is $v = 1/\sqrt{a}$. The travelling signal does not retain its original strength, it is *attenuated* by the factor $\exp(-(b/2\sqrt{a})x)$. However, the signal is "undistorted" insofar as no other signals are superimposed. In the xt-plane, the straight line (25) passes through $(0, t_0)$ having the slope $\sqrt{a}$. This line is the "world line" of the signal or boundary excitation $a_0(t_0)$ in the space-time-world, as shown in Fig. 50.

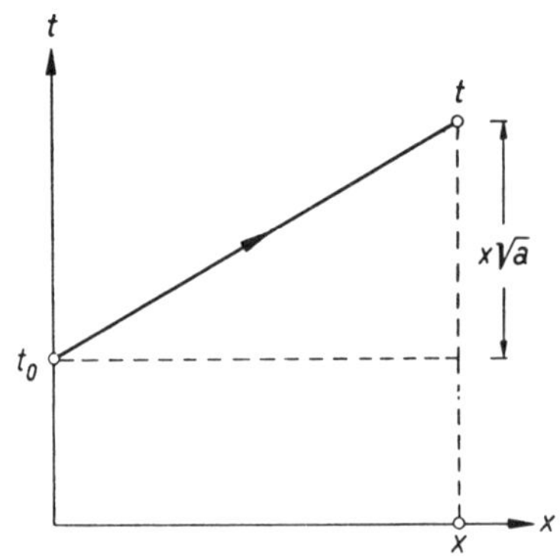

Figure 50

In the case that $d \neq 0$, we sucessfully employ a well known formula of the Bessel function:

$$J_1(z) = \sum_{n=0}^{\infty} \frac{(-1)^n}{n!\,(n+1)!} \left(\frac{z}{2}\right)^{2n+1},$$

for we have

$$e^{-x\sqrt{as^2+bs+c}} = e^{-(b/2\sqrt{a})x}\, e^{-x\sqrt{a}s} - x\sqrt{\frac{d}{a}} \int_{x\sqrt{a}}^{\infty} e^{-st}\, e^{-(b/2a)t} \frac{J_1\left(\frac{\sqrt{d}}{a}\sqrt{t^2 - a x^2}\right)}{\sqrt{t^2 - a x^2}}\, dt.$$

Defining the function $v(x, t)$ thus:

$$v(x, t) = \begin{cases} 0 & \text{for } 0 \leqq t \leqq x\sqrt{a} \\ -x\sqrt{\dfrac{d}{a}}\, e^{-(b/2a)t} \dfrac{J_1\left(\frac{\sqrt{d}}{a}\sqrt{t^2 - a x^2}\right)}{\sqrt{t^2 - a x^2}} & \text{for } t > x\sqrt{a}, \end{cases}$$

we can rewrite the above formula as follows:

$$e^{-x\sqrt{as^2+bs+c}} = e^{-(b/2\sqrt{a})x}\, e^{-x\sqrt{a}s} - \mathfrak{L}\{v(x, t)\}.$$

Upon multiplying the last equation by $A_0(s)$, we obtain on the right hand side two terms: The first term is exactly the one which was encountered in connection with the distortionless line; it may be inversely transformed with the aid of the Translation Theorem. The second term is a product of two $\mathfrak{L}$-transforms; its original function is a convolution. Accordingly, we find:

$$(26)\qquad u(x,t) = \begin{cases} 0 & \text{for } 0 \leq t < x\sqrt{a}, \\ e^{-(b/2\sqrt{a})x}\, a_0(t - x\sqrt{a}) - \int\limits_{x\sqrt{a}}^{t} a_0(t-\tau)\, v(x,\tau)\, d\tau & \text{for } t > x\sqrt{a}. \end{cases}$$

In this case, at the location x and at the time t, there arrives not merely the boundary excitation $a_0(t_0)$ with $t_0 = t - x\sqrt{a}$; instead, it is superimposed with a "distortion" which stems from all earlier excitations $(0 \leq t - \tau < t - x\sqrt{a})$ and which represents the remnants of these.[5]

Asymptotic Expansion of the Solution

The expression (26), when explicitly written, is very complicated and difficult to evaluate numerically. Moreover, it is very hard to estimate the order of magnitude of the second term. Thus, we judge here the development of a clearer representation of the solution by means of the method of asymptotic expansion far more desirable than for the equation of heat conduction.

As boundary excitation we employ, once again, the complex oscillation $a_0(t) = e^{i\omega t}$; this corresponds to the physical application of a sinusoidal voltage which, for $\omega = 0$, degenerates into a constant voltage. Accordingly, we start with the equation:

$$U_\omega(x,s) = \frac{1}{s - i\omega}\, e^{-x\sqrt{as^2+bs+c}}$$

and we represent $u_\omega(x,t)$ by means of the complex inversion integral:

$$u_\omega(x,t) = \frac{1}{2\pi i}\int\limits_{a-i\infty}^{a+i\infty} e^{ts - x\sqrt{as^2+bs+c}}\, \frac{1}{s-i\omega}\, ds \quad (a > 0).$$

We evaluate:

$$as^2 + bs + c = LCs^2 + (RC + LG)s + RG = LC\left[s^2 + \left(\frac{R}{L} + \frac{G}{C}\right)s + \frac{RG}{LC}\right]$$

$$(27)\qquad = LC\left(s + \frac{R}{L}\right)\left(s + \frac{G}{C}\right) = LC(s - \alpha_1)(s - \alpha_2),$$

where α_1 designates the algebraically larger and α_2 the smaller of the two numbers $(-R/L)$ and $(-G/C)$. The situation $\alpha_1 = \alpha_2$ can be dismissed here, since for this case condition (24) is satisfied and the solution (23) of the distortionless line is so simple that an asymptotic expansion is not needed.

---

[5] The expression $|RC - LG|/2 = \sqrt{-d}$ may be used as measure of the distortion (see (24)).

The function $U_\omega(x, s)$ has a pole at $s = i\omega$ and two branch points at the negative real values $s = \alpha_1$ and $s = \alpha_2$. We encounter here the special situation that first we must apply to the singular point with the largest real component the method for single-valued singularities developed and introduced in Chapter 35; we thus move the path of integration past the singular point whilst properly accounting for the corresponding residue. Then, at the many-valued singularity corresponding to the point with the largest real part, α_1, we must employ the method of Chapter 37.

Selecting β so that $\alpha_1 < \beta < 0$, $U_\omega(x, s)$ tends, uniformly in the strip $\beta \leqq \Re s \leqq a$ towards zero when $s \to \infty$, since $\sqrt{as^2 + bs + c}$ behaves, for large $|s|$, like $s\sqrt{a}$ and hence $\exp(-x\sqrt{as^2 + bs + c})$ remains bounded, whilst the factor $1/(s - i\omega)$ tends uniformly towards zero. Thus, we can shift the line of integration to the abscissa β and separate the residue at $i\omega$ from $u_\omega(x, t)$:[6]

$$(28) \qquad u_\omega(x, t) = e^{-x\sqrt{a(i\omega)^2 + bi\omega + c}}\, e^{i\omega t} + \frac{1}{2\pi i} \int_{\beta - i\infty}^{\beta + i\infty} e^{ts - x\sqrt{as^2 + bs + c}} \frac{1}{s - i\omega}\, ds\,.$$

Next, we investigate whether the straight line path of integration can be replaced by an angular contour $\mathfrak{W}$ centred at α_1. In all previous applications we argued the legality of the change relying on the fact that the function tends uniformly between the two paths of integration towards zero when s tends towards ∞. For the present case, this condition is not satisfied since, for large values of s, $U_\omega(x, s)$ behaves like $e^{-x\sqrt{a}\,s}/s$ and, for $x > 0$, U_ω tends towards ∞, when $\Re s$ tends towards $-\infty$. Nevertheless, the change of the path of integration can here be defended; for this purpose we separate the factor $\exp(-x\sqrt{as^2 + bs + c})$ from $U_\omega(x, s)$ and adjoin it with e^{ts}, as already indicated in the presentation (28). This function behaves, for large s, like

$$e^{ts - x\sqrt{a}\,s} = e^{(t - x\sqrt{a})\,s}\,;$$

hence, we are essentially concerned with the integral

$$\int e^{(t - x\sqrt{a})\,s} \frac{1}{s - i\omega}\, ds\,;$$

for this we may, with

$$(29) \qquad t - x\sqrt{a} > 0\,,$$

replace the straight line path of integration by $\mathfrak{W}$. The occurrence of the condition (29) is not at all surprising, for we know from (26) that $u_\omega(x, t)$ is represented by distinct analytic functions for $t - x\sqrt{a} > 0$ and for $t - x\sqrt{a} < 0$ respectively, by zero in the latter region. The $\mathfrak{W}$-integral exclusively represents functions which

[6] When using Theorem 35.1 observe that the Hypothesis 2 of the Theorem, although satisfied in our case (compare p. 237), is not really needed here, for it was then required only to develop an estimate of the "remainder integral" which is not needed for the present application.

are analytic in an angular region of the t-plane (compare Chapter 36); hence it is here only useful to represent $u_\omega(x, t)$ along the ray $x\sqrt{a} < t < \infty$ where $u_\omega(x, t)$ is analytic. The occurrence of restrictions of this type has been discussed in detail near the end of Chapter 36.

Prior to the application of Theorem 37.1 to the integral along $\mathfrak{W}$, we must expand $U_\omega(x, s)$ in a series of powers of $s - \alpha_1$. The function $U_\omega(x, s)$ has at α_1 a branch point of the type $\sqrt{s - \alpha_1}$, and it is finite at α_1; hence, we conclude that the expansion is of the form $\sum_{\nu=0}^{\infty} c_\nu (s - \alpha_1)^{\nu/2}$. Using the substitution $s - \alpha_1 = z^2$, we find the expression

$$U_\omega(x, z^2 + \alpha_1) = \sum_{\nu=0}^{\infty} c_\nu z^\nu .$$

The technique used to determine coefficients of a Taylor series may be employed here to determine the coefficients c_ν. In this manner, we find

$$c_0 = U_\omega(x, \alpha_1) = \frac{1}{\alpha_1 - i\omega} .$$

For the determination of c_1, we write $U_\omega(x, s)$ thus

$$U_\omega(x, s) = \frac{1}{s - i\omega} e^{-x\sqrt{a}\sqrt{(s-\alpha_1)(s-\alpha_2)}}$$

and we form

$$U_\omega(x, z^2 + \alpha_1) = \frac{1}{z^2 + \alpha_1 - i\omega} e^{-x\sqrt{a}z\sqrt{z^2+\alpha_1-\alpha_2}} ,$$

$$\frac{d\,U_\omega(x, z^2 + \alpha_1)}{dz}$$

$$= e^{-x\sqrt{a}z\sqrt{z^2+\alpha_1-\alpha_2}} \frac{-x\sqrt{a}\left[\sqrt{z^2 + \alpha_1 - \alpha_2} + \dfrac{z^2}{\sqrt{z^2 + \alpha_1 - \alpha_2}}\right](z^2 + \alpha_1 - i\omega) - 2z}{(z^2 + \alpha_1 - i\omega)^2} .$$

It follows that

$$c_1 = \left.\frac{dU_\omega(x, z^2 + \alpha_1)}{dz}\right|_{z=0} = -\frac{x\sqrt{a}\sqrt{\alpha_1 - \alpha_2}}{\alpha_1 - i\omega} .$$

The subsequent coefficients could be determined by the same technique; however, the first two, c_0 and c_1, suffice for the examination of the behaviour of $u_\omega(x, t)$. From the expression

$$U_\omega(x, s) = \frac{1}{\alpha_1 - i\omega} - \frac{x\sqrt{a}\sqrt{\alpha_1 - \alpha_2}}{\alpha_1 - i\omega}(s - \alpha_1)^{1/2} + \cdots$$

we derive, by Theorem 37.1, the asymptotic expansion of the second term of (28):

$$e^{\alpha_1 t}\left(-\frac{x\sqrt{a}\sqrt{\alpha_1 - \alpha_2}}{\alpha_1 - i\omega}\,\frac{1}{\Gamma(-\frac{1}{2})}\,\frac{1}{t^{3/2}} + \cdots\right) = e^{\alpha_1 t}\left(\frac{x\sqrt{a}\sqrt{\alpha_1 - \alpha_2}}{2\sqrt{\pi}(\alpha_1 - i\omega)}\,\frac{1}{t^{3/2}} + \cdots\right).$$

All terms corresponding to integer powers of $(s - \alpha_1)$ vanish in this process. Thus, for $u_\omega(x, t)$ we obtain the complete *asymptotic expansion*

$$u_\omega(x, t) \approx e^{-x\sqrt{a}\sqrt{(i\omega-\alpha_1)(i\omega-\alpha_2)}}\, e^{i\omega t} + e^{\alpha_1 t}\left(\frac{x\sqrt{a}\sqrt{\alpha_1-\alpha_2}}{2\sqrt{\pi}(\alpha_1 - i\omega)}\,\frac{1}{t^{3/2}} + \cdots\right), \tag{30}$$

as $t \to \infty$. The second term tends strongly towards zero, for growing values of t, particularly so since $\alpha_1 < 0$. Consequently, the first term alone describes the stationary state. At any location x, we encounter an oscillation having the frequency of the boundary excitation; the amplitude and the phase of the oscillation at x is determined by the modulus and the arc of the factor of the stationary state

$$e^{-x\sqrt{a}\sqrt{(i\omega-\alpha_1)(i\omega-\alpha_2)}} = e^{-x\sqrt{LC}\sqrt{(i\omega+(R/L))(i\omega+(G/C))}} \tag{31}$$

When designating the factor of $A_0(s)$ of the general solution (22) in the image space by $G(s)$,

$$G(s) = e^{-x\sqrt{as^2+bs+c}} = e^{-x\sqrt{a}\sqrt{(s-\alpha_1)(s-\alpha_2)}}, \tag{32}$$

then the *stationary state* $\tilde{u}_\omega(x, t)$ is given by[7]

$$\tilde{u}_\omega(x, t) = G(i\omega)\, e^{i\omega t}. \tag{33}$$

The stationary state may also be called *"frequency response"*. As before on p. 292, we call again attention to the complete analogy to formula (17.12). The stationary state can immediately be represented by means of the expression $G(s)$ which was generated in the image space, without inverse transformation into the original space.

For the *distortionless line* with $d = 0$, we have $\alpha_1 = \alpha_2 = -b/2a = -R/L = -G/C$, hence $\sqrt{(s - \alpha_1)(s - \alpha_2)} = (s - \alpha_1)$; α_1 is not a branch point. Therefore, the second term of (30) vanishes and $u_\omega(x, t)$ is completely determined by the first term of (30), as shown by a comparison with (23).

A closer investigation of the frequency response produces several conclusions of physical interest. When setting

$$G(i\omega) = \varrho(\omega)\, e^{-i\omega\psi(\omega)}, \text{ which produces } \tilde{u}_\omega(x, t) = \varrho(\omega)\, e^{i\omega(t-\psi(\omega))},$$

then $\varrho(\omega)$ and $\psi(\omega)$ are respectively amplitude and phase shift of $\tilde{u}_\omega(x, t)$ referred to the boundary excitation $e^{i\omega t}$. With

$$i\omega - \alpha_1 = r_1 e^{i\varphi_1}, \quad i\omega - \alpha_2 = r_2 e^{i\varphi_2}$$

[7] As remarked on p. 96, the representation (17.11) holds for an image function of the form $G(s)/(s - i\omega)$ not only when $G(s)$ is a rational function but actually for every $\mathfrak{L}$-transform which converges on the imaginary line; this formula cannot be employed here since the function $G(s)$ of (32) is not a $\mathfrak{L}$-transform.

(see Fig. 51), we find

$$G(i\omega) = \exp(-x\sqrt{a}\,(r_1 r_2)^{1/2} e^{i(\varphi_1+\varphi_2)/2}),$$

hence

$$\varrho(\omega) = e^{-x\sqrt{a}(r_1 r_2)^{1/2}\cos(\varphi_1+\varphi_2)/2},$$

(34)

$$\psi(\omega) = x\sqrt{a}\,\frac{(r_1 r_2)^{1/2}}{\omega}\sin\frac{\varphi_1+\varphi_2}{2}.$$

We wish to determine the behaviour of these quantities for large and for small frequencies ω respectively; for this purpose we study two limiting cases, $\omega = 0$ and $\omega \to \infty$.

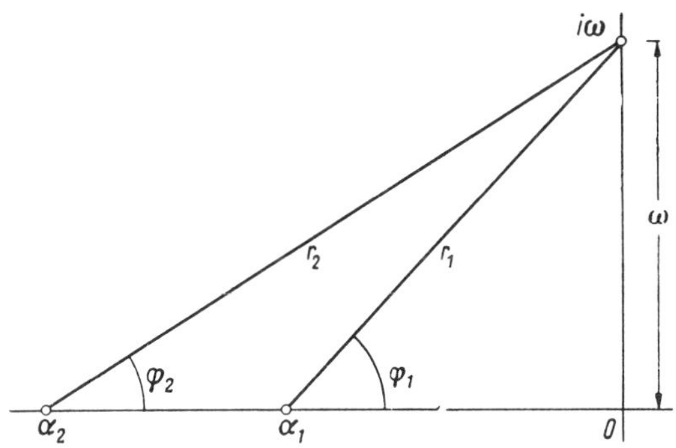

Figure 51

The Behaviour of the Amplitude

1. $\omega = 0$. The boundary excitation $a_0(t)$ becomes the unit step function $u(t)$ (constant voltage). In this case, we find $\varphi_1 = \varphi_2 = 0$, $r_1 = -\alpha_1$, $r_2 = -\alpha_2$, and

$$\varrho(0) = e^{-x\sqrt{a}\sqrt{a_1 a_2}}.$$

2. $\omega \to \infty$. We use

$$\cos^2\frac{\varphi_1+\varphi_2}{2} = \frac{1+\cos(\varphi_1+\varphi_2)}{2}$$

and

$$\cos(\varphi_1+\varphi_2) = \cos\varphi_1\cos\varphi_2 - \sin\varphi_1\sin\varphi_2 = \frac{-a_1}{r_1}\,\frac{-a_2}{r_2} - \frac{\omega}{r_1}\,\frac{\omega}{r_2},$$

hence

(35)
$$r_1 r_2 \cos^2\frac{\varphi_1+\varphi_2}{2} = \frac{1}{2}(r_1 r_2 + a_1 a_2 - \omega^2).$$

Also, we have

(36)
$$r_1 r_2 = [(\omega^2+a_1^2)(\omega^2+a_2^2)]^{1/2} = \omega^2\left(1+\frac{a_1^2}{\omega^2}\right)^{1/2}\left(1+\frac{a_2^2}{\omega^2}\right)^{1/2},$$

hence, for $\omega > |\alpha_1| > |\alpha_2|$,

$$r_1 r_2 = \omega^2\left(1+\frac{1}{2}\frac{a_1^2}{\omega^2}+\binom{1/2}{2}\frac{a_1^4}{\omega^4}+\cdots\right)\left(1+\frac{1}{2}\frac{a_2^2}{\omega^2}+\binom{1/2}{2}\frac{a_2^4}{\omega^4}+\cdots\right)$$

$$(37)\qquad = \omega^2\left[1 + \frac{a_1^2 + a_2^2}{2}\,\frac{1}{\omega^2} + O\left(\frac{1}{\omega^4}\right)\right] = \omega^2 + \frac{a_1^2 + a_2^2}{2} + O\left(\frac{1}{\omega^2}\right) \quad \text{as } \omega \to \infty .$$

Whence, one finds

$$r_1 r_2 \cos^2 \frac{\varphi_1 + \varphi_2}{2} = \frac{1}{2}\left[\frac{a_1^2 + a_2^2}{2} + a_1 a_2 + O\left(\frac{1}{\omega^2}\right)\right] \to \frac{(a_1 + a_2)^2}{4} ;$$

that is,

$$\lim_{\omega \to \infty} \varrho(\omega) = e^{-x\sqrt{a}(|a_1| + |a_2|)/2} .$$

The geometric mean of two distinct numbers $|\alpha_1|$ and $|\alpha_2|$ is always smaller than the arithmetic mean:

$$\sqrt{a_1 a_2} < \frac{|a_1| + |a_2|}{2} ;$$

hence,

$$\varrho(0) > \varrho(\infty) .$$

Forming the derivative with respect to ω of the function

$$2 r_1 r_2 \cos^2 \frac{\varphi_1 + \varphi_2}{2} = \left[\left(\omega^2 + a_1^2\right)\left(\omega^2 + a_2^2\right)\right]^{1/2} + a_1 a_2 - \omega^2 ,$$

we observe, for $\alpha_1 \neq \alpha_2$, that $\omega = 0$ is the only point where the derivative vanishes; the only extremum of the above function is at $\omega = 0$. Therefore, the function is monotonic in $(0, \infty)$. Thus, we conclude:

At every location $x > 0$, the amplitude $\varrho(\omega)$ decreases monotonically with increasing frequency; oscillations with higher frequency are more strongly attenuated than those with lower frequency. We have

$$(38)\qquad \begin{aligned} \varrho(0) &= e^{-x\sqrt{a}\sqrt{a_1 a_2}} = e^{-x\sqrt{RG}} , \\ \varrho(\infty) &= e^{-x\sqrt{a}(|a_1| + |a_2|)/2} = e^{-x(RC + LG)/2\sqrt{LC}} \end{aligned}$$

The Behaviour of the Phase Shift

We use

$$\sin^2 \frac{\varphi_1 + \varphi_2}{2} = \frac{1 - \cos(\varphi_1 + \varphi_2)}{2} ,$$

hence (compare (35))

$$(39)\qquad \frac{r_1 r_2}{\omega^2} \sin^2 \frac{\varphi_1 + \varphi_2}{2} = \frac{1}{2\,\omega^2}\left(r_1 r_2 - a_1 a_2 + \omega^2\right) .$$

Also (compare (36))

$$r_1 r_2 = a_1 a_2 \left(1 + \frac{\omega^2}{a_1^2}\right)^{1/2} \left(1 + \frac{\omega^2}{a_2^2}\right)^{1/2} ,$$

and for $\omega < |\alpha_1| < |\alpha_2|$,

$$r_1 r_2 = \alpha_1 \alpha_2 \left(1 + \frac{1}{2}\frac{\omega^2}{\alpha_1^2} + \binom{1/2}{2}\frac{\omega^4}{\alpha_1^4} + \cdots\right)\left(1 + \frac{1}{2}\frac{\omega^2}{\alpha_2^2} + \binom{1/2}{2}\frac{\omega^4}{\alpha_1^4} + \cdots\right)$$

$$= \alpha_1 \alpha_2 \left[1 + \frac{1}{2}\left(\frac{1}{\alpha_1^2} + \frac{1}{\alpha_2^2}\right)\omega^2 + O(\omega^4)\right].$$

Consequently,

$$\frac{r_1 r_2}{\omega^2}\sin^2\frac{\varphi_1+\varphi_2}{2} = \frac{1}{2\omega^2}\left[\frac{\alpha_1\alpha_2}{2}\left(\frac{1}{\alpha_1^2} + \frac{1}{\alpha_2^2}\right)\omega^2 + \omega^2 + O(\omega^4)\right]$$

$$= \frac{1}{4}\left(\frac{\alpha_2}{\alpha_1} + \frac{\alpha_1}{\alpha_2}\right) + \frac{1}{2} + O(\omega^2) \to \frac{1}{4}\left(\sqrt{\frac{\alpha_2}{\alpha_1}} + \sqrt{\frac{\alpha_1}{\alpha_2}}\right)^2$$

as $\omega \to 0$, hence

$$\lim_{\omega\to 0}\psi(\omega) = x\sqrt{a}\,\frac{1}{2}\left(\sqrt{\frac{\alpha_2}{\alpha_1}} + \sqrt{\frac{\alpha_1}{\alpha_2}}\right).$$

Moreover, (39) and (37) imply that

$$\frac{r_1 r_2}{\omega^2}\sin^2\frac{\varphi_1+\varphi_2}{2} = \frac{1}{2\omega^2}\left[2\omega^2 + \frac{\alpha_1^2+\alpha_2^2}{2} - \alpha_1\alpha_2 + O\left(\frac{1}{\omega^2}\right)\right]$$

$$= 1 + \frac{(\alpha_1-\alpha_2)^2}{4\omega^2} + O\left(\frac{1}{\omega^4}\right) \to 1 \quad \text{as } \omega \to \infty,$$

hence

$$\lim_{\omega\to\infty}\psi(\omega) = x\sqrt{a}.$$

For every positive $z \neq 1$, it is true that $(z + 1/z) > 2$; thus,

$$\frac{1}{2}\left(\sqrt{\frac{\alpha_2}{\alpha_1}} + \sqrt{\frac{\alpha_1}{\alpha_2}}\right) > 1,$$

that is

$$\psi(0^+) > \psi(\infty).$$

One can easily verify that the derivative of $\psi^2(\omega)$ vanishes only for $\omega = 0$ and for $\omega = \infty$, the only locations of extrema. Therefore, $\psi(\omega)$ is monotonic in $(0, \infty)$, and we conclude:

At every location $x > 0$, the phase shift $\psi(\omega)$ decreases monotonically with increasing frequency; oscillations with higher frequency are less phase shifted than those with lower frequency. We have:

$$\psi(0^+) = x\sqrt{a}\,\frac{1}{2}\left(\sqrt{\frac{\alpha_2}{\alpha_1}} + \sqrt{\frac{\alpha_1}{\alpha_2}}\right) = x\frac{b}{2\sqrt{c}} = \frac{1}{2}x\left(C\sqrt{\frac{R}{G}} + L\sqrt{\frac{G}{R}}\right),$$

$$\psi(\infty) = x\sqrt{a} \qquad\qquad = x\sqrt{LC}.$$

40. Integral Equations

The equations in the unknown $f(t)$ of the form

$$\int_0^t k(t,\tau)\, f(\tau)\, d\tau = g(t) \quad \text{and} \quad f(t) = g(t) + \int_0^t k(t,\tau)\, f(\tau)\, d\tau$$

are known as *Volterra linear integral equations of the first and second kind* respectively. These integral equations are of the *convolution type*, provided the kernel $k(t, \tau)$ is a function of $(t - \tau)$ only. These latter equations can be changed into algebraic equations by the $\mathfrak{L}$-transformation, invoking the Convolution Theorem. Applying the inverse Laplace transformation to the solution of the algebraic equation, one obtains the solution of the integral equation.

1. The Linear Integral Equation of the Second Kind, of the Convolution Type

One can apply the $\mathfrak{L}$-transformation to the equation

(1) $$f(t) = g(t) + \int_0^t k(t-\tau)\, f(\tau)\, d\tau\,,$$

provided the following hypothesis is satisfied: we need the convergence of

$$\mathfrak{L}\{f\} = F(s)\,, \ \mathfrak{L}\{k\} = K(s)\,, \text{ and } \mathfrak{L}\{g\} = G(s)$$

in some half-plane, where $\mathfrak{L}\{k\}$ converges absolutely. Then we can find the image equation of (1); it is, by Theorem 10,4,

(2) $$F(s) = G(s) + K(s)\, F(s)\,,$$

which has the solution

$$F(s) = \frac{G(s)}{1 - K(s)}\,.$$

The function $[1 - K(s)]^{-1}$ converges for $s \to \infty$, not towards zero, but towards 1; thus it is not a $\mathfrak{L}$-transform. It is for this reason that $F(s)$ cannot be transformed into the original space in its presented form. However, upon rewriting $F(s)$ in the form

(3) $$F(s) = G(s) + \frac{K(s)}{1 - K(s)}\, G(s)\,,$$

we can determine $f(t)$. The function

$$\frac{z}{1-z} = \sum_{n=1}^{\infty} z^n$$

is holomorphic at $z = 0$, where it has value 0, and the series representation converges in the circular disc $|z| < 1$. $\mathfrak{L}\{k\}$ converges absolutely in some half-plane $\Re s > \alpha$, hence by the Supplement of Theorem 30.4 we recognize

$$Q(s) = \frac{K(s)}{1 - K(s)} \tag{4}$$

as the $\mathfrak{L}$-transform of

$$q(t) = k(t) + \sum_{n=2}^{\infty} k(t)^{*n}, \tag{5}$$

provided $k(t)$ is a $\mathfrak{J}_0$-function. Theorem 23.7 guarantees that $|K(s)| < 1$ in some half-plane $\Re s \geqq x_0 > \alpha$, and one concludes that $\mathfrak{L}\{q\}$ converges absolutely for $\Re s \geqq x_0$. Invoking the Convolution Theorem we recognize $Q(s)\, G(s)$ as the $\mathfrak{L}$-transform of $q * g$; hence, by the Uniqueness Theorem 5.1, it follows from (3) that

$$f(t) = g(t) + q(t) * g(t) + n(t), \tag{6}$$

where $n(t)$ designates some null function. Under the specified hypothesis, there can be no other solution than (6).

The verification of this function as a solution follows from the fact that (6) implies (3) which, in turn, implies (2), and finally

$$f(t) = g(t) + k(t) * f(t) + n_1(t),$$

where $n_1(t)$ designates some null function. A null function vanishes almost everywhere, hence $f(t)$ satisfies (1) almost everywhere, in particular when we select $n(t) \equiv 0$ in (6). Thus, we have established

Theorem 40.1. *When $g(t)$ has a simply converging $\mathfrak{L}$-transform and when the $\mathfrak{J}_0$-function $k(t)$ has an absolutely converging $\mathfrak{L}$-transform, then*

$$f(t) = g(t) + \int_0^t q(t - \tau)\, g(\tau)\, d\tau, \tag{7}$$

where $q(t)$ is defined by (5), *satisfies the integral equation* (1) *almost everywhere.*

The function $q(t)$ is the to $k(t)$ *reciprocal kernel*. Eq. (4) implies that

$$Q(s) - K(s) = K(s)\, Q(s),$$

hence, almost everywhere,

$$q(t) - k(t) = \int_0^t k(\tau)\, q(t - \tau)\, d\tau.$$

Each of the two kernels, $q(t)$ and $k(t)$, can be obtained from the other as the solution of an integral equation. The representation (5) of q as the sum of *"iterated kernels"* k^{*n} is known as *Neumann series.*

Employing the principle of extension (see p. 74), we can relax the hypothesis concerning the existence of $\mathfrak{L}\{g\}$ and of $\mathfrak{L}\{k\}$ as follows: we investigate the function (7) irrespective of the manner of its derivation, and we enumerate the conditions which must be satisfied so that (7) is a solution of the integral equation (1). For this purpose we may presume that the integral equation is given only in some finite interval $0 \leqq t \leqq T$, for we encounter in the integral equation as well as in its solution only integrals over bounded intervals. (Our tacit assumption of the interval $0 \leqq t < \infty$ in the above derivation merely reflects our involvement with the $\mathfrak{L}$-transformation.)

There are several sets of conditions, each of which guarantees (7) as a solution of (1). We present here the conditions enumerated in the following theorem.

Theorem 40.2. *When in the interval* $0 \leqq t \leqq T$, $g(t)$ *is integrable and* $k(t)$ *is integrable and bounded, then the integral equation* (1) *has the solution* (7). *The function* $q(t)$ *is defined by the series* (5) *which converges absolutely and uniformly in* $0 \leqq t \leqq T$. *The functions* $[g(t) - k(t)]$ *and* $[f(t) - g(t)]$ *are continuous.*

Proof: Using $|k(t)| \leqq M$ in $0 \leqq t \leqq T$, we find

$$|k^{*2}| = \left| \int_0^t k(\tau)\, k(t-\tau)\, d\tau \right| \leqq M^2 t,$$

$$|k^{*3}| = \left| \int_0^t k^{*2}(\tau)\, k(t-\tau)\, d\tau \right| \leqq M^3 \int_0^t \tau\, d\tau = M^3 \frac{t^2}{2!},$$

....

$$|k^{*n}| = \left| \int_0^t k^{*(n-1)}(\tau)\, k(t-\tau)\, d\tau \right| \leqq M^n \int_0^t \frac{\tau^{n-2}}{(n-2)!}\, d\tau = M^n \frac{t^{n-1}}{(n-1)!};$$

hence,

$$|k(t)| + \sum_{n=2}^{\infty} |k(t)^{*n}| \leqq \sum_{n=1}^{\infty} M^n \frac{t^{n-1}}{(n-1)!} \leqq \sum_{n=1}^{\infty} M^n \frac{T^{n-1}}{(n-1)!} = M e^{MT}.$$

This demonstrates that the series representation (5) of $q(t)$ converges absolutely and uniformly on $0 \leqq t \leqq T$. Hence, we may multiply it by the bounded function $k(t-\tau)$, and then integrate the product term by term:

$$\int_0^t q(\tau)\, k(t-\tau)\, d\tau = k * q = k * \left(k + \sum_{n=2}^{\infty} k^{*n}\right) = \sum_{n=2}^{\infty} k^{*n}.$$

Defining the function $f(t)$ by

$$f = g + q * g,$$

we find

$$k * f = k * g + k * q * g = k * g + \left(\sum_{n=2}^{\infty} k^{*n}\right) * g$$

$$= q * g = f - g.$$

This shows that f satisfies (1).

The convolutions k^{*n} $(n \geqq 2)$ are continuous on $0 \leqq t \leqq T$ according to Theorems 10.2 and 10.3. The uniform convergence of the series implies continuity of

$$q(t) - k(t) = \sum_{n=2}^{\infty} k^{*n}.$$

The function $k(t)$ is bounded, hence $k * f$ is continuous on $0 \leqq t \leqq T$; it follows that $f(t) - g(t)$ too is continuous.

The immediate determination of $q(t)$ by formula (5) through evaluation of the convolution integrals is practically possible only for few problems. A more promising route is to firstly determine the powers $K^n(s)$ and subsequently return these to the original space. One can sometimes inversely transform $Q(s)$ immediately by starting with (4); this is the case, for instance, when $k(t)$ is a polynomial:

$$k(t) = a_0 + a_1 t + \cdots + a_r t^r,$$

hence

$$K(s) = \frac{a_0}{s} + \frac{1!\, a_1}{s^2} + \cdots + \frac{r!\, a_r}{s^{r+1}}$$

and, consequently,

$$Q(s) = \frac{K(s)}{1 - K(s)} = \frac{a_0 s^r + 1!\, a_1 s^{r-1} + \cdots + r!\, a_r}{s^{r+1} - a_0 s^r - 1!\, a_1 s^{r-1} - \cdots - r!\, a_r}.$$

This is a rational function, the polynomial in the numerator being of lower degree than the polynomial in the denominator. Thus, one can employ the partial fraction expansion (see p. 76), to transform $Q(s)$ into the original space. When $k(t)$ is a continuous function, then one can approximate it, arbitrarily closely, by a polynomial, and thus finally produce an approximate solution.

The integral equation

$$f(t) = g(t) + \int_0^t \left[1 - (t - \tau) + \frac{1}{2}(t - \tau)^2\right] f(\tau)\, d\tau$$

is presented here as an example. In this case, we have

$$k(t) = 1 - t + \frac{1}{2} t^2, \quad K(s) = \frac{1}{s} - \frac{1}{s^2} + \frac{1}{s^3},$$

$$Q(s) = \frac{\frac{1}{2} s^2 - s + 1}{s^3 - s^2 + s - 1} = \frac{\frac{1}{4}}{s - 1} + \frac{\frac{1}{4} s - \frac{3}{4}}{s^2 + 1},$$

$$q(t) = \frac{1}{4} e^t + \frac{1}{4} \cos t - \frac{3}{4} \sin t.$$

Thus, we find the solution:

$$f(t) = g(t) + \frac{1}{4}\int_0^t (e^\tau + \cos\tau - 3\sin\tau)\, g(t-\tau)\, d\tau.$$

In the case that $k(t)$ is a polynomial of exponential functions:

$$k(t) = a_0 e^{\alpha_0 t} + \cdots + a_r e^{\alpha_r t},$$

one finds $g(t)$ by a similar process. For those cases where $f(t)$ cannot be determined explicitly, one may resort succesfully to the technique of asymptotic expansion, which was shown in Chapters 35 and 37.

2. The Linear Integral Equation of the First Kind, of the Convolution Type

The equation in the unknown $f(t)$

$$\int_0^t k(t-\tau)\, f(\tau)\, d\tau = g(t) \tag{8}$$

requires stringent hypotheses to guarantee the existence of a solution. One may consider the left hand side of (8) as a functional transformation which transforms the function $f(t)$ into a function $g(t)$. Thus, one sees that the question whether or not (8) has a solution $f(t)$ is equivalent to the new question whether or not $g(t)$ can be represented as a transform. It is clear that, in general, the answer is negative.

The immediate application of the $\mathfrak{L}$-transformation is bound to fail, for the image equation

$$K(s)\, F(s) = G(s)$$

has the solution

$$F(s) = \frac{G(s)}{K(s)}; \tag{9}$$

however, $1/K(s)$ tends towards ∞ when $s \to \infty$, hence it certainly cannot be a $\mathfrak{L}$-transform, and the Convolution Theorem cannot be employed.

There are cases of integral equations of the first kind which can be *reduced to integral equations of the second kind*. When $k(t)$ is differentiable for $t > 0$, $k'(t)$ is a $\mathfrak{J}_0$-function, and when, moreover, $g(t)$ is differentiable for $t > 0$, and when a solution $f(t)$, continuous for $t > 0$, exists, then we have, by Theorem 10.5,

$$k(0^+)\, f(t) + \int_0^t k'(t-\tau)\, f(\tau)\, d\tau = g'(t) \quad \text{for } t > 0.$$

For $k(0^+) \neq 0$, this is an integral equation of the second kind which may be solved by the method shown above. When $k(0^+) = 0$, then possibly both $k(t)$ and $g(t)$ have $(n+1)$ derivatives $(n \geqq 1)$, and $k(0^+) = k'(0^+) = \dots k^{(n-1)}(0^+) = 0$, whilst $k^{(n)}(0^+) \neq 0$; then, one finds:

$$k^{(n)}(0^+) f(t) + \int_0^t k^{(n+1)}(t-\tau) f(\tau) \, d\tau = g^{(n+1)}(t),$$

once again, an integral equation of the second kind.

When $k'(t)$ fails to be integrable and $k(0^+)$ does not exist, a situation which is shown by the example $k(t) \equiv t^{-\alpha}$, with $0 < \alpha < 1$, then the above method cannot be applied. For such cases the following method may sometimes prove successful: Instead of the function $f(t)$, one enters its *integral*

$$\int_0^t f(\tau) \, d\tau = f * 1 = \varphi(t)$$

as the unknown in the integral equation. When f has the $\mathfrak{L}$-transform F, then φ has a $\mathfrak{L}$-transform which, by Theorem 8.1, is given by

$$\mathfrak{L}\{\varphi\} = \Phi(s) = \frac{1}{s} F(s).$$

Instead of Eq. (9) one finds

$$\Phi(s) = \frac{1}{s\,K(s)} G(s).$$

Although $1/K(s)$ is not a $\mathfrak{L}$-transform, $1/[s\,K(s)]$ may, nevertheless, be one. In this case, $\varphi(t)$ can be obtained by means of the Convolution Theorem. The sought function $f(t)$ is found (almost everywhere) by differentiation of $\varphi(t)$.

A famous example which can be solved by the above indicated substitution is

the Abel Integral Equation

which was originally encountered in physics as the generalization of the problem of the tautochrone and has the form

(10) $$\int_0^t (t-\tau)^{-a} \frac{dy(\tau)}{d\tau} \, d\tau = g(t) \qquad (0 < a < 1).$$

The substitution $y' = f$ would produce an equation of the form (8). However, we deliberately retain y' in the equation, for this is equivalent to the above suggested substitution $y = f * 1$; exactly this will enable us to produce the solution.

We assume that y' is a $\mathfrak{J}_0$-function and that its $\mathfrak{L}$-transform does exist. Consequently, by Theorem 10.2, the left hand side of (10) is continuous and it has, by Theorem 10.4, a $\mathfrak{L}$-transform. Consequently, we must assume these two properties for the function $g(t)$ on the right hand side of (10). The image equation of (10) is given by

$$\frac{\Gamma(1-\alpha)}{s^{1-\alpha}} [s\, Y(s) - y(0+)] = G(s),$$

which has the solution

$$Y(s) = \frac{y(0+)}{s} + \frac{1}{\Gamma(1-\alpha)\, s^{\alpha}} G(s).$$

The corresponding original function is

$$y(t) = y(0+) + \frac{1}{\Gamma(1-\alpha)\,\Gamma(\alpha)} t^{\alpha-1} * g(t) + \text{null function.}$$

By hypothesis, $y(t)$ is differentiable, hence it is continuous; also $t^{\alpha-1} * g(t)$ is continuous. It follows that the null function vanishes identically. Thus, we are left with

$$(11)\qquad \begin{aligned} y(t) &= y(0+) + \frac{1}{\Gamma(1-\alpha)\,\Gamma(\alpha)} t^{\alpha-1} * g(t) \\ &= y(0+) + \frac{\sin\alpha\pi}{\pi} t^{\alpha-1} * g(t). \end{aligned}$$

y' occurs in the integral equation; thus, we must be able to produce the derivative. According to Theorem 10.5, this is possible when $g(t)$ is differentiable for $t > 0$ and continuous at $t = 0$, and $g'(t)$ is a $\mathfrak{J}_0$-function. We then find:

$$(12)\qquad y'(t) = \frac{\sin\alpha\pi}{\pi} [t^{\alpha-1} * g'(t) + g(0)\, t^{\alpha-1}].$$

The above specified hypothesis that y' is a $\mathfrak{J}_0$-function and has a $\mathfrak{L}$-transform is satisfied whenever $g'(t)$ has these properties. The equations (10) to (12) can also be derived in inverse order, hence (12) and (11) respectively represent a solution of (10). This conclusion is summarized in

Theorem 40.3. *Suppose that $g(t)$ is differentiable for $t > 0$ and continuous at $t = 0$, and that $g'(t)$ is a $\mathfrak{J}_0$-function which has a $\mathfrak{L}$-transform. Then, the only solution of* (10), *the derivative of which is a $\mathfrak{J}_0$-function and has a $\mathfrak{L}$-transform, is the function* (11) *with the derivative* (12).

Once again, we can relax the hypothesis regarding the existence of the $\mathfrak{L}$-transforms, and also consider for the integral equation a finite interval instead of $0 \leq t < \infty$. For $0 < \alpha < 1$, we find:

$$t^{-\alpha} * t^{\alpha-1} \circ\!\!-\!\!\bullet \frac{\Gamma(1-\alpha)}{s^{1-\alpha}} \frac{\Gamma(\alpha)}{s^{\alpha}} = \frac{\Gamma(1-\alpha)\,\Gamma(\alpha)}{s} \bullet\!\!-\!\!\circ\, \Gamma(1-\alpha)\,\Gamma(\alpha) = \frac{\pi}{\sin\alpha\pi},$$

hence, when $g'(t)$ is a $\mathfrak{J}_0$-function, and $y'(t)$ is given by (12), then we have

$$t^{-\alpha} * y'(t) = 1 * g'(t) + g(0) = g(t)$$

in every interval where $g(t)$ is defined. Moreover, (12) is the only solution (disregarding the trivial addition of null functions), for, when two different solutions are given, then there is a solution $\not\equiv 0$ of

$$t^{-\alpha} * y' = 0;$$

this implies that

$$t^{\alpha-1} * t^{-\alpha} * y' = \frac{\pi}{\sin \alpha\pi} 1 * y' = 0;$$

that is, y' is a null function. Obviously, the function $y(t)$ is determined up to the constant $y(0+)$.

Theorem 40.4. *Suppose that $g(t)$ is differentiable in $0 < t \leqq T$ and continuous at $t = 0$; suppose, further, that $g'(t)$ is a $\mathfrak{J}_0$-function. Then the integral equation* (10) *in $0 < t \leqq T$ has the unique solution* (12) *for $y'(t)$, and $y(t)$ is given by* (11), *with arbitrary $y(0+)$. The function $y'(t)$ is continuous in $0 < t \leqq T$.*

An intuitive interpretation can be provided for the Abel integral equation and its solution. Integration of a function $f(t)$ μ times between 0 and the variable upper limit t is equivalent to μ-fold convolution with 1:

$$I^\mu f = f * 1^{*\mu}.$$

According to (11.1), we can represent $I^\mu f$ by a simple integral

(13) $$I^\mu f = \frac{1}{\Gamma(\mu)} \int_0^t f(\tau)(t-\tau)^{\mu-1}\, d\tau.$$

The last expression is meaningful not only for natural numbers $\mu = 1, 2, 3, \ldots$, but also for every real number $\mu > 0$. By means of (13), one can define *the μ-fold integral of the function f, for $\mu > 0$* ("integral" here not in the sense of primitive function, but in the sense of definite integral between 0 and t). When interpreting *μ-fold differentiation D^μ* as the *inverse* of the above process, then $D^\mu f$ is the solution $z(t)$ of the equation

(14) $$\frac{1}{\Gamma(\mu)} \int_0^t z(\tau)(t-\tau)^{\mu-1}\, d\tau = f(t).$$

Obviously, for $0 < \mu < 1$, this equation is equivalent to the Abel integral equation. Hence its solution y' (with $g(t) = \Gamma(\mu) f(t)$, $\alpha = 1 - \mu$) provides the definition of $D^\mu f$, for $0 < \mu < 1$; it is

(15) $$D^\mu f = \frac{1}{\Gamma(1-\mu)} [t^{-\mu} * f'(t) + f(0)\, t^{-\mu}] \qquad (0 < \mu < 1),$$

where one must presume that $f(t)$ is differentiable for $t > 0$ and continuous at $t = 0$, and that $f'(t)$ is a $\mathfrak{J}_0$-function.

When using expression (11) for $y(t)$ instead of the explicit formula (12) for $y'(t)$, one finds

$$(16) \qquad D^\mu f = \frac{d}{dt}\left\{\frac{t^{-\mu}}{\Gamma(1-\mu)} * f\right\} = \frac{d}{dt}\{I^{1-\mu} f\} \qquad (0 < \mu < 1).$$

Expression (16) indicates $(1-\mu)$-fold integration followed by one differentiation. One can interpret this as the annulment of one integration by one differentiation, and one is left with $(-\mu)$-fold integration. That is, D^μ appears formally as $I^{-\mu}$, a property which nicely agrees with the definition of D^μ as the inverse of I^μ.

So far we required $0 < \mu < 1$. For natural numbers $\mu = n = 1, 2, 3, \ldots$, equation (14) is equivalent to $z * 1^{*n} = f$; and one finds $z = D^n f$, almost everywhere, by n-fold differentiation of $f(t)$. $D^n f$ is identical to $f^{(n)}(t)$, provided one understands the latter as generalized differentiation in the sense of p. 41.

When μ is not a natural number, $1 \leqq n < \mu < n+1$, then we differentiate (14) n times, provided $f^{(n)}$ exists, and we obtain

$$\frac{1}{\Gamma(\mu-n)} \int_0^t z(\tau)\,(t-\tau)^{\mu-n-1}\,d\tau = f^{(n)}(t) \quad (0 < \mu - n < 1),$$

thus once again creating Eq. (14), with $0 < \mu < 1$. From Eq. (16), we find:

$$(17) \qquad D^\mu f = \frac{d}{dt}\left\{\frac{t^{-(\mu-n)}}{\Gamma(1-(\mu-n))} * f^{(n)}\right\} = \frac{d}{dt}\{I^{1-(\mu-n)} f^{(n)}\}.$$

This formula becomes intelligible, when differentiation is replaced by the operator D:

$$D^\mu f = D\, I^{1-\mu+n}\, D^n f.$$

DI and $I^n D^n$ produce the identity and, formally, there remains the equation

$$D^\mu f = I^{-\mu} f.$$

APPENDIX

Some Concepts and Theorems from the Theory of Distributions

It is presumed that the reader is familiar with the theory of distributions as introduced by L. Schwartz. The most comprehensive presentation of this theory is given by L. SCHWARTZ [1]: Théorie des distributions, Nouvelle édition. Hermann, Paris 1966, 420 pages. Shorter and less demanding expositions can be found in the books: L. SCHWARTZ [2]: Méthodes mathématiques pour les sciences physiques. Hermann, Paris 1965, 312 pages; A. H. ZEMANIAN: Distribution Theory and Transform Analysis. McGraw-Hill Book Co., New York 1965, XVIII + 371 pages.

A summary of several concepts and terms has been compiled for clarity's sake; theorems employed in the text are presented to permit brief reference.

1. All real-valued or complex-valued functions of a single real variable t are, in principle, defined on the entire real line: $-\infty < t < +\infty$, designated by R^1.

2. A *compact set* of R^1 is a closed and bounded set.

3. Given a continuous function $\varphi(t)$ on R^1, the *support* of $\varphi(t)$ is the smallest closed set which contains all points t where $\varphi(t) \neq 0$.

4. A *test function* $\varphi(t)$ is a function which has derivatives of all orders, and a compact support; hence the test function $\varphi(t)$ assumes the value zero outside some bounded interval of R^1.

5. The *space* $\mathscr{D}$ is the set of all test functions.

6. A functional maps a set of functions to the complex numbers; that is, it assigns to every function of the set one complex number. A *distribution* is a linear and continuous functional defined on $\mathscr{D}$.

7. The *space* $\mathscr{D}'$ is the set of all distributions.

8. We shall use $\langle T, \varphi\rangle$ to designate the (complex) number assigned to the test function $\varphi \in \mathscr{D}$, by the distribution T.

9. A locally integrable[1] function $f(t)$ defines a functional by means of the formula

$$\int_{-\infty}^{+\infty} f(t)\,\varphi(t)\,dt = \langle f, \varphi\rangle .$$

We shall call this distribution the *function-distribution* generated by $f(t)$, and we shall designate the function-distribution by f or by $[f]$, using the latter designation when confusion with the function f could occur. Occasionally we shall use the term function instead of function-distribution.

---

[1] All integrals are understood as Lebesgue integrals.

10. Two distributions T and U are called *equal* in some open set G of R^1 (G may be the entire real line R^1) if and only if $\langle T, \varphi\rangle = \langle U, \varphi\rangle$ for every $\varphi \in \mathcal{D}$ which has its support entirely in G. In particular, T may be equal in G to some function-distribution $[f]$, in which case we shall write $T = f$ (in G); for instance, $T = 0$ in G.

11. The *support* of some distribution T is the smallest closed subset of R^1, outside of which $T = 0$.

12. The support of the function-distribution $[f]$ generated by the continuous function f is identical to the support of the function $f(t)$.

13. The kth *distribution-derivative*[2] $D^k T$ of some distribution T is defined by means of the formula:

$$\langle D^k T, \varphi(t)\rangle = (-1)^k \langle T, \varphi^{(k)}(t)\rangle .$$

Specifically, for a function-distribution $[f]$, generated by the locally integrable function f, we find:

$$\langle D^k f, \varphi(t)\rangle = (-1)^k \int_{-\infty}^{+\infty} f(t)\, \varphi^{(k)}(t)\, dt .$$

Every locally integrable function has distribution-derivatives of all orders.

14. Suppose that the kth *derivative* $f^{(k)}(t)$ of some function $f(t)$ exists for all t in R^1 and is locally integrable; then we have

$$D^k[f] = [f^{(k)}(t)] .$$

15. We define the *unit step function* as follows:

$$u(t) = 0 \quad \text{for} \quad t \leqq 0, \ \text{and} \ u(t) = 1 \quad \text{for} \quad t > 0,$$

and we set

$$D\,u(t) = \delta .$$

The distribution δ has the property:

$$\langle \delta, \varphi\rangle = \varphi(0) .$$

We recall No. 10, and we conclude that $\delta = 0$ in both $t < 0$ and $t > 0$.

16. Yielding to the common practice used in physics, we shall make an exception by designating the *distribution-derivatives of the distribution* δ, as for derivatives of functions, by short strokes or by superscripts in round brackets. Moreover, with δ and its distribution-derivatives one often finds the variable of the real line R^1, over which these are defined, written in round brackets after the respective symbol as is customary for functions:

$$\delta = \delta(t), \qquad D\delta = \delta'(t), \qquad \ldots, \qquad D^k\delta = \delta^{(k)}(t) .$$

The distribution-derivatives of δ have the property:

$$\langle \delta', \varphi\rangle = \varphi'(0), \qquad \ldots, \qquad \langle \delta^{(k)}, \varphi\rangle = \varphi^{(k)}(0) .$$

All distributions $\delta, \delta', \ldots, \delta^{(k)}, \ldots$ have as support the single point $t = 0$.

---

[2] Another frequently used term is generalized derivative.

17. The distribution-derivative of the unit step function $u(t-a)$ having the jumping point at $t=a$ is designated in the following manner:

$$D\,u(t-a) = \delta_a .$$

However, one occasionally writes

$$D\,u(t-a) = \delta(t-a),$$

employing the notation as explained in No. 16. The distribution-derivatives are designated thus:

$$D\delta(t-a) = \delta'(t-a), \qquad \ldots, \qquad D^k\delta(t-a) = \delta^{(k)}(t-a).$$

These distributions have the property

$$\langle\delta(t-a),\varphi\rangle = \varphi(a), \quad \langle\delta'(t-a),\varphi\rangle = \varphi'(a), \quad \ldots, \quad \langle\delta^{(k)}(t-a),\varphi\rangle = \varphi^{(k)}(a).$$

18. Suppose that $\alpha(t)$ is a function which has derivatives of all orders in R^1. We define the *product* of some distribution T with $\alpha(t)$ by means of the formula:

$$\langle\alpha(t)\,T,\varphi\rangle = \langle T,\alpha\varphi\rangle.$$

19. Suppose that $\alpha(t)$ is a function which has derivatives of all orders in R^1; the *distribution-derivative of the product* of some distribution T with $\alpha(t)$ can be determined by the formal application of the same rule which is used to find the derivative of a product of two functions:

$$D[\alpha(t)\,T] = \alpha'(t)\,T + \alpha(t)\,D\,T.$$

In like manner, for the distribution-derivatives of higher order one finds:

$$D^k[\alpha(t)\,T] = \sum_{\nu=0}^{k}\binom{k}{\nu}\alpha^{(\nu)}(t)\,D^{k-\nu}T,$$

a formula which is analogous to Leibniz' formula.

20. One can generalize No. 14: If $f^{(k)}(t)$ exists for all t *with the exception of the point* $t=a$, and represents a locally integrable function, and if moreover, for both limiting processes $t\to a-0$ and $t\to a+0$ the limits

$$\begin{array}{llll} f(a^-), & f'(a^-), & \ldots, & f^{(k-1)}(a^-); \\ f(a^+), & f'(a^+), & \ldots, & f^{(k-1)}(a^+) \end{array}$$

do exist; then one finds

$$\begin{aligned} D^k[f] = [f^{(k)}(t)] &+ [f^{(k-1)}(a^+) - f^{(k-1)}(a^-)]\,\delta(t-a) \\ &+ [f^{(k-2)}(a^+) - f^{(k-2)}(a^-)]\,\delta'(t-a) \\ &+ \cdots\cdots\cdots\cdots \\ &+ [f(a^+) - f(a^-)]\,\delta^{(k-1)}(t-a). \end{aligned}$$

This theorem can be extended to a finite number of, or countably many, isolated exceptional points a_i.

21. If $f(t)$ is not locally integrable, then it may happen that the diverging integral $\int_{-\infty}^{+\infty} f(t)\varphi(t)\,dt = \langle f, \varphi\rangle$ has a "finite part" (partie finie) as defined by Hadamard, designated by Pf $\langle f, \varphi\rangle$. If the functional Pf $\langle f, \varphi\rangle$ agrees with the functional defined by some distribution T:

$$\mathrm{Pf}\,\langle f, \varphi\rangle = \langle T, \varphi\rangle,$$

then the function $f(t)$ is said to define the distribution T. One writes briefly:

$$\mathrm{Pf}\, f(t) = T.$$

We call Pf $f(t)$ a *pseudofunction-distribution*, or, more briefly, a *pseudofunction*.[3]

22. Only "right-sided" functions are of interest for the Laplace transformation; these are functions which are specified to be zero for $t < 0$ and which consequently should properly be written in the form $f(t)\,u(t)$. The most important pseudofunctions are the *powers* $t^{-\lambda} u(t)$ with $\lambda \geqq 1$. These give rise to the following distributions:[4]

(a) $\mathrm{Pf}\,[t^{-1} u(t)] = D\,[\log t \cdot u(t)]$

(b) $\mathrm{Pf}\,[t^{-2} u(t)] = -D^2\,[\log t \cdot u(t)] - \delta'(t)$

(c) $$\mathrm{Pf}\,[t^{-n} u(t)] = \frac{(-1)^{n-1}}{(n-1)!}\,\{D^n\,[\log t \cdot u(t)] + (\psi(n) + C)\,\delta^{(n-1)}(t)\}$$
$$= \frac{(-1)^{n-1}}{(n-1)!}\,D^n\,\{[\log t + \psi(n) + C]\,u(t)\}\ (n = 1, 2, \ldots)$$

with $\psi(n) + C = 1 + \frac{1}{2} + \ldots + \frac{1}{n-1}\ (n \geqq 2),\ \psi(1) = -C$
(C = Euler's constant)

(d) $$\mathrm{Pf}\,[t^{-\lambda} u(t)] = \frac{(-1)^n}{(\lambda-1)\ldots(\lambda-n)}\,D^n\,[t^{-\lambda+n}\,u(t)]$$

($\lambda > 1$ not an integer, $-\lambda + n > -1$, n an integer).

Formulae (a) and (d) are also correct in the classical analysis, provided we replace the distribution-derivatives on the right hand side by derivatives and omit the symbol Pf on the left hand side. Formulae (b) and (d) differ from the classical formulae by the occurrence of δ-distributions.

[3] The symbol Pf could be read here as abbreviation of "pseudofunction".

[4] The functions on the right hand side are locally integrable, hence the distribution-derivatives of these do exist; they are distributions.

Table of Laplace Transforms

Operations

$F(s)$	$f(t)$
$F(as)\quad (a>0)$	$\frac{1}{a} f\left(\frac{t}{a}\right)$
$F(s+\alpha)$	$e^{-\alpha t} f(t)$
$F(as+\beta)\quad (a>0,\ \beta \text{ complex})$	$\frac{1}{a} e^{-\frac{\beta}{a} t} f\left(\frac{t}{a}\right)$
$e^{-as} F(s)\quad (a \geqq 0)$	$\left\{\begin{matrix} f(t-a) \text{ for } t \geqq a \\ 0 \text{ for } t < a \end{matrix}\right\} = f(t-a)\, u(t-a)$
$F'(s)$	$-t f(t)$
$F^{(n)}(s)$	$(-t)^n f(t)$
$sF(s) - f(0^+)$	$f'(t)$
$sF(s)$	$Df(t)$
$s^n F(s) - \sum_{k=0}^{n-1} f^{(k)}(0^+)\, s^{n-k-1}$	$f^{(n)}(t)$
$s^n F(s)$	$D^n f(t)$
$\frac{1}{s} F(s)$	$\int_0^t f(\tau)\, d\tau = f(t) * 1$
$F(\sqrt{s})$	$\int_0^\infty \psi(\tau, t)\, f(\tau)\, d\tau$
$F_1(s) \cdot F_2(s)$	$\int_0^t f_1(\tau)\, f_2(t-\tau)\, d\tau = f_1(t) * f_2(t)$
$\frac{1}{2\pi i} \int_{x-i\infty}^{x+i\infty} F_1(\sigma)\, F_2(s-\sigma)\, d\sigma$	$f_1(t) \cdot f_2(t)$

Functions and Distributions

Nr.	$F(s)$	$f(t)$
1	1	$\delta(t)$
2	$s^n\ (n=1,2,\ldots)$	$\delta^{(n)}(t)$
3	$s^a\ (a>0,\ a\neq 1,2,\ldots)$	$\mathrm{Pf}\,\frac{t^{-a-1}}{\Gamma(-a)}\,u(t)$ $= -\,\mathrm{Pf}\,\frac{1}{\pi}\,\Gamma(a+1)\sin\pi a\,\frac{1}{t^{a+1}}\,u(t)$
4	$\frac{1}{s}$	$u(t)=1$ for $t>0$, $=0$ for $t<0$
5	$\frac{1}{s-\alpha}$	$e^{\alpha t}$
6	$\frac{1}{1+Ts}$	$\frac{1}{T}\,e^{-\frac{t}{T}}$
7	$\frac{1}{s^2}$	t
8	$\frac{1}{(s-\alpha)^2}$	$t\,e^{\alpha t}$
9	$\frac{1}{s(s-\alpha)}$	$\frac{1}{\alpha}\,(e^{\alpha t}-1)$
10	$\frac{1}{s(1+Ts)}$	$1-e^{-\frac{t}{T}}$
11	$\frac{1}{(s-\alpha)(s-\beta)}\ (\alpha\neq\beta)$	$\frac{e^{\alpha t}-e^{\beta t}}{\alpha-\beta}$
12	$\frac{1}{(1+Ts)^2}$	$\frac{1}{T^2}\,t\,e^{-\frac{t}{T}}$
13	$\frac{1}{(1+as)(1+bs)}\ (a\neq b)$	$\frac{e^{-\frac{t}{a}}-e^{-\frac{t}{b}}}{a-b}$
14	$\frac{a}{s^2+a^2}$	$\sin at$
15	$\frac{a}{s^2-a^2}$	$\sinh at$
16	$\frac{1}{s^2+c_1 s+c_0}\left(c_0-\frac{c_1^2}{4}=D\right)$	$\frac{1}{\sqrt{-D}}\,e^{-\frac{c_1}{2}t}\sinh\sqrt{-D}\,t\quad (D<0)$ $\frac{1}{\omega}\,e^{-\frac{c_1}{2}t}\sin\omega t\quad (D>0,\ \sqrt{-D}=i\omega)$ $t\,e^{-\frac{c_1}{2}t}\quad (D=0)$
17	$\frac{s}{(s-\alpha)^2}$	$(1+\alpha t)\,e^{\alpha t}$

Nr.	$F(s)$	$f(t)$
18	$\frac{s}{(s-\alpha)(s-\beta)}\ (\alpha \neq \beta)$	$\frac{\alpha e^{\alpha t} - \beta e^{\beta t}}{\alpha - \beta}$
19	$\frac{s}{s^2 + a^2}$	$\cos at$
20	$\frac{s}{s^2 - a^2}$	$\cosh at$
21	$\frac{s \sin b + a \cos b}{s^2 + a^2}$	$\sin(at + b)$
22	$\frac{s \cos b - a \sin b}{s^2 + a^2}$	$\cos(at + b)$
23	$\frac{1}{s^n}\ (n = 1, 2, \ldots)$	$\frac{1}{(n-1)!} t^{n-1}$
24	$\frac{1}{(s-\alpha)^n}\ (n = 1, 2, \ldots)$	$\frac{1}{(n-1)!} t^{n-1} e^{\alpha t}$
25	$\frac{1}{s}\left(1 - \frac{1}{s}\right)^n$	$L_n(t) = \frac{e^t}{n!} \frac{d^n}{dt^n}(t^n e^{-t})$ (Laguerre Polynomials)
26	$\frac{1}{s^\alpha}$ (α arbitrarily real)	$\begin{cases} \frac{t^{\alpha-1}}{\Gamma(\alpha)} u(t) & \text{for } \alpha > 0 \\ \text{Pf } \frac{t^{\alpha-1}}{\Gamma(\alpha)} u(t) & \text{for } \alpha < 0,\ \alpha \neq -1, -2, \ldots \\ \delta^{(n)}(t) & \text{for } \alpha = -n = 0, -1, -2, \ldots \end{cases}$
27	$\frac{1}{\sqrt{s}}$	$\frac{1}{\sqrt{\pi t}}$
28	$\frac{1}{\sqrt{s + \alpha}}$	$\frac{e^{-\alpha t}}{\sqrt{\pi t}}$
29	$\frac{1}{\sqrt{s} + a}$	$\frac{1}{\sqrt{\pi t}} - a e^{a^2 t} \operatorname{erfc}(a\sqrt{t})$
30	$\frac{1}{s^{3/2}}$	$2\sqrt{\frac{t}{\pi}}$
31	$\frac{1}{s^{n+\frac{1}{2}}}$	$\frac{4^n n!}{(2n)!\sqrt{\pi}} t^{n-\frac{1}{2}}$
32	$\frac{1}{\sqrt{s^2 + a^2}}$	$J_0(at)$
33	$\frac{1}{(s^2 + a^2)^{\nu+\frac{1}{2}}}\ (\Re \nu > -\frac{1}{2})$	$\frac{\sqrt{\pi}}{\Gamma(\nu + \frac{1}{2})} \left(\frac{t}{2a}\right)^\nu J_\nu(at)$
34	$\left(\sqrt{s^2 + \alpha^2} - s\right)^\nu\ (\Re \nu > 0)$	$\frac{\nu \alpha^\nu}{t} J_\nu(\alpha t)$

Nr.	$F(s)$	$f(t)$
35	$\dfrac{(\sqrt{s^2+\alpha^2}-s)^\nu}{\sqrt{s^2+\alpha^2}} \quad (\Re\nu > -1)$	$\alpha^\nu J_\nu(\alpha t)$
36	$\sqrt{s-\alpha}-\sqrt{s-\beta}$	$\dfrac{1}{2t\sqrt{\pi t}}(e^{\beta t}-e^{\alpha t})$
37	$\dfrac{\log s}{s}$	$-\log t - C$
38	$\log\dfrac{s+a}{s-a}$	$\dfrac{2}{t}\sinh at$
39	$\log\dfrac{s-a}{s-b}$	$\dfrac{e^{bt}-e^{at}}{t}$
40	$\log\dfrac{s^2+a^2}{s^2+b^2}$	$\dfrac{2}{t}(\cos bt-\cos at)$
41	$\log s + C$	$-\operatorname{Pf}\dfrac{1}{t}u(t)$
42	$s(\log s + C - 1)$	$\operatorname{Pf}\dfrac{1}{t^2}u(t)$
43	$s^n(\log s - \psi(n+1))$ $(n = 0, 1, \ldots)$	$-(-1)^n n!\operatorname{Pf}\dfrac{1}{t^{n+1}}u(t)$
44	$\log\sqrt{s^2+a^2}+C$	$\operatorname{Pf}\dfrac{\cos at}{t}u(t)$
45	$e^{-Ts}\ (T>0)$	$\delta(t-T)$
46	$\dfrac{e^{-Ts}}{s}\ (T>0)$	$u(t-T)$
47	$\dfrac{1}{s}e^{-\frac{a^2}{4s}}$	$J_0(a\sqrt{t})$
48	$\dfrac{1}{s^{\nu+1}}e^{-\frac{a^2}{4s}}\ (\Re\nu > -1)$	$\left(\dfrac{2}{a}\right)^\nu t^{\frac{\nu}{2}} J_\nu(a\sqrt{t})$
49	$e^{-a\sqrt{s}} \quad (a>0)$	$\psi(a,t) = \dfrac{a}{2\sqrt{\pi}\,t^{3/2}}e^{-\frac{a^2}{4t}}$
50	$\dfrac{1}{\sqrt{s}}e^{-a\sqrt{s}} \quad (a \geqq 0)$	$\chi(a,t) = \dfrac{1}{\sqrt{\pi t}}e^{-\frac{a^2}{4t}}$
51	$\psi(a,s)$	$\dfrac{\sin a\sqrt{t}}{\pi}$
52	$\chi(a,s)$	$\dfrac{\cos a\sqrt{t}}{\pi\sqrt{t}}$
53	$\dfrac{1}{s}e^{-a\sqrt{s}} \quad (a \geqq 0)$	$\operatorname{erfc}\dfrac{a}{2\sqrt{t}} = \dfrac{2}{\sqrt{\pi}}\displaystyle\int_{a/2\sqrt{t}}^{\infty} e^{-u^2}du$

Nr.	$F(s)$	$f(t)$
54	$\frac{1}{\sqrt{s}} \sin \frac{a}{s}$	$\frac{\sinh \sqrt{2at} \sin \sqrt{2at}}{\sqrt{\pi t}}$
55	$\frac{1}{\sqrt{s}} \cos \frac{a}{s}$	$\frac{\cosh \sqrt{2at} \cos \sqrt{2at}}{\sqrt{\pi t}}$
56	$\frac{1}{\sqrt{s}} e^{-\sqrt{as}} \sin \sqrt{as}$	$\frac{1}{\sqrt{\pi t}} \sin \frac{a}{2t}$
57	$\frac{1}{\sqrt{s}} e^{-\sqrt{as}} \cos \sqrt{as}$	$\frac{1}{\sqrt{\pi t}} \cos \frac{a}{2t}$
58	$e^{\frac{s^2}{4}} \operatorname{erfc} \frac{s}{2}$	$\frac{2}{\sqrt{\pi}} e^{-t^2}$
59	$\operatorname{arctg} \frac{a}{s}$	$\frac{\sin at}{t}$
60	$\frac{\sinh x \sqrt{s}}{\sinh l \sqrt{s}} \quad (\lvert x \rvert < l)$	$\frac{1}{l} \frac{\partial}{\partial x} \vartheta_3 \left(\frac{l-x}{2l}, \frac{t}{l^2} \right)$ $= \sum_{n=-\infty}^{\infty} \psi(2nl + l - x, t)$ $= -\frac{2\pi}{l^2} \sum_{n=1}^{\infty} (-1)^n n \sin n \frac{\pi}{l} x \ e^{-n^2 \frac{\pi^2}{l^2} t}$

INDEX

A

B

C

D

E

F

G

H

I

J

K

L

M

N

O

P

Q

R

S

T

U

V

W

The following books by the same author appeared in Birkhäuser Verlag Basel und Stuttgart

1. Einführung in Theorie und Anwendung der Laplace-Transformation
Ein Lehrbuch für Studierende der Mathematik, Physik und Ingenieurwissenschaft
351 Seiten mit 51 Figuren (1970)
Zweite, neubearbeitete und erweiterte Auflage
ISBN 3-7643-0086-8
(Mathematische Reihe, Band 24 – Sammlung LMW)

2. Handbuch der Laplace-Transformation
In 3 Bänden
(Mathematische Reihe, Bände 14, 15 und 19 – Sammlung LMW)

Band I: Theorie der Laplace-Transformation
Verbesserter Nachdruck der ersten Auflage. 1971
581 Seiten mit 40 Figuren, ISBN 3-7643-0083-3

Band II: Anwendungen der Laplace-Transformation
Verbesserter Nachdruck der ersten Auflage. 1972
436 Seiten mit 48 Figuren, ISBN 3-7643-0653-X

Band III: Anwendungen der Laplace-Transformation
Verbesserter Nachdruck der ersten Auflage. 1973
300 Seiten mit 23 Figuren, ISBN 3-7643-0674-4

3. Voelker, Doetsch
Die zweidimensionale Laplace-Transformation
Eine Einführung in ihre Anwendung zur Lösung von Randwertproblemen nebst Tabellen von Korrespondenzen
259 Seiten mit 17 Figuren und vielen Tabellen (1950)
ISBN 3-7643-0394-8
(Mathematische Reihe, Band 12 – Sammlung LMW)